Cognitive Technologies

T0137491

For further volumes:
http://www.springer.com/series/5216

Dov M. Gabbay · Karl Schlechta

Logical Tools for Handling Change in Agent-Based Systems

With 36 Figures and 8 Tables

Springer

Authors
Prof. Dov M. Gabbay
Department of Computer Science
King's College London, Strand
London WC2R 2LS, UK
dov.gabbay@kcl.ac.uk

and

Department of Computer Science
Bar-Ilan University
52900 Ramat-Gan
Israel

and

Computer Science and Communications
Faculty of Sciences
University of Luxembourg
6 rue Coudenhove-Kalergi
1359 Luxembourg

Prof. Karl Schlechta
Laboratoire d'Informatique Fondamentale de Marseille
UMR 6166
CNRS

and

Université de Provence, CMI
39, rue Joliot-Curie
13453 Marseille Cedex 13
France
ks@cmi.univ-mrs.fr
karl.schlechta@web.de

Managing Editors
Prof. Dov M. Gabbay
Augustus De Morgan Professor of Logic
Department of Computer Science
King's College London
Strand, London WC2R 2LS, UK

Prof. Dr. Jörg Siekmann
Forschungsbereich Deduktions- und
Multiagentensysteme, DFKI
Stuhlsatzenweg 3, Geb. 43
66123 Saarbrücken, Germany

Cognitive Technologies ISSN 1611-2482
ISBN 978-3-642-26187-9 e-ISBN 978-3-642-04407-6
DOI 10.1007/978-3-642-04407-6
Springer Heidelberg Dordrecht London New York

ACM Computing Classification (1998): I.2.3, F.4.1

Cover design: KünkelLopka GmbH, Heidelberg

Printed on acid-free paper

Springer is part of Springer Science+Business Media (www.springer.com)

Preface

Nonmonotonic logics were created in 1970s by John McCarthy ([McC80]) and others in order to model more closely common sense everyday reasoning. Such logics avoid explicit mentioning of a very large number of exceptional cases. Thus, formal nonmonotonic systems are logics which are closer to human reasoning. In subsequent years, a multitude of formalisms were invented, some presented from the proof theoretic point of view and some from the semantic point of view.

Axiomatic systems for nonmonotonic consequence relations was one approach ([Gab85]) and tinkering and changing existing monotonic logics were other approaches (many authors). Some tinkered with the semantics and some with the proof theory. The result was a wild landscape of diverse logics driven by applications and a multitude of formal relationships among them.

Several observations helped bring some order into this chaotic netscape. On the syntactical side, it was observed by Delgrande ([Del87]) that a nonmonotonic consequence relation corresponds to first-level conditionals (no embedded conditional implications). Later it was shown by Gabbay that if we bring (fibre) a nonmonotonic consequence relation from the metalevel into the object level, we get conditionals ([Gab98]). Thus we have a strong connection between nonmonotonic consequence and conditionals. Also, Boutilier ([Bou90a]) made a connection with modal logic. Gardenfors and Makinson ([MG91]) made a connection with the rich area of AGM revision theory.

On the semantical side, it was observed by Shoham ([Sho87b]) and later systematically developed by Kraus, Lehmann, and Magidor, ([KLM90]) that preferential models can yield many axiomatic nonmonotonic systems. This approach was studied mathematically by Schlechta (summarized in [Sch04]) for a number of completeness and incompleteness results. In a certain way, most of the semantical work can be seen as continuing the tradition begun by Kripke (e.g. [Kri59]), to create semantics by adding structure to a set of classical models. However, despite all the above observations, numerous connections and representations results, the landscape of nonmonotonic logics still was not properly organised.

The present research monograph goes further and begins to unite more logics, not only nonmonotonic ones but also modal logic, counterfactuals, and others, using a novel tool, that of reactive diagrams. Reactive diagrams allow "higher" nonmonotonicity, as the arrows of preferential structures, cancelling models as representing non-normal states, can now be cancelled themselves.

Thus, this book is a first step towards properly and fundamentally organising the field. It systematizes logical properties and separates them from underlying abstract semantical properties. It bases many of them on simple rules of working with a notion of size in an abstract semantics and shows the power of reactivity in structural semantics, breaking the strong coherence conditions of preferential models.

There are however problems and differences:

- We show the limitations of traditional size semantics in our investigation of defeasible inheritance; ongoing research will strive to close this gap, using again the idea of reactivity.
- We underline the differences between the semantics of what is (e.g. in "normal worlds") and of what should be (e.g. in the "best worlds" of deontic logic).
- We emphasize the importance of closure conditions of the domain, in particular of closure under finite unions, a problem largely overlooked up to now.

We wish to thank D. Makinson for many very interesting discussions.

London, UK — Dov M. Gabbay
Marseille, France — Karl Schlechta
August 2009

Contents

Chapter 1
Introduction and Motivation

Throughout, unless said otherwise, we will work in propositional logic.

1.1 Programme

The human agent in his daily activity has to deal with many situations involving change. Chief among them are the following:

(1) Common sense reasoning from available data. This involves prediction of what unavailable data is supposed to be (nonmonotonic deduction) but it is a defeasible prediction, geared towards immediate change. This is formally known as nonmonotonic reasoning and is studied by the nonmonotonic community.
(2) Belief revision, studied by a very large community. The agent is unhappy with the totality of his beliefs which he finds internally unacceptable (usually logically inconsistent but not necessarily so) and needs to change/revise it.
(3) Receiving and updating his data, studied by the update community.
(4) Making morally correct decisions, studied by the deontic logic community.
(5) Dealing with hypothetical and counterfactual situations. This is studied by a large community of philosophers and AI researchers.
(6) Considering temporal future possibilities; this is covered by modal and temporal logic.
(7) Dealing with properties that persist through time in the near future and with reasoning that is constructive. This is covered by intuitionistic logic.

All the above types of reasoning exist in the human mind and are used continuously and coherently every hour of the day. The formal modelling of these types is done by diverse communities which are largely distinct with no significant communication or cooperation. The formal models they use are very similar and arise from a more general theory, what we might call

"Reasoning with information bearing binary relations".

D.M. Gabbay, K. Schlechta, *Logical Tools for Handling Change in Agent-Based Systems*, Cognitive Technologies, DOI 10.1007/978-3-642-04407-6_1,

1.2 Short Overview of the Different Logics

We will discuss the semantics of the propositional logic situation only. In all cases except the last (i.e. inheritance), the semantics consist of a set of classical models for the underlying language, with an additional structure, usually a binary relation (sometimes relative to a point of origin). This additional structure is not unique, and the result of the reasoning based on this additional structure will largely depend on the specific choice of this structure. The laws which are usually provided (as axioms or rationality postulates) are those which hold for any such additional structure.

1.2.1 Nonmonotonic Logics

Nonmonotonic logics (NML) were created to deal with principled reasoning about "normal" situation. Thus, "normal" birds will (be able to) fly, but there are many exceptions, like penguins and roasted chickens, and it is usually difficult to enumerate all exceptions, so they will be treated in bulk as "abnormal" birds.

The standard example is – as we began to describe already – that "normal" birds will (be able to) fly, that there are exceptions, like penguins, that "normal" penguins will not fly, but that there might be exceptions to the exceptions, that some abnormal penguin might be able to fly – due to a jet pack on its back, some genetic tampering, etc. Then, if we know that some animal is a bird, call it "Tweety" as usual, and if we want to keep it as a pet, we should make sure that its cage has a roof, as it might fly away otherwise. If, however, we know that Tweety is not only a bird but also a penguin, then its cage need not have a roof.

Note that this reasoning is nonmonotonic: From the fact "Tweety is a bird", we conclude that it will (normally) fly, but from the facts that "Tweety is a bird" and "Tweety is a penguin", we will not conclude that it will (normally) fly any more; we will even conclude the contrary, that it will (normally) not fly.

We can also see here a general principle at work: more specific information (Tweety is a penguin) and its consequences (Tweety will not fly) will usually be considered more reliable than the more general information (Tweety is a bird) and its consequences (Tweety will fly). Then, NML can also be considered as a principled treatment of information of different quality or reliability. The classical information is the best one, and the conjecture that the case at hand is a normal one is less reliable.

Note that normality is absolute here in the following sense: normal birds will be normal with respect to all "normal" properties of birds, i.e. they will fly, lay eggs, build nests, etc. In this treatment, there are no birds normal with respect to flying, but not laying eggs, etc.

It is sometimes useful to introduce a generalized quantifier ∇. In a first-order (FOL) setting $\nabla x\phi(x)$ will mean that $\phi(x)$ holds almost everywhere; in a propositional setting $\nabla\phi$ will mean that in almost all models ϕ holds. Of course, this "almost everywhere" or "almost all" has to be made precise, e.g. by a filter over the FOL universe, or the set of all propositional models.

Inheritance systems will be discussed separately below.

- Formal semantics by preferential systems
  The semantics for preferential logics are preferential structures, a set of classical models with an arbitrary binary relation. This relation need not be transitive, nor does it need to have any other of the usual properties. If $m \prec m'$, then m is considered more normal (or less abnormal) than m'. m is said to be minimal in a set of models M iff there is no $m' \in M$, $m' \prec m$ – a word of warning: there might be $m' \prec m$, but $m' \notin M$! This defines a semantic consequence relation as follows: we say $\phi \mid\sim \psi$ iff ψ holds in all minimal models of ϕ.

  As a model m might be minimal in $M(\phi)$ – the set of models of ϕ – but not minimal in $M(\psi)$, where $\models \phi \rightarrow \psi$ classically, this consequence relation $\sim$ | is nonmonotonic. Non-flying penguins are normal (= minimally abnormal) penguins, but all non-flying birds are abnormal birds.
  Minimal models of ϕ need not exist, even if ϕ is consistent – there might be cycles or infinite descending chains. We will write $M(\phi)$ for the set of ϕ -models and $\mu(\phi)$ or $\mu(M(\phi))$ for the set of minimal models of ϕ. If there is some set X and some $x' \in X$ s.t. $x' \prec x$, we say that x' minimizes x, likewise that X minimizes x. We will be more precise in Chap. 4 (p. 73).

  One can impose various restrictions on $\prec$, they will sometimes change the resulting logic. The most important one is perhaps rankedness: If m and m' are $\prec$ – incomparable, then for all m'', $m'' \prec m$ iff $m'' \prec m'$ and also $m \prec m''$ iff $m' \prec m''$. We can interpret the fact that m and m' are $\prec$– incomparable by putting them at the same distance from some imaginary point of maximal normality. Thus, if m is closer to this point than m'' is, then so will be m', and if m is farther away from this point than m'' is, then so will be m'. (The also very important condition, smoothness, is more complicated, so the reader is referred to Chap. 4 (p. 73) for discussion.)

  Preferential structures are presented and discussed in Chap. 4 (p. 73) and Chap. 5 (p. 119).

1.2.2 Theory Revision

The problem of theory revision is to "integrate" some new information ϕ into an old body of knowledge K such that the result is consistent, even if K together with ϕ (i.e. the union $K \cup \{\phi\}$) is inconsistent. (We will assume that K and ϕ are consistent separately.)

The best examined approach was first published in [AGM85], and is known by the initials of its authors as the AGM approach. The formal presentation of this approach (and more) is given in Sect. 8.2 (p. 227). This problem is well known in juridical thinking, where a new law might be inconsistent with the old set of laws, and the task is to "throw away" enough, but not too many, of the old laws, so we can incorporate the new law into the old system in a consistent way.

We can take up the example for NML and modify it slightly. Suppose our background theory K is that birds fly, in the form: Blackbirds fly, ravens fly, penguins fly, robins fly, ..., and that the new information is that penguins don't fly. Then, of course, the minimal change to the old theory is to delete the information that penguins fly and replace it with the new information.

Often, however, the situation is not so simple. K might be that ψ holds, and so does $\psi \to \rho$. The new information might be that $\neg\rho$ holds. The radical – and usually excessive – modification will be to delete all information from K and just take the new information. More careful modifications will be to delete either ψ or $\psi \to \rho$, but not both. But there is a decision problem here: which of the two do we throw out? Logic alone cannot tell us and we will need more information to take this decision.

- Formal semantics

  In many cases, revising K by ϕ is required to contain ϕ; thus, if $*$ denotes the revision operation, then $K * \phi \vdash \phi$ (classically). Dropping this requirement does not change the underlying mechanism enormously; we will uphold it.

  Speaking semantically, $K * \phi$ will then be defined by some subset of $M(\phi)$. If we choose all of ϕ, then any influence of K is forgotten. A good way to capture this influence seems to choose those models of ϕ, which are closest to the K-models, in some way, and with respect to some distance d. We thus choose those ϕ-models m such that there is $n \in M(K)$ with $d(n, m)$ minimal among all $d(n', m')$, $n' \in M(K)$, $m' \in M(\phi)$. (We assume again that the minimal distance exists, i.e. that there are no infinite descending chains of distances, without any minimum.) Of course, the choice of the distance is left open and will influence the outcome. For instance, choosing as d the trivial distance, i.e. $d(x, y) = 1$ iff $x \neq y$, and 0 otherwise, will give us just ϕ – if K is inconsistent with ϕ.

  This semantic approach corresponds well to the classical, syntactic AGM revision approach in the following sense: When we fix K, this semantics corresponds exactly to the AGM postulates (which leave K fixed). When we allow K to change, we can also treat iterated revision, i.e. something like $(K * \phi) * \psi$, thus go beyond the AGM approach (but pay the price of arbitrarily long axioms). This semantics leaves the order (or distance) untouched, and is thus fundamentally different from, e.g. Spohn's ordinal conditional functions, see [Spo88].

1.2.3 Theory Update

Theory update is the problem of "guessing" the results of optimal developments.

Consider the following situation: There is a house, at time 0, the light is on, and so is the deep freezer. At time 1, the light is off. Problem: Is the deep freezer still on? The probably correct answer depends on circumstances. Suppose in situation

A, there is someone in the house and weather conditions are normal. In situation B, there is no one in the house and there is a very heavy thunderstorm going on. Then, in situation A, we will conjecture that the people in the house have switched the light off, but left the deep freezer on. In situation B, we might conjecture a general power failure and that the deep freezer is now off, too.

We can describe the states at time 0 and 1 by a triple: light on/off, freezer on/off, and power failure yes/no.

In situation A, we will consider the development (light on, freezer on, no power failure) to (light off, freezer on, no power failure) as the most likely (or normal) one.

In situation B, we will consider the development (light on, freezer on, no power failure) to (light off, freezer off, power failure) as the most likely (or normal) one.

Often, we will assume a general principle of inertia: things stay the way as they are, unless they are forced to change. Thus, when the power failure is repaired, freezer and light will go on again.

- Formal semantics

  In the general case, we will consider a set of fixed length sequences of classical models, say $m = \langle m_0, m_1, \ldots, m_n \rangle$, which represent developments considered possible. Among this set, we have some relation $\prec$, which is supposed to single out the most natural or probable ones. We then look at some coordinate, say i, and try to find the most probable situation at coordinate i. For example, we have a set S of sequences and look at the theory defined by the information at coordinate i of the most probable sequences of $S : Th(\{m_i : m \in \mu(S)\})$ – where $\mu(S)$ is the set of most probable sequences of S and $Th(X)$ is the set of formulas which hold in all models $x \in X$.

  Looking back at our above intuitive example, S will be the set of sequences consisting of
  $\langle (l, f, -p), (l, f, p) \rangle$,
  $\langle (l, f, -p), (l, f, -p) \rangle$,
  $\langle (l, f, -p), (l, -f, p) \rangle$,
  $\langle (l, f, -p), (l, -f, -p) \rangle$,
  $\langle (l, f, -p), (-l, f, p) \rangle$,
  $\langle (l, f, -p), (-l, f, -p) \rangle$,
  $\langle (l, f, -p), (-l, -f, p) \rangle$,
  $\langle (l, f, -p), (-l, -f, -p) \rangle$,
  where "l" stands for "light on", "f" for "freezer on", "p" for "power failure", etc. The "best" sequence in situation A will be $\langle (l, f, -p), (-l, f, -p) \rangle$, and in situation B $\langle (l, f, -p), (-l, -f, p) \rangle$. Thus, in situation A, the result is defined by $-$l, f, $-$p, etc. – the theory of the second coordinate.

  Thus, again, the choice of the actual distance has an enormous influence on the outcome.

1.2.4 Deontic Logic

Deontic logic treats (among other things) the moral acceptability of situations or acts. For instance, when driving a car, you should not cause accidents and hurt someone. So, in all "good" driving situations, there are no accidents and no victims. Yet, accidents unfortunately happen. And if you have caused an accident, you should stop and help the possibly injured. Thus, in the "morally bad" situations where you have caused an accident, the morally best situations are those where you help the victims, if there are any.

The parallel to above example for NML is obvious, and, as a matter of fact, the first preferential semantics was given for deontic and not for nonmonotonic logics – see Sect. 7.1 (p. 175). There is, however, an important difference to be made. Preferential structures for NML describe what *holds* in the normally best models, those for deontic logic what holds in "morally" best models. But obligations are not supposed to say what holds in the morally best worlds, but are supposed to distinguish in some way the "good" from the "bad" models. This problem is discussed in extenso in Sect. 7.1 (p. 175).

- Formal semantics

  As said already, preferential structures as defined above for NML were given as a semantics for deontic logics, before NML came into existence.

  A word of warning: Here, the morally optimal models describe "good" situations, and not directly actions to take. This is already obvious by the law of weakening, which holds for all such structures: If ϕ holds in all minimal models and $\vdash \phi \rightarrow \psi$ (classically), then so does ψ. But if one should be kind, then it does not follow that one should be kind or kill one's grandmother. Of course, we can turn this reasoning into advice for action: act the way that the outcome of your actions assures you to be in a morally good situation.

1.2.5 Counterfactual Conditionals

A counterfactual conditional states an implication, where the antecedent is (at the moment, at least) wrong. "If it were to rain, he would open his umbrella." This is comprehensible, the person has an umbrella, and if it were to start to rain now, he would open the umbrella. If, however, the rain would fall in the midst of a hurricane, then opening the umbrella would only lead to its destruction. Thus, if, at the moment the sentence was uttered, there was no hurricane, and no hurricane announced, then the speaker was referring to a situation which was different from the present situation only insofar as it is raining, or, in other words, minimally different from the actual situation, but with rain falling. If, however, there was a hurricane in sight at the moment of uttering the sentence, we might doubt the speaker's good sense and point the problem out to him/her. We see here again a reasoning about minimal change or normal situations.

- Formal semantics
  Stalnaker and Lewis first gave a minimal distance semantics in the following way:

  If we are in the actual situation m, then $\phi > \psi$ (read: if ϕ were the case, then ψ would also hold) holds in m, iff in all ϕ-models which are closest to m, ψ also holds. Thus, there might well be ϕ-models where ψ fails, but these are not among the ϕ-models closest to m. The distance will, of course, express the difference between the situation m and the models considered. Thus, in the first scenario, situations where it rains and there is no extreme wind condition are closer to the original one than those where a hurricane blows.

  In the original approach, distances from each possible actual situation are completely independent. It can, however, be shown that we can achieve the same results with one uniform distance over the whole structure, see [SM94].

1.2.6 Modal Logic

Modal logic reasons about the possible or necessary. If we are in the midwest of the United States and it is a hurricane season, then the beautiful sunny weather might turn into a hurricane over the next hours. Thus, the weather need not necessarily stay the way it is, but it might become a very difficult situation. Note that we reason here not about what is likely, or normal, but about what is considered possible at all. We are not concerned only about what might happen in time $t + 1$, but about what might happen in some (foreseeable, reasonable) future – and not about what will be the case at the end of the developments considered possible either. Just everything which might be the case sometime in the near future.

"Necessary" and "possible" are dual: if ϕ is necessarily the case, this means that it will always hold in all situations evolving from the actual situation, and if ϕ is possibly the case, this means that it is not necessary that $\neg\phi$ holds, i.e. there is at least some situation into which the present can evolve and where ϕ holds.

- Formal semantics
  Kripke gave a semantics for modal logic by possible worlds, i.e. a set of classical models, with an additional binary relation, expressing accessibility.

  If m is in relation R with n, mRn, then m can possibly become n, is a possibility seen from m, or whatever one might want to say. Again, R can be any binary relation. The necessity operator is essentially a universal quantifier, $\Box\phi$ holds in m iff ϕ holds in all n accessible via R from m. Likewise, the possibility operator is an existential quantifier, $\Diamond\phi$ holds in m iff there is at least one n accessible from m where ϕ holds.

  Again, it is interesting to impose additional postulates on the relation R, like reflexivity and transitivity.

1.2.7 Intuitionistic Logic

Intuitionistic logic is (leaving philosophy aside) reasoning about performed constructions and proofs in mathematics or development of (certain) knowledge. We may have a conjecture – or, simply, any statement –, and a proof for it, a proof for its contrary, or neither. Proofs are supposed to be correct, so what is considered a proof will stay one forever. Knowledge can only be won, but not lost. If we have neither a proof nor a refutation, then we might one day have one or the other, or we might stay ignorant forever.

- Formal semantics
  Intuitionistic logic can also be given a semantics in the style of the one for modal logics. There are two, equivalent, variants. The one closer to modal logic interprets intuitionistic statements (in the above sense that a construction or proof has been performed) as preceded by the necessity quantifier. Thus, it is possible that in m neither $\Box\phi$ nor $\Box\neg\phi$ holds, as we have a proof neither for ϕ nor for its contrary, and might well find one for one or the other in some future, possible, situation. Progressing along R may, if the relation is transitive, only make more statements of the form $\Box\phi$ true, as we quantify then over less situations.

1.2.8 Inheritance Systems

Inheritance systems or diagrams are directed acyclic graphs with two types of arrows, positive and negative ones. Roughly, nodes stand for sets of objects, like birds and penguins or properties like "able to fly". A positive arrow $a \to b$ stands for "(almost) all $x \in a$ are also in b" – so it admits exceptions. A negative arrow $a \to b$ stands for "(almost) all $x \in a$ are *not* in b" – so it also admits exceptions. Negation is thus very strong. The problem is to find the valid paths (concatenations of arrows) in a given diagram, considering contradictions and specificity. See Chap. 9 (p. 251) for a deeper explanation and formal definitions.

- Formal semantics
  They are still the subject of ongoing research.

1.2.9 A Summarizing Table for the Semantics

Table 1.1 summarizes important properties of the semantics of some of the logics discussed.

where "cpm" stands for "classical propositional model", "CPM" for a set of such;
"T." means number of types of nodes/arrows;
"Counterfactuals": counterfactual conditionals;

Table 1.1 Various semantics

	Meaning, properties, use							
Logic	Nodes		Arrows (relations)				Remarks	Relativization
	T.	Meaning	T.	Meaning	Properties	Use		
Modal	1	cpm	1	Poss. development?	Transitive reflexive,...	Collect (reachable) nodes		What is reachable from one point Summary over all points (monotonic)
Intuitionistic	1	cpm	1	Epist. change: progress of knowl.	Transitive	Collect		What is reachable from one point Summary over all points (monotonic)
Preferential	1	cpm	1	Epist. change: conjecture	Ranked smooth,...	Best nodes or Λ		What is optimal below one point Summary over all points (monotonic)
Deontic	1	cpm	1	Moral (value) change	Ranked...	Best/Λ		What is optimal below one point Summary over all points (monotonic)
Counterfactuals	1	cpm	1	Change in world to make prereq. real	Ranked (distance)	Individ. best/ Λ		What is closest to one point Summary over all points (monotonic)
Full TR	1	cpm	1	Epist. change: incorp. new info	Ranked (distance)	Global best/ Λ	Global best needs distance for comparison	$A \mid B$ is all of B not more distant from A than m is from A If m is closer to A than B is, then $\emptyset$
Fixed K TR	2	CPM cpm	1	Epist. change: incorp. new info	Ranked? (AGM tradition)	Global best/ Λ		Same as full TR
Update	1	CPM	1	Change of world, evolving sit.	Ranked? distance?	Individ. best/ Λ		What is optimal in all threads through a point satisfying start condition Summary over all points (monotonic)
Inheritance	1	CPM	2	Normally, ϕ's are ψ's (or $\neg\psi$'s)	Acyclic?	Collect	Comparison of paths	–
Argumentation	1	CPM	2	"Soft" implication	Acyclic?	Collect		
Causation	1	CPM	2	"Normally causes" "Normally inhibits"	Acyclic?	Collect		

"Λ" means limit version in the absence of minimal/closest or best points, see Sect. 5.5 (p. 145);
"Collect" means all points in relation are collected;
"Relativization" means and some reasonable ways to make the logic dependent on a fixed model as starting point. (This is evident for modal logic and less evident for some other logics.)

1.3 A Discussion of Concepts

The aim of this section is to describe the concepts and roles of models and operators in various propositional logics.

Notation 1.3.1 We will use ♠ for the global universal modal quantifier: ♠ϕ holds in a model iff ϕ holds everywhere – it is the dual of consistency.
$\Box$ and $\Diamond$ are the usual universal and existential modal quantifiers; recall that ∇ is, respectively some normality quantifier, see, e.g. Chap. 3 (p. 53).

1.3.1 Basic Semantic Entities, Truth Values, and Operators

1.3.1.1 The Levels of Language and Semantics

We have several levels:

(1) The language and the truth values.
(2) The basic semantical entities, e.g. classical models and maximal consistent sets of modal formulas.
(3) Abstract or algebraic semantics, which describe the interpretation of the operators of the language in set-theoretic terms, like the interpretation of $\wedge$ by $\cap$ and ∇ ("the normal cases of") by μ (choice of minimal elements). These semantics do not indicate any mechanism which generates these abstract operators.
(4) Structural semantics, like Kripke structures, preferential structures, which give such mechanisms and generate the abstract behaviour of the operators. They are or should be the intuitive basis of the whole enterprise.

(In analogue, we have a structural, an abstract, and a logical limit – see Sect. 5.5 (p. 145).)

1.3.1.2 Language and Truth Values

A language has

- variable parts, like propositional variables and
- constant parts, like
  - operators, e.g. $\wedge$, ∇, and $\Box$. and
  - relations like a consequence relation $\mid\!\sim$.

Operators may have a

- unique interpretation, like $\wedge$, which is always interpreted by $\cap$ or
- only restrictions on the interpretation, like ∇, $\Box$.

Operators and relations may be

- nested, like $\wedge$, ∇, or
- only flat (perhaps on the top level), like $\mid\!\sim$.

The truth values are part of the overall framework. For the moment, we will tacitly assume that there is only TRUE and FALSE. This restriction is unimportant for our purposes.

1.3.1.3 Basic Semantical Entities

The language speaks about the basic semantic entities. Note that the language will usually NOT speak about the relation of a Kripke or preferential structures, etc., only about the resulting function, resp. the operator which is interpreted by the function, as part of formulas – we do not speak directly about operators, but only as part of formulas.

For the same language, there may be different semantic entities. The semantic entities are (perhaps consistent, perhaps complete wrt the logic) sets of formulas of the language. They are descriptions – in the language – of situations. They are NOT objects (in FOL we have names for objects), nor situations, but only descriptions, even if it may help to consider them as such objects (but unreal ones, just as mannequins in shop windows are unreal – they are only there to exhibit the garments).

An example for different semantic entities is intuitionistic logics, where we may take

- knowledge states, which may be incomplete (this forces the relation in Kripke structures to be monotonic) or
- classical models, where $\Box$ codes knowledge, and, automatically, its growth.

(Their equivalence is a mathematical result; the former approach is perhaps the philosophically better one, the second one easier to handle, as intuitionistic formulas are distinct by the preceding $\Box$.)

The entities need not contain all formulas with all operators; perhaps they are only sets of propositional variables, with no operators at all; perhaps they are all consistent sets of formulas of some sublanguage. For classical logic, we can take as basic entities either just sets of propositional variables or maximal consistent sets of formulas.

In the case of maximal consistent formula sets in the full language, perhaps the simplest way to find all semantic entities is by the following ways:

- Any set of formulas is a candidate for a semantic entity.
- If we want them to be complete, eliminate those who are not.

- Eliminate all which are contradictory under the operators and the logic which governs their behaviour – e.g. p, q, and $\neg p \vee \neg q$ together cannot hold in classical models.

In this approach, the language determines all situations, the logic those which are possible. We thus have a clean distinction between the work of the language and that of the logic.

For preferential reasoning (with some relation $\mid\!\sim$ outside the "core language"), we may again take all classical models – however defined – and introduce the interpretation of $\mid\!\sim$ in the algebraic or abstract superstructure (see below), but we may also consider a normality operator ∇ directly in the language, and all consistent sets of such formulas (see, e.g. [SGMRT00]). Our picture is large enough to admit both possibilities.

In modal logic, we may again consider classical models, or, as is mostly done (maximal consistent), sets of formulas of the full language. The choice of these entities is a philosophical decision, not dictated by the language, but it has some consequences. See below (nested preferential operators) for details. We call these entities models or basic models.

1.3.1.4 Several Truth Values in Basic Semantic Entities

When we consider sets of formulas as basic models, we assume implicitly two truth values: everything which is in the set has truth value TRUE, the rest truth value FALSE (or undecided – context will tell). Of course, we can instead consider (partial) functions from the set of formulas into any set of truth values – this does not change the overall approach. For example, in an information state, we might have been informed that ϕ holds with some reliability r, in another one with reliability r', etc., so r, r', etc. may be truth values, and even pairs $\{r, r'\}$ when we have been informed with reliability r that ϕ, and with reliability r' that $\neg\phi$. Whatever is reasonable in the situation considered should be admitted as truth value.

1.3.2 Algebraic and Structural Semantics

We now make a major conceptual distinction between an "algebraic" and a "structural" semantics, which can best be illustrated by an example.

Consider nonmonotonic logics as discussed above. In preferential structures, we only consider the minimal elements, say $\mu(X)$, if X is a set of models. Abstractly, we thus have a choice function μ, defined on the power set of the model set, and μ has certain properties, e.g. $\mu(X) \subseteq X$. More important is the following property: $X \subseteq Y \rightarrow \mu(Y) \cap X \subseteq \mu(X)$. (The proof is trivial: suppose there were $x \in \mu(Y) \cap X$, $x \notin \mu(X)$. Then there must be $x' \prec x$, $x' \in X \subseteq Y$, but then x cannot be minimal in Y.) Thus, all preferential structures generate μ functions with certain properties, and once we have a complete list, we can show that any arbitrary model choice function with these properties can be generated by an appropriate preferential structure.

Note that we do not need here the fact that we have a relation between models, just any relation on an arbitrary set suffices. It seems natural to call the complete list of properties of such μ-functions an algebraic semantics, forgetting that the function itself was created by a preferential structure, which is the structural semantics.

This distinction is very helpful; it incites us not only to separate the two semantics conceptually, but also to split completeness proof into two parts: One part, where we show correspondence between the logical side and the algebraic semantics, and a second one, where we show the correspondence between the algebraic and the structural semantics. The latter part will usually be more difficult, but any result obtained here is independent from logics itself, and can thus often be re-used in other logical contexts. On the other hand, there are often some subtle problems for the correspondence between the logics and the algebraic semantics (see definability preservation, in particular the discussion in [Sch04]), which we can then more clearly isolate, identify, and solve.

1.3.2.1 Abstract or Algebraic Semantics

In all cases, we see that the structural semantics define a set operator, and thus an algebraic semantics:

- In nonmonotonic logics (and deontic logic), the function chooses the minimal (morally best) models, a subset, $\mu(X) \subseteq X$.
- In (distance based) theory revision, we have a binary operator, say $|$ which chooses the ϕ-models closest to the set of K-models: $M(K) \mid M(\phi)$.
- In theory update, the operator chooses the ith coordinate of all best sequences.
- In the logic of counterfactual conditionals, we have again a binary operator $m \mid M(\phi)$ which chooses the ϕ-models closest to m or when we consider a whole set X of models as starting points $X \mid M(\phi) = \bigcup\{m \mid M(\phi) : m \in X\}$.
- In modal and intuitionistic logic, seen from some model m, we choose a subset of all the models (thus not a subset of a more restricted model set), those which can be reached from m.

Thus, in each case, the structure "sends" us to another model set, and this expresses the change from the original situation to the "most plausible", "best", "possible", etc. situations. It seems natural to call all such logics "generalized modal logics", as they all use the idea of a model choice function. (Note again that we have neglected here the possibility that there are no best or closest models (or sequences), but only ever better ones.)

Abstract semantics are interpretations of the operators of the language (all, flat, top level or not) by functions (or relations in the case of $\mid\sim$), which assign to sets of models, $\mathcal{O} : \mathcal{P}(\mathcal{M}) \to \mathcal{P}(\mathcal{M})$ – $\mathcal{P}$ the power set operator, $\mathcal{M}$ the set of basic models, or binary functions for binary operators, etc. These functions are determined or restricted by the laws for the corresponding operators. For example, in classical, preferential, or modal logic, $\wedge$ is interpreted by $\cap$, etc.; in preferential logic ∇ by μ; in modal logic, we interpret $\Box$, etc.

Operators may be truth-functional or not. $\neg$ is truth-functional. It suffices to know the truth value of ϕ at some point, to know that of $\neg\phi$ at the same point. $\Box$ is not truth-functional: ϕ and ψ may hold, and $\Box\phi$, but not $\Box\psi$, all at the same point (= base model), we have to look at the full picture, not only at some model.

We consider first those operators, which have a unique possible interpretation, like $\wedge$, which is interpreted by $\cap$, $\neg$ by $\mathbf{C}$, the set-theoretic complement, etc. ∇ (standing for "most", "the important", etc.) e.g., has only restrictions to its interpretation, like $\mu(X) \subseteq X$. Given a set of models without additional structure, we do not know its exact form; we know it only once we have fixed the additional structure (the relation in this case).

If the models contain already the operator, the function will respect it, i.e. we cannot have ϕ and $\neg\phi$ in the same model, as $\neg$ is interpreted by $\mathbf{C}$. Thus, the functions can, at least in some cases, control consistency. If, e.g., the models contain $\wedge$, then we have two ways to evaluate $\phi \wedge \psi$: we can first evaluate ϕ, then ψ, and use the function for $\wedge$ to evaluate $\phi \wedge \psi$. Alternatively, we can look directly at the model for $\phi \wedge \psi$ – provided we considered the full language in constructing the models.

As we can apply one function to the result of the other, we can evaluate complicated formulas, using the functions on the set of models. Consequently, if $\mid\sim$ or ∇ is evaluated by μ, we can consider $\mu(\mu(X))$, etc.; thus, the machinery for the flat case gives immediately an interpretation for nested formulas too – whether we looked for it or not.

As far as we see, our picture covers the usual presentations of classical logic, preferential, intuitionist, and modal logic, and also of linear logic (where we have more structure on the set of basic models, a monoid, with a distinct set $\perp$, plus some topology for! and? – see below) and quantum logic a la Birkhoff/von Neumann.

We can introduce new truth-functional operators into the language as follows: Suppose we have a distinct truth value TRUE, then we may define $\mathcal{O}_X(\phi) = TRUE$ iff the truth value of ϕ is an element of X. This might sometimes be helpful. Making the truth value explicit as element of the object language may facilitate the construction of an accompanying proof system – experience will tell whether this is the case. In this view, $\neg$ has now a double meaning in the classical situation: It is an operator for the truth value "false" and an operator on the model set, and corresponds to the complement. "Is true" is the identical truth-functional operator, $is - true(\phi)$ and ϕ have the same truth value.

If the operators have a unique interpretation, this might be all there is to say in this abstract framework. (This does not mean that it is impossible to introduce new operators which are independent from any additional structure, and based only on the set of models for the basic language. We can, for instance, introduce a "CON" operator, saying that ϕ is consistent, and $CON(\phi)$ will hold everywhere iff ϕ is consistent, i.e. holds in at least one model, or, for a more bizarre example, a 3 operator, which says that ϕ has at least three models (which is then dependent on the language). We can also provide exactly one additional structure, e.g. in the following way: Introduce a ranked order between models as follows: At the bottom, put the single model which makes all propositional variables true, on the next level those which make exactly one propositional variable true, then two, etc., with the model making all false on top. So there is room to play; if one can find many useful examples is another question.)

If the operator has no unique interpretation (like ∇, $\Box$ which are only restricted, e.g. by $\spadesuit(\phi \to \psi) \to \spadesuit(\nabla\psi \wedge \phi \to \nabla\phi)$, the situation seems more complicated and is discussed below in Sect. 1.3.3 (p. 15)).

It is sometimes useful to consider the abstract semantics as a (somehow coherent) system of filters. For instance, in preferential structures, $\mu(X) \subseteq X$ can be seen as the basis of a principal filter. Thus, $\phi \mid\sim \psi$ iff ψ holds in all minimal models of ϕ, iff there is a "big" subset of $M(\phi)$ where ψ holds, recalling that a filter is an abstraction of size – sets in the filter are big, their complements are small, and the other sets have medium size. Thus, the "normal" elements form the smallest big subset. Rules like $X \subseteq Y \to \mu(Y) \cap X \subseteq \mu(X)$ form the coherence between the individual filters; we cannot choose them totally independently. Particularly for preferential structures, the reasoning with small and big subsets can be made very precise and intuitively appealing, and we will come back to this point later. We can also introduce a generalized quantifier, say ∇, with the same meaning, i.e. $\phi \mid\sim \psi$ iff $\nabla(\phi) \cdot \psi$, i.e. "almost everywhere" or "in the important cases" where ϕ holds, so will ψ. This is then the syntactic analogue of the semantical filter system. These aspects are discussed in detail in Chap. 3 (p. 53).

1.3.2.2 Structural Semantics

Structural semantics generate the abstract or algebraic semantics, i.e. the behaviour of the functions or relations (and of the operators in the language when we work with "rich" basic models). Preferences between models generate corresponding μ-functions, relations in Kripke structures generate the functions corresponding to $\Box$-operators, etc.

Ideally, structural semantics capture the essence of what we want to reason and speak about (beyond classical logic), they come, or should come, first. Next, we try to see the fundamental ingredients and laws of such structures, code them in an algebraic semantics and the language, i.e. extract the functions and operators, and their laws. In a backward movement, we make the roles of the operators (or relations) precise (should they be nested or not? etc.), and define the basic models and the algebraic operators. This may result in minor modifications of the structural semantics (like introduction of copies), but should still be close to the point of outset. In this view, the construction of a logic is a back-and-forth movement.

1.3.3 Restricted Operators and Relations

We discuss only operators; relations seem to be similar. The discussion applies as well to abstract as to structural semantics. An operator, which is only restricted in its behaviour, but not fully defined, has to be interpreted by a unique function. Thus, the interpretation will be more definite than the operator. It seems that the problem has no universal solution.

(1) If there is tacitly a "best choice", it seems natural to make this choice. At the same time, such a best choice may also serve to code our ignorance, without enumerating all possible cases among which we do not know how to decide.

    For instance, in reasoning about normality (preferential structures), the interpretation which makes nothing more normal than explicitly required – corresponding to a set-theoretically minimal relation – seems a natural choice. This will NOT always give the same result as a disjunction over all possible interpretations: e.g. if the operator is in the language, and we have finitely many possibilities, we can express them by "or", and this need not be the same as considering the unique minimal solution. (Note that this will usually force us to consider "copies" in preferential structures – see below.)
(2) We can take all possible interpretations, and consider them separately, and take as result only those facts which hold in all possibilities. Bookkeeping seems difficult, especially when we have nested operators, which have all to be interpreted in various ways. In the second step, we can unite all possibilities in one grand picture (a universal structure, as it permits to find exactly all consequences in one construction, and not in several ones as is done for classical logic), essentially by a disjoint union – this was done (more or less) by the authors [SGMRT00] for preferential structures.
(3) We can work with basic models already in the full language and capture the different possibilities already on the basic level. The interpretation of the operator will then be on the big set of models for the full language, which serve essentially as bookkeeping device for different possibilities of interpretation – again a universal structure. This is done in the usual completeness proofs for modal logic.

1.3.4 Copies in Preferential Models

Copies in preferential structures (variant (2) in Sect. 1.3.3 (p. 15)) thus seem to serve to construct universal structures or code our ignorance, i.e. we know that x is minimized by X, but we do not know by which element of X, they are in this view artificial. But they have an intuitive justification too: They allow minimization by sets of other elements only. We may consider an element m only abnormal in the presence of several other elements together. For example, considering penguins, nightingales, woodpeckers, ravens, they all have some exceptional qualities, so we may perhaps not consider a penguin more abnormal than a woodpecker, etc., but seeing all these birds together, the penguin stands out as the most abnormal one. But we cannot code minimization by a set, without minimization by its elements, without the use of copies. Copies will then code the different aspects of abnormality.

1.3.5 Further Remarks on Universality of Representation Proofs

There is a fundamental difference between considering all possible ramifications and coding ignorance. For instance, if we know that $\{a, b, c\}$ is minimized by $\{b, c\}$, we can create two structures: one, where a is minimized by b, the other, where a is minimized by c. These are all possible ramifications (if $\mu(\{a\}) \neq \emptyset$). Or, we can code our ignorance with copies of a, as is done in our completeness constructions. $((\{a, b\} \mid\sim b)$ or $(\{a, c\} \mid\sim c))$ is different from $\{a, b, c\} \mid\sim \{b, c\}$, and if the language is sufficient, we can express this. In a "directly ignorant" structure, none of the two disjoints hold, so the disjunction will fail.

Our proofs try to express ignorance directly. Note that a representing structure can be optimal in several ways: (1) optimal, or universal, as it expresses exactly the logic, (2) optimal, as it has exactly the required properties, but not more. For instance, a smooth structure can be optimal, as it expresses exactly the logic it is supposed to code, or optimal, as it preserves exactly smoothness, there is no room to move left. Usually, one seeks only the first variant. If both variants disagree, then structure and model do not coincide exactly, the structure still has more space to move than necessary for representation.

In our constructions, all possibilities are coded into the choice functions (the indices), but as they are not directly visible to the language, they are only seen together, so there is no way to analyse the "or" of the different possibilities separately.

1.3.6 $\mid\sim$ in the Object Language?

It is tempting to try and put a consequence relation $\mid\sim$ into the object language by creating a new modal operator ∇ (expressing "most" or so), with the translation $\phi \mid\sim \psi$ iff $\vdash \nabla\phi \rightarrow \psi$. We examine now this possibility and the resulting consequences.

We suppose that $\rightarrow$ will be interpreted in the usual way, i.e. by the subset relation. The aim is then to define the interpretation of ∇ s.t. $\phi \mid\sim \psi$ iff $M(\nabla\phi) \subseteq M(\psi)$ – $M(\phi)$ the set of models of ϕ.

It need not be the case – but it will be desirable – that ∇ is insensitive to logical equivalence. $\nabla\phi$ and $\nabla\phi'$ may well be interpreted by different model sets, even if ϕ and ϕ' are interpreted by the same model sets. Thus, we need not leave logical equivalence – whatever the basic logic is. On the right hand side, as we define $\mid\sim$ via $\rightarrow$, and $\rightarrow$ via model subsets, we will have closure under semantic consequence (even infinite closure, if this makes a difference). It seems obvious that this is the only property a relation $\mid\sim$ has to have to be translatable into object language via "classical" $\rightarrow$, or, better, subsets of models.

Note that the interpretation of $\nabla\phi$ can be equivalent to a formula, a theory, or a set of models logically equivalent to a formula or theory, even if some models are missing (lack of definability preservation). Likewise, we may also consider ∇T for a full theory T. The standard solution will, of course, be $M(\nabla\phi) := \bigcap\{M(\psi) : \phi \mid\sim \psi\}$.

1.3.6.1 Possibilities and Problems of External $\mid\!\sim$ vs. Internal ∇

We first enumerate a small number of various differences, before we turn to a few deeper ones.

- The translation of $\mid\!\sim$ into the object language – whenever possible – introduces contraposition into the logic, see also [Sch04] for a discussion.
- It may also help to clarify questions like validity of the deduction theorem (we see immediately that one half is monotony), Cut, situations like $\nabla\phi \to \psi$ and $\nabla\psi \to \phi$, etc.
- It is more usual to consider full theories outside the language, even if, a priori, there is no problem to define a formula ∇T using a theory T in the inductive step. Thus, the external variant can be more expressive in one respect.
- At least in principle, having ∇ inside the language makes it more amenable to relativize it to different viewpoints, as in modal logic. It seems at least more usual to write $m \models \nabla\phi \to \psi$ than to say that in m, $\phi \mid\!\sim \psi$ holds.

1.3.6.2 Disjunction in Preferential and Modal Structures

In a modal Kripke model, $\Box\phi \vee \Box\psi$ may hold everywhere, but neither $\Box\phi$ nor $\Box\psi$; may hold everywhere. This is possible, as formulas are evaluated locally¡??¿ and at one point m, we may make $\Box\phi$ hold, at another m', $\Box\psi$.

This is not the case for the globally evaluated modal operator ♠. Then, in *one* structure, if ♠$(\phi) \vee$ ♠(ψ) holds, either ♠(ϕ) or ♠(ψ) holds, but a structure where ♠(ϕ) (or ♠(ψ)) holds has more information than a structure in which ♠$(\phi) \vee$ ♠(ψ) holds. Consequently, one Kripke structure is not universal any more for ♠ (and $\vee$); we need several such structures to represent disjunctions of ♠.

The same is the case for preferential structures. If we put $\mid\!\sim$ into the language as ∇, to represent disjunctions, we may need several preferential structures: $(\nabla\phi \to \psi) \vee (\nabla\phi \to \psi')$ will only be representable as $\nabla\phi \to \psi$ or as $\nabla\phi \to \psi'$, but in one structure, only one of them may hold. Thus, again, *one* structure will say more than the original formula $(\nabla\phi \to \psi)\vee(\nabla\phi \to \psi')$, and to express this formula faithfully, we will need again two structures. Thus, putting $\mid\!\sim$ into the object language destroys the universality of preferential structures by its richer language.

Remark 1.3.1 rational monotony is, semantically, also universally quantified, $\alpha \mid\!\sim \gamma \to \alpha \wedge \beta \mid\!\sim \gamma$ or *everywhere* $\alpha \mid\!\sim \neg\beta$.

1.3.6.3 Iterated ∇ in Preferential Structures

Once we have ∇ in the object language, we can form $\nabla\nabla\phi$, etc. When we consider preferential structures in the standard way, it is obvious that $\nabla\nabla\phi = \nabla\phi$, etc. will hold. But, even in a preferential context, it is not obvious that this has to hold; it suffices to interpret the relation slightly differently. Instead of setting $\mu(X)$ the set of minimal elements of X, it suffices to define $\mu(X)$ the set of non-worst elements of X, i.e. everything except the upper layer. (One of the authors once had a similar discussion with M. Magidor.) (Note that, in this interpretation, for instance, $\nabla\nabla\phi$

may seem to be the same as $\nabla\phi$, but $\nabla\nabla\nabla\phi$ clearly is different from $\nabla\nabla\phi$. In going from $\nabla\phi$ to $\nabla\nabla\phi$, we lose some models, but not enough to be visible by logics – a problem of definability preservation. The loss becomes visible in the next step.)

But, we can have a similar result even in the usual interpretation of "normal" by "best":

Consider for any $X \subseteq \omega$ the logic $\mid\!\sim_X$ defined by the axioms $\{A_i : i \in X\}$, where $A_i := \nabla^{i+1}\phi \leftrightarrow \nabla^i\phi$, and $\nabla^i\phi$ is, of course, i many ∇'s, followed by ϕ, etc., plus the usual axioms for preferential logics. This defines 2^ω many logics, and we show that they are all different. The semantics we will give show at the same time that the usual axioms for preferential logics do not entail $\nabla\nabla\phi \leftrightarrow \nabla\phi$.

For simplicity, we first show that the ω many logics defined by the axioms $B_i = \nabla^{i+1}\phi \leftrightarrow \nabla^i\phi$ for arbitrary ϕ are different. We consider sequences of ω many preferential structures over some infinite language, s.t. we choose exactly at place i the same structure twice, and all the other times different structures. Let S_i be the structure which minimizes $\neg p_i$ to p_i, i.e. every p_i-model m is smaller than its opposite m', which is like m, only $m' \models \neg p_i$. It is obvious that the associated μ-functions μ_i will give different result on many sets X (if a model $x \in X$ is minimized at all in X, it will be minimized in different ways). Consider now, e.g. a formula $\nabla\nabla\phi$. We start to evaluate at S_0, evaluating the leftmost ∇ by μ_0. The second ∇ will be evaluated at S_1 by μ_1. If, for instance, ϕ is a tautology, we eliminate in the first step the $\neg p_0$-models, in the second step the $\neg p_1$-models, so $\nabla\nabla$ is not equivalent to ∇. If, instead of taking at position $i+1$ S_{i+1}, we just take S_i again, then axiom B_i will hold, but not the other B_j, $i \neq j$. Thus, the logics B_i are really different.

In the first case, for the A_i, we repeat S_j for all $j \in X$, instead of taking S_{j+1}, and start evaluation again at S_0. Again, the different sequences of structures will distinguish the different logics, and we are done.

Yet the standard interpretation of ∇ in one single preferential structure does not allow to distinguish between all these logics, as sequences of ∇'s will always be collapsed to one ∇. As long as a (preferential) consequence relation satisfies this, it will hold in the standard interpretation; if not, it will fail.

This is another example which shows that putting $\mid\!\sim$ as ∇ in the object language can increase substantially the expressiveness of the logics.

Compare the situation to modal logic and Kripke semantics. In Kripke semantics, any evaluation of $\Box$ takes us to different points in the structure, and, if the relation is not transitive, what is beyond the first step is not visible from the origin. In preferential structures, this "hiding" is not possible, by the very definition of μ; we have to follow any chain as far as it goes.

(One of the authors had this idea when looking at the very interesting article "Mathematical modal logic: a view of its evolution" by Robert Goldblatt [Gol03] and begun to mull a bit over the short remark it contains on the continuum many modal logics indistinguishable by Kripke structures.)

1.3.7 Various Considerations on Abstract Semantics

We can think of preferential structures as elements attacking each other: if $x \prec y$, then x attacks y, so y cannot be minimal any more. Thus the nonmonotonicity.

In topology, e.g. things are different. Consider the open interval $(0, 1)$ and its closure $[0, 1]$. Sequences converging to 0 or 1 "defend" 0 or 1 – thus, the more elements there are, the bigger the closure will be, monotony. The open core has a similar property. The same is true for, e.g. Kripke structures for modal logic: If y is in relation R with x, then any time y is there, x can be reached: y defends x.

A neuron may have inhibitory (attackers) and excitatory (defending) inputs. In preferential models, we may need many attackers to destroy minimality of one model, provided this model occurs in several copies. Of course, in a neuron system, there are usually many attackers and many defenders, so we have here a rather complicated system. Abstractly, both defense and attack are combined in Gabbay's reactive diagrams; see [Gab04] and Sect. 1.4 (p. 21).

1.3.7.1 Set Functions

We can make a list of possible formal properties of such functions, which might include (U is the universe or base set)

$f(X) = X$, $f(X) \subseteq X$, $X \subseteq f(X)$
$f(X) = f(f(X))$, $f(X) \subseteq f(f(X))$, $f(f(X)) \subseteq f(X)$
$X \neq \emptyset \rightarrow f(X) \neq \emptyset$
$X \neq U \rightarrow f(X) \cap X = \emptyset$
$X \neq \emptyset \rightarrow f(X) \cap X \neq \emptyset$
$f(\mathbf{C}(X)) = \mathbf{C}(f(X))$
$f(X \cup Y) = f(X) \cup f(Y)$
$f(X \cup Y) = f(X)$ or $f(Y)$ or $f(X) \cup f(Y)$ (true in ranked structures)
$X \subseteq Y \rightarrow f(Y) \cap X \subseteq f(X)$ (attack, basic law of preferential structures)
$X \subseteq Y \rightarrow f(X) \subseteq f(Y)$ (defense, e.g. modal logic)
$f(X) \subseteq Y \subseteq X \rightarrow f(X) = f(Y)$ (smoothness), holds also for the open core of a set in topology
$X \subseteq Y \subseteq f(X) \rightarrow f(X) = f(Y)$ counterpart, holds for topological closure

(Note that the last two properties will also hold in all other situations where one chooses the biggest subset or smallest superset from a set of candidates.)

$$f\left(\bigcup \mathcal{X}\right) = \bigcup \{f(X) : X \in \mathcal{X}\}$$

etc.

In general, distance-based revision is a two-argument set function (in the AGM approach, it has only one argument):

$$M(K) \mid M(\phi) \subseteq M(\phi)$$

This is nonmonotonic in both arguments, as the result depends more on the "shape" of both model sets than on their size.

Counterfactuals (and update) are also two-argument set functions:

$M(\phi) \uparrow M(\psi)$ is defined as the set of ψ-models closest to some individual ϕ-model; here the function is monotonic in the first argument and nonmonotonic in the second; – we collect for all $m \in M(\phi)$ the closest ψ-models.

1.3.8 A Comparison with Reiter Defaults

The meaning of Reiter defaults differs from that of preferential structures in a number of aspects.

(1) The simple (Reiter) default "normally ϕ" does not only mean that in normal cases ϕ holds, but also that if ψ holds, then normally also $\phi \wedge \psi$ holds. It thus inherits "normally ϕ" down on subsets.
(2) Of course, this is itself a default rule, as we might have that for ψ-cases, normally $\neg\phi$ holds. But this is a meta-default.
(3) Defaults can also be concatenated, if normally ϕ holds, and normally, if ϕ holds, then also ψ holds; we conclude that normally $\phi \wedge \psi$ holds. Again, this is a default rule.

Thus, Reiter defaults give us (at least) three levels of certainty: classical information, the information directly expressed by defaults, and the information concluded by the usual treatment of defaults.

1.4 IBRS

1.4.1 Definition and Comments

Definition 1.4.1

(1) An *information bearing binary relation frame IBR* has the form $(S, \mathfrak{R})$, where S is a non empty set and $\mathfrak{R}$ is a subset of S_ω, where S_ω is defined by induction as follows:

(1.1) $S_0 := S$.
(1.2) $S_{n+1} := S_n \cup (S_n \times S_n)$.
(1.3) $S_\omega = \bigcup\{S_n : n \in \omega\}$.

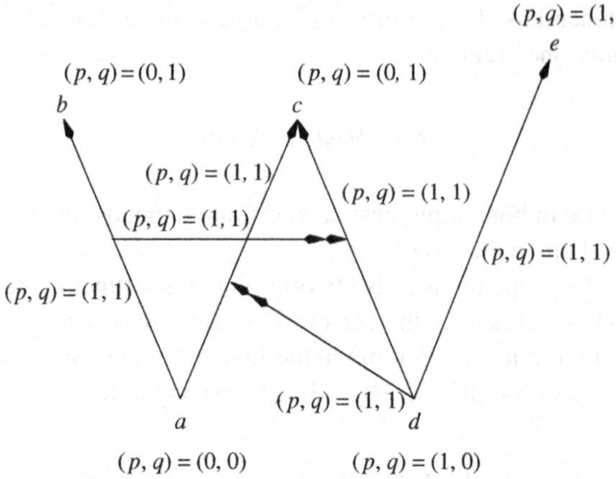

Diagram 1.4.1 A simple example of an information bearing system

We call elements from S *points* or *nodes* and elements from $\mathfrak{R}$ *arrows*. Given $(S, \mathfrak{R})$, we also set $\boldsymbol{P}((S, \mathfrak{R})) := S$ and $\boldsymbol{A}((S, \mathfrak{R})) := \mathfrak{R}$.

If α is an arrow, the origin and destination of α are defined as usual, and we write $\alpha : x \to y$ when x is the origin and y the destination of the arrow α. We also write $o(\alpha)$ and $d(\alpha)$ for the origin and destination of α.

(2) Let Q be a set of atoms and $\boldsymbol{L}$ be a set of labels (usually $\{0, 1\}$ or $[0, 1]$). An *information assignment* h on $(S, \mathfrak{R})$ is a function $h : Q \times \mathfrak{R} \to \boldsymbol{L}$.

(3) An *information bearing system* $IBRS$ has the form $(S, \mathfrak{R}, h, Q, \boldsymbol{L})$, where $S, \mathfrak{R}, h, Q, \boldsymbol{L}$ are as above.

See Diagram 1.4.1 (p. 23) for an illustration.

We have here

$S = \{a, b, c, d, e\}$.
$\mathfrak{R} = S \cup \{(a, b), (a, c), (d, c), (d, e)\} \cup \{((a, b), (d, c)), (d, (a, c))\}$.
$Q = \{p, q\}$.

The values of h for p and q are as indicated in the diagram. For example, $h(p, (d, (a, c))) = 1$.

Comment 1.4.1 The elements in Diagram 1.4.1 (p. 22) can be interpreted in many ways, depending on the area of application.

(1) The points in S can be interpreted as possible worlds, or as nodes in an argumentation network or nodes in a neural net or states, etc.
(2) The direct arrows from nodes to nodes can be interpreted as accessibility relation, attack or support arrows in an argumentation networks, connection in a neural nets, a preferential ordering in a nonmonotonic model, etc.

(3) The labels on the nodes and arrows can be interpreted as fuzzy values in the accessibility relation or weights in the neural net or strength of arguments and their attack in argumentation nets, or distances in a counterfactual model, etc.
(4) The double arrows can be interpreted as feedback loops to nodes or to connections, or as reactive links changing the system which are activated as we pass between the nodes.

Thus, IBRS can be used as a source of information for various logics based on the atoms in Q. We now illustrate by listing several such logics.

Modal Logic

One can consider the figure as giving rise to two modal logic models. One with actual world a and one with d, these being the two minimal points of the relation. Consider a language with $\Box q$. How do we evaluate $a \models \Box q$?

The modal logic will have to give an algorithm for calculating the values.

Say, we choose algorithm $\mathcal{A}_1$ for $a \models \Box q$, namely
$[\mathcal{A}_1(a, \Box q) = 1]$ iff for all $x \in S$ such that $a = x$ or $(a, x) \in \mathfrak{R}$ we have $h(q, x) = 1$. According to $\mathcal{A}_1$ we get that $\Box q$ is false at a. $\mathcal{A}_1$ gives rise to a T-modal logic. Note that the reflexivity is not anchored at the relation $\mathfrak{R}$ of the network but in the algorithm $\mathcal{A}_1$ in the way we evaluate. We say $(S, \mathfrak{R}, \ldots) \models \Box q$ iff $\Box q$ holds in all minimal points of $(S, \mathfrak{R})$.

For orderings without minimal points we may choose a subset of distinguished points.

Nonmonotonic Deduction

We can ask whether $p \mid\!\sim q$ according to algorithm $\mathcal{A}_2$ defined below. $\mathcal{A}_2$ says that $p \mid\!\sim q$ holds iff q holds in all minimal models of p. Let us check the value of $\mathcal{A}_2$ in this case:

Let $S_p = \{s \in S \mid h(p, s) = 1\}$. Thus $S_p = \{d, e\}$.
The minimal points of S_p are $\{d\}$. Since $h(q, d) = 0$, we have that $p \not\mid\!\sim q$.

Note that in the cases of modal logic and nonmonotonic logic, we ignored the arrows $(d, (a, c))$ (i.e. the double arrow from d to the arrow (a, c)) and the h values to arrows. These values do not play a part in the traditional modal or nonmonotonic logic. They do play a part in other logics. The attentive reader may already suspect that we have here an opportunity for generalization of, say, nonmonotonic logic, by giving a role to arrow annotations.

Argumentation Nets

Here the nodes of S are interpreted as arguments. The atoms $\{p, q\}$ can be interpreted as types of arguments and the arrows, e.g. $(a, b) \in \mathfrak{R}$ as indicating that the argument a is attacking the argument b.

So, for example, let

a = We must win votes.
b = Death sentence for murderers.
c = We must allow abortion for teenagers.
d = Bible forbids taking of life.
q = The argument is a social argument.
p = The argument is a religious argument.
$(d, (a, c))$ = There should be no connection between winning votes and abortion.
$((a, b), (d, c))$ = If we attack the death sentence in order to win votes, then we must stress (attack) that there should be no connection between religion (Bible) and social issues.

Thus we have according to this model that supporting abortion can lose votes. The argument for abortion is a social one and the argument from the Bible against it is a religious one.

We can extract information from this IBRS using two algorithms. The modal logic one can check whether, for example, every social argument is attacked by a religious argument. The answer is no, since the social argument b is attacked only by a which is not a religious argument.

We can also use algorithm $\mathcal{A}_3$ (following Dung) to extract the winning arguments of this system. The arguments a and d are winning since they are not attacked. d attacks the connection between a and c (i.e. stops a attacking c). The attack of a on b is successful and so b is out. However, the arrow (a, b) attacks the arrow (d, c). So c is not attacked at all as both arrrows leading into it are successfully eliminated. So c is in. e is out because it is attacked by d. So the winning arguments are $\{a, c, d\}$.

In this model, we ignore the annotations on arrows. To be consistent in our mathematics, we need to say that h is a partial function on $\mathfrak{R}$. The best way is to give more specific definition on IBRS to make it suitable for each logic. See also [Gab08c] and [BGW05].

Counterfactuals

The traditional semantics for counterfactuals involves closeness of worlds. The clause $y \models p \hookrightarrow q$, where $\hookrightarrow$ is a counterfactual implication is that q holds in all worlds y' "near enough" to y in which p holds. So if we interpret the annotation on arrows as distances, then we can define "near" as distance ≤ 2, we get: $a \models p \hookrightarrow q$ iff in all worlds of p-distance ≤ 2 if p holds so does q. Note that the distance depends on p.

In this case, we get that $a \models p \hookrightarrow q$ holds. The distance function can also use the arrows from arrows to arrows, etc. There are many opportunities for generalization in our IBRS setup.

Intuitionistic Persistence

We can get an intuitionistic Kripke model out of this IBRS by letting, for $t, s \in S$, $t\rho_0 s$ iff $t = s$ or $[tRs \wedge \forall q \in Q(h(q,t) \leq h(q,s))]$. We get that

$$[r_0 = \{(y, y) \mid y \in S\} \cup \{(a, b), (a, c), (d, e)\}.]$$

Let ρ be the transitive closure of ρ_0. Algorithm $\mathcal{A}_4$ evaluates $p \Rightarrow q$ in this model, where $\Rightarrow$ is intuitionistic implication.

$\mathcal{A}_4$: $p \Rightarrow q$ holds at the IBRS iff $p \Rightarrow q$ holds intuitionistically at every ρ-minimal point $of(S, \rho)$.

1.4.2 The Power of IBRS

We now show how a number of logics fit into our general picture of IBRS.

(1) Nonmonotonic logics in the form of preferential logics

There are only arrows from nodes to nodes, and they are unlabelled. The nodes are classical models, and as such all propositional variables of the base language are given a value from $\{0, 1\}$.

The structure is used as described above, i.e. the R-minimal models of a formula or theory are considered.

(2) Theory revision

In the full case, i.e. where the left hand side can change, nodes are again classical models, arrows exist only between nodes, and express by their label the distance between nodes. Thus, there is just one (dummy) p and a real value as label. In the AGM situation, where the left hand side is fixed, nodes are classical models (on the right) or sets thereof (on the left), arrows go from sets of models to models, and express again distance from a set (the K-models in AGM notation) to a model (of the new formula ϕ).

The structure is used by considering the closest ϕ-models.

The framework is sufficiently general to express revision also differently: Nodes are pairs of classical models, and arrows express that in pair (a, b) the distance from a to b is smaller than the distance in the pair (a', b').

(3) Theory update

As developments of length 2 can be expressed by a binary relation and the distance associated, we can – at least in the simple case – proceed analogously to the first revision situation. It seems, however, more natural to consider as nodes threads of developments, i.e. sequences of classical models, as arrows comparisons between such threads, i.e. unlabelled simple arrows only, expressing that one thread is more natural or likely than another.

The evaluation is then by considering the "best" threads under above comparison and taking a projection on the desired coordinate (i.e. classical model). The result is then the theory defined by these projections.

(4) Deontic logic

Just as for preferential logics.

(5) The logic of counterfactual conditionals

Again, we can compare pairs (with same left element) as above, or, alternatively, compare single models with respect to distance from a fixed other model. This would give arrows with indices, which stand for this other model.

Evaluation will then be as usual, taking the closest ϕ-models, and examining whether ψ holds in them.

(6) Modal logic

Nodes are classical models, and thus have the usual labels, arrows are unlabelled, and only between nodes, and express reachability.

For evaluation, starting from some point, we collect all reachable other models, perhaps adding the point of departure.

(7) Intuitionistic logic

Just as for modal logic.

(8) Inheritance systems

Nodes are properties (or sets of models), arrows come in two flavours, positive and negative, and exist between nodes only.

The evaluation is relatively complicated, and the subject of ongoing discussion.

(9) Argumentation theory

There is no unique description of an argumentation system as an IBRS. For instance, an inheritance system is an argumentation system, so we can describe such a system as detailed above. But an argument can also be a deontic statement, as we saw in the first part of this introduction, and a deontic statement can be described as an IBRS itself. Thus, a node can be, under finer granularity, itself an IBRS. Labels can describe the type of argument (social, etc.) or its validity, etc.

1.4.3 Abstract Semantics for IBRS and Its Engineering Realization

1.4.3.1 Introduction

We give here a rough outline of a formal semantics for IBRS. It consists more of some hints where difficulties are, than of a finished construction. Still, the authors feel that this is sufficient to complete the work, and, on the other hand, our remarks might be useful to those who intend to finalize the construction.

(1) Nodes and arrows

As we may have counterarguments not only against nodes, but also against arrows, they must be treated basically the same way, i.e. in some way there has to be a positive, but also a negative influence on both. So arrows cannot just be concatenation between the contents of nodes.

We will differentiate between nodes and arrows by labelling arrows in addition with a time delay. We see nodes as situations, where the output is computed

instantaneously from the input, whereas arrows describe some "force" or "mechanism" which may need some time to "compute" the result from the input. Consequently, if α is an arrow and β an arrow pointing to α, then it should point to the input of α, i.e. before the time lapse. Conversely, any arrow originating in α should originate after the time lapse.
Apart from this distinction, we will treat nodes and arrows the same way, so the following discussion will apply to both – which we call just "objects".

(2) Defeasibility
The general idea is to code each object, say X, by $I(X) : U(X) \to C(X)$: If $I(X)$ holds then, unless $U(X)$ holds, consequence $C(X)$ will hold. (We adopted Reiter's notation for defaults, as IBRS have common points with the former.)
The situation is slightly more complicated, as there can be several counterarguments, so $U(X)$ really is an "or". Likewise, there can be several supporting arguments, so $I(X)$ also is an "or".
A counterargument must not always be an argument against a specific supporting argument, but it can be. Thus, we should admit both possibilities. As we can use arrows to arrows, the second case is easy to treat (as is the dual, a supporting argument can be against a specific counterargument). How do we treat the case of unspecific pro- and counterarguments? Probably the easiest way is to adopt Dung's idea: an object is in, if it has at least one support and no counterargument – see [Dun95]. Of course, other possibilities may be adopted, counting, use of labels, etc., but we just consider the simple case here.

(3) Labels
In the general case, objects stand for some kind of defeasible transmission. We may in some cases see labels as restricting this transmission to certain values. For instance, if the label is $p = 1$ and $q = 0$, then the p-part may be transmitted and the q-part not. Thus, a transmission with a label can sometimes be considered as a family of transmissions, and which ones are active is indicated by the label.

Example 1.4.1 In fuzzy Kripke models, labels are elements of $[0, 1]$. $p = 0.5$ as label for a node m' which stands for a fuzzy model means that the value of p is 0.5. $p = 0.5$ as label for an arrow from m to m' means that p is transmitted with value 0.5. Thus, when we look from m to m', we see p with value $0.5 \times 0.5 = 0.25$. So, we have $\Diamond p$ with value 0.25 at m – if, e.g. m, m' are the only models.

(4) Putting things together
If an arrow leaves an object, the object's output will be connected to the (only) positive input of the arrow. (An arrow has no negative inputs from objects it leaves.) If a positive arrow enters an object, it is connected to one of the positive inputs of the object, analogously for negative arrows and inputs.
When labels are present, they are transmitted through some operation.

In slightly more formal terms, we have

Definition 1.4.2 In the most general case, objects of IBRS have the form: $(\langle I_1, L_1\rangle, \ldots, \langle I_n, L_n\rangle) : (\langle U_1, L'_1\rangle, \ldots, \langle U_n, L'_n\rangle)$, where the L_i, L'_i are labels and the I_i, U_i

might be just truth values, but can also be more complicated, a (possibly infinite) sequence of some values. Connected objects have, of course, corresponding such sequences. In addition, the object X has a criterion for each input, whether it is valid or not (in the simple case, this will just be the truth value "true"). If there is at least one positive valid input I_i, and no valid negative input U_i, then the output $C(X)$ and its label are calculated on the basis of the valid inputs and their labels. If the object is an arrow, this will take some time, t, otherwise, this is instantaneous.

Evaluating a Diagram

An evaluation is relative to a fixed input, i.e. some objects will be given certain values, and the diagram is left to calculate the others. It may well be that it oscillates, i.e. shows a cyclic behaviour. This may be true for a subset of the diagram or the whole diagram. If it is restricted to an unimportant part, we might neglect this. Whether it oscillates or not can also depend on the time delays of the arrows (see Example 1.4.2 (p. 29)).

We therefore define for a diagram Δ

$\alpha \mid\!\sim_{\Delta} \beta$ iff

(a) α is a (perhaps partial) input – where the other values are set "not valid"
(b) β is a (perhaps partial) output
(c) after sometime, β is stable, i.e. all still possible oscillations do not affect β
(d) the other possible input values do not matter, i.e. whatever the input, the result is the same.

In the cases examined here more closely, all input values will be defined.

1.4.3.2 A Circuit Semantics for Simple IBRS Without Labels

It is natural to implement IBRS by (modified) logical circuits. The nodes will be implemented by subcircuits which can store information and the arcs by connections between them. As connections can connect connections, connections will not just be simple wires. The objective of the present discussion is to warn the reader that care has to be taken about the temporal behaviour of such circuits, especially when feedback is allowed.

All details are left to those interested in a realization.

Background: It is standard to implement the usual logical connectives by electronic circuits. These components are called gates. Circuits with feedback sometimes show undesirable behaviour when the initial conditions are not specified. (When we switch a circuit on, the outputs of the individual gates can have arbitrary values.) The technical realization of these initial values shows the way to treat defaults. The initial values are set via resistors (in the order of 1 kΩ) between the point in the circuit we want to initialize and the desired tension (say 0 V for false, 5 V for true). They are called pull-down or pull-up resistors (for default 0 or 5 V). When a "real" result comes in, it will override the tension applied via the resistor.

Table 1.2 Oscillating variant

	$In1$	$In2$	$A1$	$A2$	$A3$	$A4$	$Out1$	$Out2$	
1:	T	F	F	F	F	F	F	F	
2:	T	F	F	F	F	F	T	T	
3:	T	F	T	F	T	T	T	T	
4:	T	F	T	F	T	T	F	F	
5:	T	F	F	F	T	F	F	F	Oscillation starts
6:	T	F	F	F	F	F	F	T	
7:	T	F	F	F	T	F	T	T	
8:	T	F	T	F	T	T	F	T	
9:	T	F	F	F	T	F	F	F	Back to start of oscillation

Table 1.3 No oscillation

	$In1$	$In2$	$A1$	$A2$	$A3$	$A4$	$Out1$	$Out2$	
1:	T	F	F	F	F	F	F	F	
2:	T	F	F	F	F	F	T	T	
3:	T	F	F	F	T	T	T	T	
4:	T	F	T	F	T	T	F	F	
5:	T	F	T	F	T	F	F	F	
6:	T	F	F	F	T	F	F	T	Stable state reached
7:	T	F	F	F	T	F	F	T	
8:	T	F	F	F	T	F	F	T	

Closer inspection reveals that we have here a three-level default situation: The initial value will be the weakest, which can be overridden by any "real" signal, but a positive argument can be overridden by a negative one. Thus, the biggest resistor will be for the initialization, the smaller one for the supporting arguments, and the negative arguments have full power. Technical details will be left to the experts.

We now give an example which shows that the delays of the arrows can matter. In one situation, a stable state is reached; in another, the circuit begins to oscillate (Tables 1.2 and 1.3).

Example 1.4.2 (In engineering terms, this is a variant of a JK flip-flop with $R * S = 0$, a circuit with feedback.)

We have eight measuring points.

$In1, In2$ are the overall input, $Out1, Out2$ the overall output, $A1, A2, A3, A4$ are auxiliary internal points. All points can be true or false.

The logical structure is as follows:

$$A1 = In1 \wedge Out1, \ A2 = In2 \wedge Out2,$$
$$A3 = A1 \vee Out2, \ A4 = A2 \vee Out1,$$
$$Out1 = \neg A3, \ Out2 = \neg A4.$$

Thus, the circuit is symmetrical, with $In1$ corresponding to $In2$, $A1$ to $A2$, $A3$ to $A4$, $Out1$ to $Out2$.

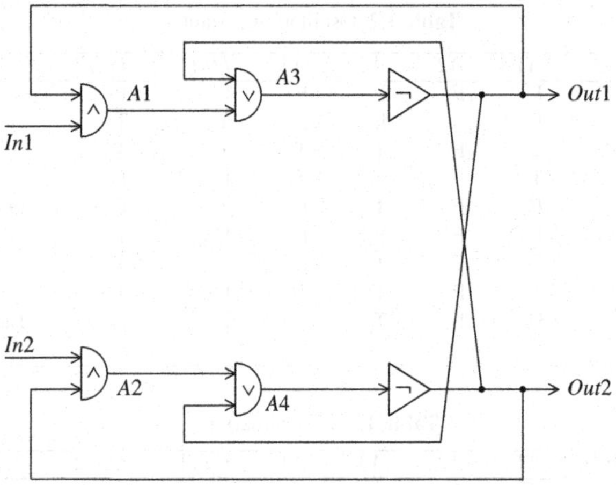

Diagram 1.4.2 Gate semantics

The input is held constant. See Diagram 1.4.2 (p. 30).

We suppose that the output of the individual gates is present n time slices after the input was present. In the first circuit n will be equal to 1 for all gates; in the second circuit equal to 1 for all but the AND gates, which will take two time slices. Thus, in both cases, e.g. $Out1$ at time t will be the negation of $A3$ at time $t-1$. In the first case, $A1$ at time t will be the conjunction of $In1$ and $Out1$ at time $t-1$, and in the second case the conjunction of $In1$ and $Out1$ at time $t-2$.

We initialize $In1$ as true, all others as false. (The initial value of $A3$ and $A4$ does not matter, the behaviour is essentially the same for all such values.)

The first circuit will oscillate with a period of 4; the second circuit will go to a stable state.

We have the following transition tables (time slice shown at left):

Circuit 1, $delay = 1$ everywhere:
Circuit 2, $delay = 1$ everywhere, except for AND with $delay = 2$:

(Thus, $A1$ and $A2$ are held at their initial value up to time 2, then they are calculated using the values of time $t-2$.)

Note that state 6 of circuit 2 is also stable in circuit 1, but it is never reached in that circuit.

Chapter 2
Basic Definitions and Results

2.1 Algebraic Definitions

Notation 2.1.1 We use sometimes FOL as abbreviation for first-order logic and NML for nonmonotonic logic. To avoid Latex complications in bigger expressions, we replace $\widetilde{xxxxx}$ by $\widehat{xxxxx}$.

Definition 2.1.1

(1) We use $\mathcal{P}$ to denote the power set operator, $\Pi\{X_i : i \in I\} := \{g : g : I \rightarrow \bigcup\{X_i : i \in I\}, \forall i \in I, g(i) \in X_i\}$ is the general cartesian product, $card(X)$ shall denote the cardinality of X, and V the set-theoretic universe we work in – the class of all sets. Given a set of pairs $\mathcal{X}$, and a set X, we denote by $\mathcal{X}\lceil X := \{\langle x, i\rangle \in \mathcal{X} : x \in X\}$. When the context is clear, we will sometimes simply write X for $\mathcal{X}\lceil X$. (The intended use is for preferential structures, where x will be a point (intention: a classical propositional model) and i an index, permitting copies of logically identical points.)

(2) $A \subseteq B$ will denote that A is a subset of B or equal to B, and $A \subset B$ that A is a proper subset of B, likewise for $A \supseteq B$ and $A \supset B$.
Given some fixed set U we work in, and $X \subseteq U$, then $\boldsymbol{C}(X) := U - X$.

(3) If $\mathcal{Y} \subseteq \mathcal{P}(X)$ for some X, we say that $\mathcal{Y}$ satisfies
($\cap$) iff it is closed under finite intersections,
($\bigcap$) iff it is closed under arbitrary intersections,
($\cup$) iff it is closed under finite unions,
($\bigcup$) iff it is closed under arbitrary unions,
($\boldsymbol{C}$) iff it is closed under complementation,
($-$) iff it is closed under set difference.

(4) We will sometimes write $A = B \parallel C$ for $A = B$, or $A = C$, or $A = B \cup C$.

We make ample and tacit use of the axiom of choice.

Definition 2.1.2 $\prec^*$ will denote the transitive closure of the relation $\prec$. If a relation $<$, $\prec$, or similar is given, $a \perp b$ will express that a and b are $<-$ (or $\prec -$) incomparable – context will tell. Given any relation $<$, $\leq$ will stand for $<$ or $=$; conversely, given $\leq$, $<$ will stand for $\leq$, but not $=$, similarly for $\prec$, etc.

D.M. Gabbay, K. Schlechta, *Logical Tools for Handling Change in Agent-Based Systems*, Cognitive Technologies, DOI 10.1007/978-3-642-04407-6_2,

Definition 2.1.3 A child (or successor) of an element x in a tree t will be a direct child in t. A child of a child, etc. will be called an indirect child. Trees will be supposed to grow downwards, so the root is the top element.

Definition 2.1.4 A subsequence $\sigma_i : i \in I \subseteq \mu$ of a sequence $\sigma_i : i \in \mu$ is called cofinal, iff for all $i \in \mu$ there is $i' \in I$ $i \leq i'$. Given two sequences σ_i and τ_i of the same length, their Hamming distance is the quantity of i where they differ.

Definition 2.1.5 Let $\mathcal{Y} \subseteq \mathcal{P}(Z)$ be given and closed under arbitrary intersections.

(1) For $A \subseteq Z$, let $\overbrace{A} := \bigcap\{X \in \mathcal{Y} : A \subseteq X\}$.
(2) For $B \in \mathcal{Y}$, we call $A \subseteq B$ a small subset of B iff there is no $X \in \mathcal{Y}$ such that $B - A \subseteq X \subset B$.

(Context will disambiguate from other uses of "small".)

Intuitively, Z is the set of all models for $\mathcal{L}$, $\mathcal{Y}$ is $\mathbf{D}_{\mathcal{L}}$, and $\overbrace{A} = M(Th(A))$ is the intended application – $Th(A)$ is the set of formulas which hold in all $a \in A$ and $M(Th(A))$ is the set of models of $Th(A)$. Note that then $\overbrace{\emptyset} = \emptyset$.

Fact 2.1.1

(1) If $\mathcal{Y} \subseteq \mathcal{P}(Z)$ is closed under arbitrary intersections and finite unions, $Z \in \mathcal{Y}$, $X, Y \subseteq Z$, then the following hold:
$(Cl\cup)$ $\overbrace{X \cup Y} = \overbrace{X} \cup \overbrace{Y}$,
$(Cl\cap)$ $\overbrace{X \cap Y} \subseteq \overbrace{X} \cap \overbrace{Y}$, but usually not conversely,
$(Cl-)$ $\overbrace{A} - \overbrace{B} \subseteq \overbrace{A - B}$,
$(Cl=)$ $X = Y \Rightarrow \overbrace{X} = \overbrace{Y}$, but not conversely,
$(Cl\subseteq 1)$ $\overbrace{X} \subseteq Y \Rightarrow X \subseteq Y$, but not conversely,
$(Cl\subseteq 2)$ $X \subseteq \overbrace{Y} \Rightarrow \overbrace{X} \subseteq \overbrace{Y}$.
(2) If, in addition, $X \in \mathcal{Y}$ and $\boldsymbol{C}X := Z - X \in \mathcal{Y}$, then the following two properties hold too:
$(Cl\cap+)$ $\overbrace{A} \cap X = \overbrace{A \cap X}$,
$(Cl-+)$ $\overbrace{A} - X = \overbrace{A - X}$.
(3) In the intended application, i.e. $\overbrace{A} = M(Th(A))$, the following hold:
(3.1) $Th(X) = Th(\overbrace{X})$,
(3.2) Even if $A = \overbrace{A}$, $B = \overbrace{B}$, it is not necessarily true that $\overbrace{A - B} \subseteq \overbrace{A} - \overbrace{B}$.

Proof $(Cl=)$, $(Cl\subseteq 1)$, $(Cl\subseteq 2)$, and (3.1) are trivial.

$(Cl\cup)$ Let $\mathcal{Y}(U) := \{X \in \mathcal{Y} : U \subseteq X\}$. If $A \in \mathcal{Y}(X \cup Y)$, then $A \in \mathcal{Y}(X)$ and $A \in \mathcal{Y}(Y)$, so $\overbrace{X \cup Y} \supseteq \overbrace{X} \cup \overbrace{Y}$. If $A \in \mathcal{Y}(X)$ and $B \in \mathcal{Y}(Y)$, then $A \cup B \in \mathcal{Y}(X \cup Y)$, so $\overbrace{X \cup Y} \subseteq \overbrace{X} \cup \overbrace{Y}$.

($Cl\cap$) Let $X', Y' \in \mathcal{Y}$, $X \subseteq X'$, $Y \subseteq Y'$, then $X \cap Y \subseteq X' \cap Y'$, so $\widehat{X \cap Y} \subseteq \widehat{X} \cap \widehat{Y}$. For the converse, set $X := M_{\mathcal{L}} - \{m\}$, $Y := \{m\}$ in Example 2.2.1 (p. 35). ($M_{\mathcal{L}}$ is the set of all models of the language $\mathcal{L}$.)

($Cl-$) Let $A - B \subseteq X \in \mathcal{Y}$, $B \subseteq Y \in \mathcal{Y}$, so $A \subseteq X \cup Y \in \mathcal{Y}$. Let $x \not\in \widehat{B}$ $\Rightarrow \exists Y \in \mathcal{Y}(B \subseteq Y, x \not\in Y)$, $x \not\in \widehat{A - B} \Rightarrow \exists X \in \mathcal{Y}(A - B \subseteq X, x \not\in X)$, so $x \not\in X \cup Y$, $A \subseteq X \cup Y$, so $x \not\in \widehat{A}$. Thus, $x \not\in \widehat{B}$, $x \not\in \widehat{A - B} \Rightarrow x \not\in \widehat{A}$ or $x \in \widehat{A} - \widehat{B} \Rightarrow x \in \widehat{A - B}$.

($Cl\cap+$) $\widehat{A} \cap X \supseteq \widehat{A \cap X}$ by ($Cl\cap$). For "$\subseteq$": Let $A \cap X \subseteq A' \in \mathcal{Y}$, then by closure under ($\cup$), $A \subseteq A' \cup \mathbf{C}X \in \mathcal{Y}$, $(A' \cup \mathbf{C}X) \cap X \subseteq A'$. So $\widehat{A} \cap X \subseteq \widehat{A \cap X}$.

($Cl-+$) $\widehat{A - X} = \widehat{A \cap \mathbf{C}X} = \widehat{A} \cap \mathbf{C}X = \widehat{A} - X$ by ($Cl\cap+$).

(3.2) Set $A := M_{\mathcal{L}}$, $B := \{m\}$ for $m \in M_{\mathcal{L}}$ arbitrary, $\mathcal{L}$ infinite. So $A = \widehat{A}$, $B = \widehat{B}$, but $\widehat{A - B} = A \neq A - B$. □

Definition 2.1.6 Given $x, y \in \Sigma$, a set of sequences over an index set I, the Hamming distance comes in two flavours:

$d_s(x, y) := \{i \in I : x(i) \neq y(i)\}$, the set variant and

$d_c(x, y) := card(d_s(x, y))$, the counting variant.

We define $d_s(x, y) \leq d_s(x', y')$ iff $d_s(x, y) \subseteq d_s(x', y')$. Thus, s-distances are not always comparable.

There are straightforward generalizations of the counting variant:

We can also give different importance to different i in the counting variant, so, e.g. $d_c(\langle x, x' \rangle, \langle y, y' \rangle)$ might be 1 if $x \neq y$ and $x' = y'$, but 2 if $x = y$ and $x' \neq y'$.

If the $x \in \Sigma$ may have more than two different values, then a varying individual distance may also reflect to the distances in Σ. So, if $d(x(i), x'(i)) < d(x(i), x''(i))$, then (the rest being equal), we may have $d(x, x') < d(x, x'')$.

Fact 2.1.2

(1) If the $x \in \Sigma$ have only two values, say TRUE and FALSE, then $d_s(x, y) = \{i \in I : x(i) = TRUE\} \triangle \{i \in I : y(i) = TRUE\}$, where $\triangle$ is the symmetric set difference.

(2) d_c has the normal addition, set union takes the role of addition for d_s, $\emptyset$ takes the role of 0 for d_s, both are distances in the following sense:

(2.1) $d(x, y) = 0$ iff $x = y$,

(2.2) $d(x, y) = d(y, x)$,

(2.3) the triangle inequality holds for the set variant in the form $d_s(x, z) \subseteq d_s(x, y) \cup d_s(y, z)$.

Proof (2.3) If $i \not\in d_s(x, y) \cup d_s(y, z)$, then $x(i) = y(i) = z(i)$, so $x(i) = z(i)$ and $i \not\in d_s(x, z)$.

The others are trivial. □

2.2 Basic Logical Definitions

Definition 2.2.1

(1) We work here in a classical propositional language $\mathcal{L}$, a theory T will be an arbitrary set of formulas. Formulas will often be named ϕ, ψ, etc., theories T, S, etc.
$v(\mathcal{L})$ will be the set of propositional variables of $\mathcal{L}$.
$M_{\mathcal{L}}$ will be the set of (classical) models for $\mathcal{L}$, $M(T)$ or M_T is the set of models of T, likewise $M(\phi)$ for a formula ϕ.

(2) $\boldsymbol{D}_{\mathcal{L}} := \{M(T) : T \text{ a theory in } \mathcal{L}\}$, the set of *definable* model sets.
Note that, in classical propositional logic, $\emptyset, M_{\mathcal{L}} \in \boldsymbol{D}_{\mathcal{L}}$, $\boldsymbol{D}_{\mathcal{L}}$ contains singletons, is closed under arbitrary intersections and finite unions.
An operation $f : \mathcal{Y} \to \mathcal{P}(M_{\mathcal{L}})$ for $\mathcal{Y} \subseteq \mathcal{P}(M_{\mathcal{L}})$ is called *definability preserving*, (dp) or (μdp) in short, iff for all $X \in \boldsymbol{D}_{\mathcal{L}} \cap \mathcal{Y}$ $f(X) \in \boldsymbol{D}_{\mathcal{L}}$.
We will also use (μdp) for binary functions $f : \mathcal{Y} \times \mathcal{Y} \to \mathcal{P}(M_{\mathcal{L}})$ – as needed for theory revision – with the obvious meaning.

(3) $\vdash$ will be classical derivability, and
$\overline{T} := \{\phi : T \vdash \phi\}$, the closure of T under $\vdash$.

(4) $Con(.)$ will stand for classical consistency, so $Con(\phi)$ will mean that ϕ is classical consistent, likewise for $Con(T)$. $Con(T, T')$ will stand for $Con(T \cup T')$, etc.

(5) Given a consequence relation $\mid\!\sim$, we define
$\overline{\overline{T}} := \{\phi : T \mid\!\sim \phi\}$.
(There is no fear of confusion with $\overline{T}$, as it is just not useful to close twice under classical logic.)

(6) $T \vee T' := \{\phi \vee \phi' : \phi \in T, \phi' \in T'\}$.

(7) If $X \subseteq M_{\mathcal{L}}$, then $Th(X) := \{\phi : X \models \phi\}$, likewise for $Th(m)$, $m \in M_{\mathcal{L}}$. ($\models$ will usually be classical validity.)

We recollect and note:

Fact 2.2.1 Let $\mathcal{L}$ be a fixed propositional language, $\boldsymbol{D}_{\mathcal{L}} \subseteq X$, $\mu : X \to \mathcal{P}(M_{\mathcal{L}})$, for a $\mathcal{L}$-theory T. Suppose $\overline{\overline{T}} = Th(\mu(M_T))$, let T, T' be arbitrary theories, then:

(1) $\mu(M_T) \subseteq M_{\overline{\overline{T}}}$,
(2) $M_T \cup M_{T'} = M_{T \vee T'}$ and $M_{T \cup T'} = M_T \cap M_{T'}$,
(3) $\mu(M_T) = \emptyset \Leftrightarrow \bot \in \overline{\overline{T}}$.

If μ is definability preserving or $\mu(M_T)$ is finite, then the following also hold:

(4) $\mu(M_T) = M_{\overline{\overline{T}}}$,
(5) $T' \vdash \overline{\overline{T}} \Leftrightarrow M_{T'} \subseteq \mu(M_T)$,
(6) $\mu(M_T) = M_{T'} \Leftrightarrow \overline{T'} = \overline{\overline{T}}$.

Fact 2.2.2 Let $A, B \subseteq M_{\mathcal{L}}$.
Then $Th(A \cup B) = Th(A) \cap Th(B)$.

Proof $\phi \in Th(A \cup B) \Leftrightarrow A \cup B \models \phi \Leftrightarrow A \models \phi$ and $B \models \phi \Leftrightarrow \phi \in Th(A)$ and $\phi \in Th(B)$. □

Fact 2.2.3 Let $X \subseteq M_{\mathcal{L}}$, ϕ, ψ formulas.

(1) $X \cap M(\phi) \models \psi$ iff $X \models \phi \rightarrow \psi$.
(2) $X \cap M(\phi) \models \psi$ iff $M(Th(X)) \cap M(\phi) \models \psi$.
(3) $Th(X \cap M(\phi)) = \overline{Th(X) \cup \{\phi\}}$.
(4) $X \cap M(\phi) = \emptyset \Leftrightarrow M(Th(X)) \cap M(\phi) = \emptyset$.
(5) $Th(M(T) \cap M(T')) = \overline{T \cup T'}$.

Proof

(1) "$\Rightarrow$": $X = (X \cap M(\phi)) \cup (X \cap M(\neg\phi))$. If both parts hold $\neg\phi \vee \psi$, so $X \models \phi \rightarrow \psi$. "$\Leftarrow$": Trivial.
(2) $X \cap M(\phi) \models \psi$ (by (1)) iff $X \models \phi \rightarrow \psi$ iff $M(Th(X)) \models \phi \rightarrow \psi$ iff (again by (1)) $M(Th(X)) \cap M(\phi) \models \psi$.
(3) $\psi \in Th(X \cap M(\phi)) \Leftrightarrow X \cap M(\phi) \models \psi \Leftrightarrow_{(2)} M(Th(X) \cup \{\phi\}) = M(Th(X)) \cap M(\phi) \models \psi \Leftrightarrow Th(X) \cup \{\phi\} \vdash \psi$.
(4) $X \cap M(\phi) = \emptyset \Leftrightarrow X \models \neg\phi \Leftrightarrow M(Th(X)) \models \neg\phi \Leftrightarrow M(Th(X)) \cap M(\phi) = \emptyset$.
(5) $M(T) \cap M(T') = M(T \cup T')$. □

Fact 2.2.4 If $X = M(T)$, then $M(Th(X)) = X$.

Proof $X \subseteq M(Th(X))$ is trivial. $Th(M(T)) = \overline{T}$ is trivial by classical soundness and completeness. So $M(Th(M(T)) = M(\overline{T}) = M(T) = X$. □

Example 2.2.1 If $v(\mathcal{L})$ is infinite, and m any model for $\mathcal{L}$, then $M := M_{\mathcal{L}} - \{m\}$ is not definable by any theory T. (Proof: Suppose it were the case, and let ϕ hold in M, but not in m, so in m $\neg\phi$ holds, but as ϕ is finite, there is a model m' in M which coincides on all propositional variables of ϕ with m, so in m' $\neg\phi$ holds too, a contradiction.) Thus, in the infinite case, $\mathcal{P}(M_{\mathcal{L}}) \neq \boldsymbol{D}_{\mathcal{L}}$.

(There is also a simple cardinality argument, which shows that almost no model sets are definable, but it is not constructive and thus less instructive than above argument. We give it nonetheless: Let $\kappa := card(v(\mathcal{L}))$. Then there are κ many formulas, so 2^κ many theories, and thus 2^κ many definable model sets. But there are 2^κ many models, so $(2^\kappa)^\kappa$ many model sets.)

2.3 Basic Definitions and Results for Nonmonotonic Logics

The numbers in the first column "Corr.", meaning "Correspondence", refer to Proposition 2.3.1 (p. 45), published as Proposition 21 in [GS08c] Tables 2.1 and 2.2., those in the second column "Corr." to Proposition 3.2.1 (p. 69).

Sometimes, supplementary conditions are noted in the "Correspondence" columns. $-(\mu dp)$ stands for "without (μdp)", $T = \phi$ (or $T' = \phi$) stands for "T (or T') is equivalent to a formula".

Table 2.1 Logical rules, definitions, and connections I

Logical rule		Corr.	Model set	Corr.	Size rules
Basics					
(SC) Supraclassicality $\alpha \vdash \beta \Rightarrow \alpha \mid\sim \beta$ (REF) Reflexivity $T \cup \{\alpha\} \mid\sim \alpha$	(SC) $\overline{T} \subseteq \overline{\overline{T}}$	$\Rightarrow (4.1)$ $\Leftarrow (4.2)$	$(\mu \subseteq)$ $f(X) \subseteq X$	Trivial	(Opt)
(LLE) Left logical equivalence $\vdash \alpha \leftrightarrow \alpha', \alpha \mid\sim \beta \Rightarrow$ $\alpha' \mid\sim \beta$	(LLE) $\overline{T} = \overline{T'} \Rightarrow \overline{\overline{T}} = \overline{\overline{T'}}$				
(RW) Right weakening $\alpha \mid\sim \beta, \vdash \beta \to \beta' \Rightarrow$ $\alpha \mid\sim \beta'$	(RW) $T \mid\sim \beta, \vdash \beta \to \beta' \Rightarrow$ $T \mid\sim \beta'$			Trivial	(iM)
(wOR) $\alpha \mid\sim \beta, \alpha' \vdash \beta \Rightarrow$ $\alpha \vee \alpha' \mid\sim \beta$	(wOR) $\overline{\overline{T}} \cap \overline{T'} \subseteq \overline{\overline{T \vee T'}}$	$\Rightarrow (3.1)$ $\Leftarrow (3.2)$	(μwOR) $f(X \cup Y) \subseteq f(X) \cup Y$	$\Leftrightarrow (1)$	$(eM\mathcal{I})$
$(disjOR)$ $\alpha \vdash \neg\alpha', \alpha \mid\sim \beta,$ $\alpha' \mid\sim \beta \Rightarrow \alpha \vee \alpha' \mid\sim \beta$	$(disjOR)$ $\neg Con(T \cup T') \Rightarrow$ $\overline{\overline{T}} \cap \overline{\overline{T'}} \subseteq \overline{\overline{T \vee T'}}$	$\Rightarrow (2.1)$ $\Leftarrow (2.2)$	$(\mu disjOR)$ $X \cap Y = \emptyset \Rightarrow$ $f(X \cup Y) \subseteq f(X) \cup f(Y)$	$\Leftrightarrow (4)$	$(I \cup disj)$
(CP) Consistency preservation $\alpha \mid\sim \bot \Rightarrow \alpha \vdash \bot$	(CP) $T \mid\sim \bot \Rightarrow T \vdash \bot$	$\Rightarrow (5.1)$ $\Leftarrow (5.2)$	$(\mu\emptyset)$ $f(X) = \emptyset \Rightarrow X = \emptyset$	Trivial	(I_1)
			$(\mu\emptyset fin)$ $X \neq \emptyset \Rightarrow f(X) \neq \emptyset$ for finite X		(I_1)
	(AND_1) $\alpha \mid\sim \beta \Rightarrow \alpha \not\mid\sim \neg\beta$				(I_2)

Table 2.1 (continued)

Logical rule		Corr.	Model set	Corr.	Size rules
Basics					
	(AND_n) $\alpha \mid\sim \beta_1, \ldots, \alpha \mid\sim \beta_{n-1} \Rightarrow$ $\alpha \not\mid\sim (\neg\beta_1 \vee \ldots \vee \neg\beta_{n-1})$				(I_n)
(AND) $\alpha \mid\sim \beta, \alpha \mid\sim \beta' \Rightarrow$ $\alpha \mid\sim \beta \wedge \beta'$	(AND) $T \mid\sim \beta, T \mid\sim \beta' \Rightarrow$ $T \mid\sim \beta \wedge \beta'$			Trivial	(I_ω)
(CCL) Classical closure	(CCL) $\overline{\overline{T}}$ classically closed			Trivial	$(iM)+(I_\omega)$
(OR) $\alpha \mid\sim \beta, \alpha' \mid\sim \beta \Rightarrow$ $\alpha \vee \alpha' \mid\sim \beta$	(OR) $\overline{\overline{T}} \cap \overline{\overline{T'}} \subseteq \overline{\overline{T \vee T'}}$	$\Rightarrow (1.1)$ $\Leftarrow (1.2)$	(μOR) $f(X \cup Y) \subseteq f(X) \cup f(Y)$	$\Leftrightarrow (2)$	$(eM\mathcal{I})+(I_\omega)$
$\overline{\overline{\alpha \wedge \alpha'}} \subseteq \overline{\overline{\overline{\alpha}}} \cup \{\alpha'\}$	(PR) $\overline{\overline{T \cup T'}} \subseteq \overline{\overline{\overline{T}}} \cup T'$	$\Rightarrow (6.1)$ $\Leftarrow (\mu dp)+(\mu\subseteq)$ (6.2) $\not\Leftarrow -(\mu dp)$ (6.3) $\Leftarrow (\mu\subseteq)$ (6.4) $T' = \phi$	(μPR) $X \subseteq Y \Rightarrow$ $f(Y) \cap X \subseteq f(X)$	$\Leftrightarrow (3)$	$(eM\mathcal{I})+(I_\omega)$
		$\Leftarrow (6.5)$ $T' = \phi$	$(\mu PR')$ $f(X) \cap Y \subseteq f(X \cap Y)$		
(CUT) $T \mid\sim \alpha; T \cup \{\alpha\} \mid\sim \beta \Rightarrow$ $T \mid\sim \beta$	(CUT) $T \subseteq \overline{T'} \subseteq \overline{\overline{T}} \Rightarrow$ $\overline{\overline{T'}} \subseteq \overline{\overline{T}}$	$\Rightarrow (7.1)$ $\Leftarrow (7.2)$	(μCUT) $f(X) \subseteq Y \subseteq X \Rightarrow$ $f(X) \subseteq f(Y)$	$\Leftarrow (8.1)$ $\not\Leftarrow (8.2)$	$(eM\mathcal{I})+(I_\omega)$

Table 2.2 Logical rules, definitions and connections II

Logical rule		Corr.	Model set	Corr.	Size-rule
Cumulativity					
(wCM) $\alpha \mid\!\sim \beta, \alpha' \vdash \alpha, \alpha \wedge \beta \vdash \alpha' \Rightarrow$ $\alpha' \mid\!\sim \beta$				Trivial	$(e\mathcal{MF})$
(CM_2) $\alpha \mid\!\sim \beta, \alpha \mid\!\sim \beta' \Rightarrow \alpha \wedge \beta \not\mid\!\sim \neg\beta'$					(I_2)
(CM_n) $\alpha \mid\!\sim \beta_1, \ldots, \alpha \mid\!\sim \beta_n \Rightarrow$ $\alpha \wedge \beta_1 \wedge \ldots \wedge \beta_{n-1} \not\mid\!\sim \neg\beta_n$					(I_n)
(CM) Cautious monotony $\alpha \mid\!\sim \beta, \alpha \mid\!\sim \beta' \Rightarrow$ $\alpha \wedge \beta \mid\!\sim \beta'$	(CM) $T \subseteq \overline{T'} \subseteq \overline{\overline{T}} \Rightarrow$ $\overline{\overline{T}} \subseteq \overline{\overline{T'}}$	$\Rightarrow$ (8.1) $\Leftarrow$ (8.2)	(μCM) $f(X) \subseteq Y \subseteq X \Rightarrow$ $f(Y) \subseteq f(X)$	$\Leftrightarrow$ (5)	$(\mathcal{M}^+_\omega)(4)$
or $(ResM)$ Restricted monotony $T \mid\!\sim \alpha, \beta \Rightarrow T \cup \{\alpha\} \mid\!\sim \beta$		$\Rightarrow$ (9.1) $\Leftarrow$ (9.2)	$(\mu ResM)$ $f(X) \subseteq A \cap B \Rightarrow$ $f(X \cap A) \subseteq B$		
(CUM) Cumulativity $\alpha \mid\!\sim \beta \Rightarrow$ $(\alpha \mid\!\sim \beta' \Leftrightarrow \alpha \wedge \beta \mid\!\sim \beta')$	(CUM) $T \subseteq \overline{T'} \subseteq \overline{\overline{T}} \Rightarrow$ $\overline{\overline{T}} = \overline{\overline{T'}}$	$\Rightarrow$ (11.1) $\Leftarrow$ (11.2)	(μCUM) $f(X) \subseteq Y \subseteq X \Rightarrow$ $f(Y) = f(X)$	$\Leftarrow$ (9.1) $\not\Rightarrow$ (9.2)	$(e\mathcal{MI}) + (I_\omega) + (\mathcal{M}^+_\omega)(4)$
	$(\subseteq\supseteq)$ $T \subseteq \overline{\overline{T'}}, T' \subseteq \overline{\overline{T}} \Rightarrow$ $\overline{\overline{T'}} = \overline{\overline{T}}$	$\Rightarrow$ (10.1) $\Leftarrow$ (10.2)	$(\mu \subseteq\supseteq)$ $f(X) \subseteq Y, f(Y) \subseteq X \Rightarrow$ $f(X) = f(Y)$	$\Leftarrow$ (10.1) $\not\Rightarrow$ (10.2)	$(e\mathcal{MI}) + (I_\omega) + (e\mathcal{MF})$

Table 2.2 (continued)

Logical rule		Corr.	Model set	Corr.	Size-rule
Rationality					
$(RatM)$ Rational monotony	$(RatM)$	$\Rightarrow$ (12.1)	$(\mu RatM)$	$\Leftrightarrow$ (6)	$(\mathcal{M}^{++})$
$\alpha \mid\sim \beta, \alpha \mid\not\sim \neg\beta' \Rightarrow$	$Con(T \cup \overline{\overline{\overline{T'}}}), T \vdash T' \Rightarrow$	$\Leftarrow (\mu dp)$ (12.2)	$X \subseteq Y, X \cap f(Y) \neq \emptyset \Rightarrow$		
$\alpha \wedge \beta' \mid\sim \beta$	$\overline{\overline{\overline{T}}} \supseteq \overline{\overline{\overline{\overline{\overline{\overline{T'}}}} \cup T}}$	$\not\Leftarrow -(\mu dp)$ (12.3)	$f(X) \subseteq f(Y) \cap X$		
		$\Leftarrow T = \phi$ (12.4)			
	$(RatM =)$	$\Rightarrow$ (13.1)	$(\mu =)$		
	$Con(T \cup \overline{\overline{\overline{T'}}}), T \vdash T' \Rightarrow$	$\Leftarrow (\mu dp)$ (13.2)	$X \subseteq Y, X \cap f(Y) \neq \emptyset \Rightarrow$		
	$\overline{\overline{\overline{T}}} = \overline{\overline{\overline{\overline{\overline{\overline{T'}}}} \cup T}}$	$\not\Leftarrow -(\mu dp)$ (13.3)	$f(X) = f(Y) \cap X$		
		$\Leftarrow T = \phi$ (13.4)			
	$(Log =')$	$\Rightarrow$ (14.1)	$(\mu =')$		
	$Con(\overline{\overline{\overline{T'}}} \cup T) \Rightarrow$	$\Leftarrow (\mu dp)$ (14.2)	$f(Y) \cap X \neq \emptyset \Rightarrow$		
	$\overline{\overline{\overline{T \cup T'}}} = \overline{\overline{\overline{\overline{\overline{\overline{T'}}}} \cup T}}$	$\not\Leftarrow -(\mu dp)$ (14.3)	$f(Y \cap X) = f(Y) \cap X$		
		$\Leftarrow T = \phi$ (14.4)			
(DR)	$(Log \parallel)$	$\Rightarrow$ (15.1)	$(\mu \parallel)$		
$\alpha \vee \beta \mid\sim \gamma \Rightarrow$	$\overline{\overline{\overline{T \vee T'}}}$ is one of	$\Leftarrow$ (15.2)	$f(X \cup Y)$ is one of		
$\alpha \mid\sim \gamma$ or $\beta \mid\sim \gamma$	$\overline{\overline{\overline{T}}}$, or $\overline{\overline{\overline{T'}}}$, or $\overline{\overline{\overline{T}}} \cap \overline{\overline{\overline{T'}}}$ (by (CCL))		$f(X)$, $f(Y)$, or $f(X) \cup f(Y)$		
	$(Log\cup)$	$\Rightarrow (\mu \subseteq) + (\mu =)$ (16.1)	$(\mu\cup)$		
	$Con(\overline{\overline{\overline{T'}}} \cup T), \neg Con(\overline{\overline{\overline{T'}}} \cup \overline{\overline{\overline{T}}}) \Rightarrow$	$\Leftarrow (\mu dp)$ (16.2)	$f(Y) \cap (X - f(X)) \neq \emptyset \Rightarrow$		
	$\neg Con(\overline{\overline{\overline{T \vee T'}}} \cup T')$	$\not\Leftarrow -(\mu dp)$ (16.3)	$f(X \cup Y) \cap Y = \emptyset$		
	$(Log\cup')$	$\Rightarrow (\mu \subseteq) + (\mu =)$ (17.1)	$(\mu\cup')$		
	$Con(\overline{\overline{\overline{T'}}} \cup T), \neg Con(\overline{\overline{\overline{T'}}} \cup \overline{\overline{\overline{T}}}) \Rightarrow$	$\Leftarrow (\mu dp)$ (17.2)	$f(Y) \cap (X - f(X)) \neq \emptyset \Rightarrow$		
	$\overline{\overline{\overline{T \vee T'}}} = \overline{\overline{\overline{T}}}$	$\not\Leftarrow -(\mu dp)$ (17.3)	$f(X \cup Y) = f(X)$		
			$(\mu \in)$		
			$a \in X - f(X) \Rightarrow$		
			$\exists b \in X.a \notin f(\{a, b\})$		

The proof of the following fact – together with the subsequent examples – requires some knowledge of preferential structures, which will be introduced "officially" only in Chap. 4 (p. 73). We chose to give those results here, so the reader will have immediately a global picture, and can come back later, if desired, and read the proofs and (counter) examples.

Fact 2.3.1 The following table (Table 2.3) is to be read as follows: If the left hand side holds for some function $f : \mathcal{Y} \rightarrow \mathcal{P}(U)$, and the auxiliary properties noted in the middle also hold for f or $\mathcal{Y}$, then the right hand side will hold too – and conversely.
"sing." stand for "$\mathcal{Y}$ contains singletons".

Table 2.3 Interdependencies of algebraic rules

		Basics	
(1.1)	(μPR)	$\Rightarrow (\cap) + (\mu \subseteq)$	$(\mu PR')$
(1.2)		$\Leftarrow$	
(2.1)	(μPR)	$\Rightarrow (\mu \subseteq)$	(μOR)
(2.2)		$\Leftarrow (\mu \subseteq) + (-)$	
(2.3)		$\Rightarrow (\mu \subseteq)$	(μwOR)
(2.4)		$\Leftarrow (\mu \subseteq) + (-)$	
(3)	(μPR)	$\Rightarrow$	(μCUT)
(4)	$(\mu \subseteq) + (\mu \subseteq\supseteq) + (\mu CUM)$ $+(\mu RatM) + (\cap)$	$\not\Rightarrow$	(μPR)
		Cumulativity	
(5.1)	(μCM)	$\Rightarrow (\cap) + (\mu \subseteq)$	$(\mu ResM)$
(5.2)		$\Leftarrow$ (infin.)	
(6)	$(\mu CM) + (\mu CUT)$	$\Leftrightarrow$	(μCUM)
(7)	$(\mu \subseteq) + (\mu \subseteq\supseteq)$	$\Rightarrow$	(μCUM)
(8)	$(\mu \subseteq) + (\mu CUM) + (\cap)$	$\Rightarrow$	$(\mu \subseteq\supseteq)$
(9)	$(\mu \subseteq) + (\mu CUM)$	$\not\Rightarrow$	$(\mu \subseteq\supseteq)$
		Rationality	
(10)	$(\mu RatM) + (\mu PR)$	$\Rightarrow$	$(\mu =)$
(11)	$(\mu =)$	$\Rightarrow$	$(\mu PR) + (\mu RatM)$
(12.1)	$(\mu =)$	$\Rightarrow (\cap) + (\mu \subseteq)$	$(\mu =')$
(12.2)		$\Leftarrow$	
(13)	$(\mu \subseteq) + (\mu =)$	$\Rightarrow (\cup)$	$(\mu \cup)$
(14)	$(\mu \subseteq) + (\mu \emptyset) + (\mu =)$	$\Rightarrow (\cup)$	$(\mu \parallel)$, $(\mu \cup')$, (μCUM)
(15)	$(\mu \subseteq) + (\mu \parallel)$	$\Rightarrow (-)$ of $\mathcal{Y}$	$(\mu =)$
(16)	$(\mu \parallel) + (\mu \in) + (\mu PR) +$ $(\mu \subseteq)$	$\Rightarrow (\cup)$ + sing.	$(\mu =)$
(17)	$(\mu CUM) + (\mu =)$	$\Rightarrow (\cup)$ + sing.	$(\mu \in)$
(18)	$(\mu CUM) + (\mu =) + (\mu \subseteq)$	$\Rightarrow (\cup)$	$(\mu \parallel)$
(19)	$(\mu PR) + (\mu CUM) + (\mu \parallel)$	$\Rightarrow$ sufficient, e.g. true in $\boldsymbol{D}_{\mathcal{L}}$	$(\mu =)$.
(20)	$(\mu \subseteq) + (\mu PR) + (\mu =)$	$\not\Rightarrow$	$(\mu \parallel)$
(21)	$(\mu \subseteq) + (\mu PR) + (\mu \parallel)$	$\not\Rightarrow$ (without $(-)$)	$(\mu =)$
(22)	$(\mu \subseteq) + (\mu PR) + (\mu \parallel) +$ $(\mu =) + (\mu \cup)$	$\not\Rightarrow$	$(\mu \in)$ (thus not representable by ranked structures)

Proof All sets are to be in $\mathcal{Y}$.

(1.1) $(\mu PR)+(\cap)+(\mu\subseteq)\Rightarrow(\mu PR')$:

By $X\cap Y\subseteq X$ and (μPR), $f(X)\cap X\cap Y\subseteq f(X\cap Y)$. By $(\mu\subseteq)$ $f(X)\cap Y=f(X)\cap X\cap Y$.

(1.2) $(\mu PR')\Rightarrow(\mu PR)$:

Let $X\subseteq Y$, so $X=X\cap Y$, so by $(\mu PR')$ $f(Y)\cap X\subseteq f(X\cap Y)=f(X)$.

(2.1) $(\mu PR)+(\mu\subseteq)\Rightarrow(\mu OR)$:

$f(X\cup Y)\subseteq X\cup Y$ by $(\mu\subseteq)$, so $f(X\cup Y)=(f(X\cup Y)\cap X)\cup(f(X\cup Y)\cap Y)$ $\subseteq f(X)\cup f(Y)$.

(2.2) $(\mu OR)+(\mu\subseteq)+(-)\Rightarrow(\mu PR)$:

Let $X\subseteq Y$, $X':=Y-X$. $f(Y)\subseteq f(X)\cup f(X')$ by (μOR), so $f(Y)\cap X\subseteq(f(X)\cap X)\cup(f(X')\cap X)=_{(\mu\subseteq)}f(X)\cup\emptyset=f(X)$.

(2.3) $(\mu PR)+(\mu\subseteq)\Rightarrow(\mu wOR)$:

Trivial by (2.1).

(2.4) $(\mu wOR)+(\mu\subseteq)+(-)\Rightarrow(\mu PR)$:

Let $X\subseteq Y$, $X':=Y-X$. $f(Y)\subseteq f(X)\cup X'$ by (μwOR), so $f(Y)\cap X\subseteq(f(X)\cap X)\cup(X'\cap X)=_{(\mu\subseteq)}f(X)\cup\emptyset=f(X)$.

(3) $(\mu PR)\Rightarrow(\mu CUT)$:

$f(X)\subseteq Y\subseteq X\Rightarrow f(X)\subseteq f(X)\cap Y\subseteq f(Y)$ by (μPR).

(4) $(\mu\subseteq)+(\mu\subseteq\supseteq)+(\mu CUM)+(\mu RatM)+(\cap)\not\Rightarrow(\mu PR)$:

This is shown in Example 2.3.2 (p. 43).

(5.1) $(\mu CM)+(\cap)+(\mu\subseteq)\Rightarrow(\mu ResM)$:

Let $f(X)\subseteq A\cap B$, so $f(X)\subseteq A$, so by $(\mu\subseteq)$ $f(X)\subseteq A\cap X\subseteq X$, so by (μCM) $f(A\cap X)\subseteq f(X)\subseteq B$.

(5.2) $(\mu ResM)\Rightarrow(\mu CM)$:

We consider here the infinitary version, where all sets can be model sets of infinite theories. Let $f(X)\subseteq Y\subseteq X$, so $f(X)\subseteq Y\cap f(X)$, so by $(\mu ResM)$ $f(Y)=f(X\cap Y)\subseteq f(X)$.

(6) $(\mu CM)+(\mu CUT)\Leftrightarrow(\mu CUM)$:

Trivial.

(7) $(\mu\subseteq)+(\mu\subseteq\supseteq)\Rightarrow(\mu CUM)$:

Suppose $f(D)\subseteq E\subseteq D$. So by $(\mu\subseteq)$ $f(E)\subseteq E\subseteq D$, so by $(\mu\subseteq\supseteq)$ $f(D)=f(E)$.

(8) $(\mu\subseteq)+(\mu CUM)+(\cap)\Rightarrow(\mu\subseteq\supseteq)$:

Let $f(D)\subseteq E$, $f(E)\subseteq D$, so by $(\mu\subseteq)$ $f(D)\subseteq D\cap E\subseteq D$, $f(E)\subseteq D\cap E\subseteq E$. As $f(D\cap E)$ is defined, so $f(D)=f(D\cap E)=f(E)$ by (μCUM).

(9) $(\mu\subseteq)+(\mu CUM)\not\Rightarrow(\mu\subseteq\supseteq)$:

This is shown in Example 2.3.1 (p. 43).

(10) $(\mu RatM)+(\mu PR)\Rightarrow(\mu=)$:

Trivial.

(11) $(\mu=)$ entails (μPR) and $(\mu RatM)$:

Trivial.

(12.1) $(\mu=)\Rightarrow(\mu=')$:

Let $f(Y)\cap X\neq\emptyset$, we have to show $f(X\cap Y)=f(Y)\cap X$. By $(\mu\subseteq)$ $f(Y)\subseteq Y$, so $f(Y)\cap X=f(Y)\cap(X\cap Y)$, so by $(\mu=)$ $f(Y)\cap X=f(Y)\cap(X\cap Y)=f(X\cap Y)$.

(12.2) $(\mu =') \Rightarrow (\mu =)$:
Let $X \subseteq Y$, $f(Y) \cap X \neq \emptyset$, then $f(X) = f(Y \cap X) = f(Y) \cap X$.
(13) $(\mu \subseteq)$, $(\mu =) \Rightarrow (\mu \cup)$:
If not, $f(X \cup Y) \cap Y \neq \emptyset$, but $f(Y) \cap (X - f(X)) \neq \emptyset$. By (11), (μPR) holds, so $f(X \cup Y) \cap X \subseteq f(X)$, so $\emptyset \neq f(Y) \cap (X - f(X)) \subseteq f(Y) \cap (X - f(X \cup Y))$, so $f(Y) - f(X \cup Y) \neq \emptyset$, so by $(\mu \subseteq)$ $f(Y) \subseteq Y$ and $f(Y) \neq f(X \cup Y) \cap Y$. But by $(\mu =)$ $f(Y) = f(X \cup Y) \cap Y$, a contradiction.
(14) $(\mu \subseteq)$, $(\mu\emptyset)$, $(\mu =) \Rightarrow (\mu \parallel)$:
If X or Y or both are empty, then this is trivial. Assume then $X \cup Y \neq \emptyset$, so by $(\mu\emptyset)$ $f(X \cup Y) \neq \emptyset$. By $(\mu \subseteq)$ $f(X \cup Y) \subseteq X \cup Y$, so $f(X \cup Y) \cap X = \emptyset$ and $f(X \cup Y) \cap Y = \emptyset$ together are impossible. Case 1, $f(X \cup Y) \cap X \neq \emptyset$ and $f(X \cup Y) \cap Y \neq \emptyset$: By $(\mu =)$ $f(X \cup Y) \cap X = f(X)$ and $f(X \cup Y) \cap Y = f(Y)$, so by $(\mu \subseteq)$ $f(X \cup Y) = f(X) \cup f(Y)$. Case 2, $f(X \cup Y) \cap X \neq \emptyset$ and $f(X \cup Y) \cap Y = \emptyset$: So by $(\mu =)$ $f(X \cup Y) = f(X \cup Y) \cap X = f(X)$. Case 3, $f(X \cup Y) \cap X = \emptyset$ and $f(X \cup Y) \cap Y \neq \emptyset$: Symmetrical.
$(\mu \subseteq)$, $(\mu\emptyset)$, $(\mu =) \Rightarrow (\mu \cup')$:
Let $f(Y) \cap (X - f(X)) \neq \emptyset$. If $X \cup Y = \emptyset$, then $f(X \cup Y) = f(X) = \emptyset$ by $(\mu \subseteq)$. So suppose $X \cup Y \neq \emptyset$. By (13), $f(X \cup Y) \cap Y = \emptyset$, so $f(X \cup Y) \subseteq X$ by $(\mu \subseteq)$. By $(\mu\emptyset)$, $f(X \cup Y) \neq \emptyset$, so $f(X \cup Y) \cap X \neq \emptyset$ and $f(X \cup Y) = f(X)$ by $(\mu =)$.
$(\mu \subseteq)$, $(\mu\emptyset)$, $(\mu =) \Rightarrow (\mu CUM)$:
Let $f(Y) \subseteq X \subseteq Y$. If $Y = \emptyset$, this is trivial by $(\mu \subseteq)$. If $Y \neq \emptyset$, then by $(\mu\emptyset)$ – which is crucial here – $f(Y) \neq \emptyset$, so by $f(Y) \subseteq X$ $f(Y) \cap X \neq \emptyset$, so by $(\mu =)$ $f(Y) = f(Y) \cap X = f(X)$.
(15) $(\mu \subseteq) + (\mu \parallel) \Rightarrow (\mu =)$:
Let $X \subseteq Y$, $X \cap f(Y) \neq \emptyset$, and consider $Y = X \cup (Y{-}X)$. Then $f(Y) = f(X) \parallel f(Y{-}X)$. As $f(Y) \cap X \neq \emptyset$, $f(Y) = f(Y{-}X)$ is impossible. Otherwise, $f(X) = f(Y) \cap X$, and we are done.
(16) $(\mu \parallel) + (\mu \in) + (\mu PR) + (\mu \subseteq) \Rightarrow (\mu =)$:
Suppose $X \subseteq Y$, $x \in f(Y) \cap X$, we have to show $f(Y) \cap X = f(X)$. "$\subseteq$" is trivial by (μPR). "$\supseteq$": Assume $a \notin f(Y)$ (by $(\mu \subseteq)$), but $a \in f(X)$. By $(\mu \in)$ $\exists b \in Y.a \notin f(\{a, b\})$. As $a \in f(X)$, by (μPR), $a \in f(\{a, x\})$. By $(\mu \parallel)$, $f(\{a, b, x\}) = f(\{a, x\}) \parallel f(\{b\})$. As $a \notin f(\{a, b, x\})$, $f(\{a, b, x\}) = f(\{b\})$, so $x \notin f(\{a, b, x\})$, contradicting (μPR), as $a, b, x \in Y$.
(17) $(\mu CUM) + (\mu =) \Rightarrow (\mu \in)$:
Let $a \in X - f(X)$. If $f(X) = \emptyset$, then $f(\{a\}) = \emptyset$ by (μCUM). If not: Let $b \in f(X)$, then $a \notin f(\{a, b\})$ by $(\mu =)$.
(18) $(\mu CUM) + (\mu =) + (\mu \subseteq) \Rightarrow (\mu \parallel)$:
By (μCUM), $f(X \cup Y) \subseteq X \subseteq X \cup Y \Rightarrow f(X) = f(X \cup Y)$ and $f(X \cup Y) \subseteq Y \subseteq X \cup Y \Rightarrow f(Y) = f(X \cup Y)$. Thus, if $(\mu \parallel)$ were to fail, $f(X \cup Y) \not\subseteq X$, $f(X \cup Y) \not\subseteq Y$, but then by $(\mu \subseteq)$ $f(X \cup Y) \cap X \neq \emptyset$, so $f(X) = f(X \cup Y) \cap X$ and $f(X \cup Y) \cap Y \neq \emptyset$, so $f(Y) = f(X \cup Y) \cap Y$ by $(\mu =)$. Thus, $f(X \cup Y) = (f(X \cup Y) \cap X) \cup (f(X \cup Y) \cap Y) = f(X) \cup f(Y)$.
(19) $(\mu PR) + (\mu CUM) + (\mu \parallel) \Rightarrow (\mu =)$:

Suppose $(\mu =)$ does not hold. So, by (μPR), there are X, Y, y s.t. $X \subseteq Y$, $X \cap f(Y) \neq \emptyset$, $y \in Y - f(Y)$, $y \in f(X)$. Let $a \in X \cap f(Y)$. If $f(Y) = \{a\}$, then by (μCUM) $f(Y) = f(X)$, so there must be $b \in f(Y)$, $b \neq a$. Take now Y', Y'' s.t. $Y = Y' \cup Y''$, $a \in Y'$, $a \notin Y''$, $b \in Y''$, $b \notin Y'$, $y \in Y' \cap Y''$. Assume now $(\mu \parallel)$ to hold, we show a contradiction. If $y \notin f(Y'')$, then by (μPR) $y \notin f(Y'' \cup \{a\})$. But $f(Y'' \cup \{a\}) = f(Y'') \parallel f(\{a, y\})$, so $f(Y'' \cup \{a\}) = f(Y'')$, contradicting $a \in f(Y)$. If $y \in f(Y'')$, then by $f(Y) = f(Y') \parallel f(Y'')$, $f(Y) = f(Y')$, *contradiction* as $b \notin f(Y')$.

(20) $(\mu \subseteq) + (\mu PR) + (\mu =) \not\Rightarrow (\mu \parallel)$:
See Example 2.3.3 (p. 43).

(21) $(\mu \subseteq) + (\mu PR) + (\mu \parallel) \not\Rightarrow (\mu =)$:
See Example 2.3.4 (p. 44).

(22) $(\mu \subseteq) + (\mu PR) + (\mu \parallel) + (\mu =) + (\mu \cup) \not\Rightarrow (\mu \in)$:
See Example 2.3.5 (p. 45).

Thus, by Fact 4.2.7 (p. 92), the conditions do not assure representability by ranked structures. □

Remark 2.3.1 Note that $(\mu =')$ is very close to $(RatM)$: $(RatM)$ says: $\alpha \mid\sim \beta$, $\alpha \not\mid\sim \neg\gamma \Rightarrow \alpha \wedge \gamma \mid\sim \beta$. Or, $f(A) \subseteq B$, $f(A) \cap C \neq \emptyset \Rightarrow f(A \cap C) \subseteq B$ for all A, B, C. This is not quite, but almost: $f(A \cap C) \subseteq f(A) \cap C$ (it depends how many B there are, if $f(A)$ is some such B, the fit is perfect).

Example 2.3.1 We show here $(\mu \subseteq) + (\mu CUM) \not\Rightarrow (\mu \subseteq\supseteq)$.
Consider $X := \{a, b, c\}$, $Y := \{a, b, d\}$, $f(X) := \{a\}$, $f(Y) := \{a, b\}$, $\mathcal{Y} := \{X, Y\}$. (If $f(\{a, b\})$ were defined, we would have $f(X) = f(\{a, b\}) = f(Y)$, *contradiction*.) Obviously, $(\mu \subseteq)$ and (μCUM) hold, but not $(\mu \subseteq\supseteq)$.

Example 2.3.2 We show here $(\mu \subseteq) + (\mu \subseteq\supseteq) + (\mu CUM) + (\mu RatM) + (\cap) \not\Rightarrow (\mu PR)$.

Let $U := \{a, b, c\}$. Let $\mathcal{Y} = \mathcal{P}(U)$. So $(\cap)$ is trivially satisfied. Set $f(X) := X$ for all $X \subseteq U$ except for $f(\{a, b\}) = \{b\}$. Obviously, this cannot be represented by a preferential structure and (μPR) is false for U and $\{a, b\}$. But it satisfies $(\mu \subseteq)$, (μCUM), and $(\mu RatM)$. $(\mu \subseteq)$ is trivial. (μCUM): Let $f(X) \subseteq Y \subseteq X$. If $f(X) = X$, we are done. Consider $f(\{a, b\}) = \{b\}$. If $\{b\} \subseteq Y \subseteq \{a, b\}$, then $f(Y) = \{b\}$, so we are done again. It is shown in Fact 2.3.1 (p. 40), (8) that $(\mu \subseteq\supseteq)$ follows. $(\mu RatM)$: Suppose $X \subseteq Y$, $X \cap f(Y) \neq \emptyset$, we have to show $f(X) \subseteq f(Y) \cap X$. If $f(Y) = Y$, the result holds for $X \subseteq Y$, so it does if $X = Y$. The only remaining case is $Y = \{a, b\}$, $X = \{b\}$, and the result holds again.

Example 2.3.3 The example shows that $(\mu \subseteq) + (\mu PR) + (\mu =) \not\Rightarrow (\mu \parallel)$.
Consider the following structure without transitivity: $U := \{a, b, c, d\}$, c and d have ω many copies in descending order $c_1 \succeq c_2 \ldots$, etc. a, b have one single copy each. $a \succeq b$, $a \succeq d_1$, $b \succeq a$, $b \succeq c_1$. $(\mu \parallel)$ does not hold: $f(U) = \emptyset$, but $f(\{a, c\}) = \{a\}$, $f(\{b, d\}) = \{b\}$. (μPR) holds as in all preferential structures. $(\mu =)$ holds: If it were to fail, then for some $A \subseteq B$, $f(B) \cap A \neq \emptyset$, so $f(B) \neq \emptyset$. But the only possible cases for B are now $(a \in B,\ b, d \notin B)$ or $(b \in B,\ a, c \notin B)$. Thus, B

can be $\{a\}$, $\{a, c\}$, $\{b\}$, $\{b, d\}$ with $f(B) = \{a\}$, $\{a\}$, $\{b\}$, $\{b\}$. If $A = B$, then the result will hold trivially. Moreover, A has to be $\neq \emptyset$. So the remaining cases of B where it might fail are $B = \{a, c\}$ and $\{b, d\}$, and by $f(B) \cap A \neq \emptyset$, the only cases of A where it might fail are $A = \{a\}$ or $\{b\}$, respectively. So the only cases remaining are $B = \{a, c\}$, $A = \{a\}$ and $B = \{b, d\}$, $A = \{b\}$. In the first case, $f(A) = f(B) = \{a\}$; in the second case $f(A) = f(B) = \{b\}$, but $(\mu =)$ holds in both.

Example 2.3.4 The example shows that $(\mu \subseteq) + (\mu PR) + (\mu \parallel) \not\Rightarrow (\mu =)$.

Work in the set of theory definable model sets of an infinite propositional language. Note that this is not closed under set difference, and closure properties will play a crucial role in the argumentation. Let $U := \{y, a, x_{i<\omega}\}$, where $x_i \to a$ in the standard topology. For the order, arrange s.t. y is minimized by any set iff this set contains a cofinal subsequence of x_i; this can be done by the standard construction. Moreover, let all x_i kill themselves, i.e. with ω many copies $x_i^1 \succeq x_i^2 \succeq \ldots$. There are no other elements in the relation. Note that if $a \not\in \mu(X)$, then $a \not\in X$, and X cannot contain a cofinal subsequence of x_i, as X is closed in the standard topology. (A short argument: Suppose X contains such a subsequence, but $a \not\in X$. Then the theory of a $Th(a)$ is inconsistent with $Th(X)$, so already a finite subset of $Th(a)$ is inconsistent with $Th(X)$, but such a finite subset will finally hold in a cofinal sequence converging to a.) Likewise, if $y \in \mu(X)$, then X cannot contain a cofinal subsequence of x_i.

Obviously, $(\mu \subseteq)$ and (μPR) hold, but $(\mu =)$ does not hold. Set $B := U$, $A := \{a, y\}$. Then $\mu(B) = \{a\}$, $\mu(A) = \{a, y\}$, contradicting $(\mu =)$.

It remains to show that $(\mu \parallel)$ holds.

$\mu(X)$ can only be $\emptyset$, $\{a\}$, $\{y\}$, $\{a, y\}$. As $\mu(A \cup B) \subseteq \mu(A) \cup \mu(B)$ by (μPR),

Case 1: $\mu(A \cup B) = \{a, y\}$ is settled.

Note that if $y \in X - \mu(X)$, then X will contain a cofinal subsequence, and thus $a \in \mu(X)$.

Case 2: $\mu(A \cup B) = \{a\}$.

Case 2.1: $\mu(A) = \{a\}$ – we are done.

Case 2.2: $\mu(A) = \{y\}$: A does not contain a, nor a cofinal subsequence. If $\mu(B) = \emptyset$, then $a \not\in B$, so $a \not\in A \cup B$, a contradiction. If $\mu(B) = \{a\}$, we are done. If $y \in \mu(B)$, then $y \in B$, but B does not contain a cofinal subsequence, so $A \cup B$ does not either, so $y \in \mu(A \cup B)$, *contradiction*.

Case 2.3: $\mu(A) = \emptyset$: A cannot contain a cofinal subsequence. If $\mu(B) = \{a\}$, we are done. $a \in \mu(B)$ does have to hold, so $\mu(B) = \{a, y\}$ is the only remaining possibility. But then B does not contain a cofinal subsequence, and neither does $A \cup B$, so $y \in \mu(A \cup B)$, *contradiction*.

Case 2.4: $\mu(A) = \{a, y\}$: A does not contain a cofinal subsequence. If $\mu(B) = \{a\}$, we are done. If $\mu(B) = \emptyset$, B does not contain a cofinal subsequence (as $a \not\in B$), so neither does $A \cup B$, so $y \in \mu(A \cup B)$, *contradiction*. If $y \in \mu(B)$, B does not contain a cofinal subsequence, and we are done again.

Case 3: $\mu(A \cup B) = \{y\}$: To obtain a contradiction, we need $a \in \mu(A)$ or $a \in \mu(B)$. But in both cases $a \in \mu(A \cup B)$.

Case 4: $\mu(A \cup B) = \emptyset$: Thus, $A \cup B$ contains no cofinal subsequence. If, e.g. $y \in \mu(A)$, then $y \in \mu(A \cup B)$, if $a \in \mu(A)$, then $a \in \mu(A \cup B)$, so $\mu(A) = \emptyset$.

Example 2.3.5 The example show that $(\mu \subseteq) + (\mu PR) + (\mu \parallel) + (\mu =) + (\mu\cup) \not\Rightarrow (\mu \in)$.

Let $U := \{y, x_{i<\omega}\}$, x_i a sequence, each x_i kills itself, $x_i^1 \succeq x_i^2 \succeq \ldots$ and y is killed by all cofinal subsequences of x_i. Then for any $X \subseteq U$ $\mu(X) = \emptyset$ or $\mu(X) = \{y\}$.

$(\mu \subseteq)$ and (μPR) hold obviously.

$(\mu \parallel)$: Let $A \cup B$ be given. If $y \notin X$, then for all $Y \subseteq X$ $\mu(Y) = \emptyset$. So, if $y \notin A \cup B$, we are done. If $y \in A \cup B$ and $\mu(A \cup B) = \emptyset$, one of A, B must contain a cofinal sequence, it will have $\mu = \emptyset$. If not, then $\mu(A \cup B) = \{y\}$, and this will also hold for the one y is in.

$(\mu =)$: Let $A \subseteq B$, $\mu(B) \cap A \neq \emptyset$, show $\mu(A) = \mu(B) \cap A$. But now $\mu(B) = \{y\}$, $y \in A$, so B does not contain a cofinal subsequence, neither does A, so $\mu(A) = \{y\}$.

$(\mu\cup)$: $(A - \mu(A)) \cap \mu(A') \neq \emptyset$, so $\mu(A') = \{y\}$, so $\mu(A \cup A') = \emptyset$, as $y \in A - \mu(A)$.

But $(\mu \in)$ does not hold: $y \in U - \mu(U)$, but there is no x s.t. $y \notin \mu(\{x, y\})$.

We turn to interdependencies of the different μ-conditions. Again, we will sometimes use preferential structures in our arguments.

Fact 2.3.2 $(\mu wOR) + (\mu \subseteq) \Rightarrow f(X \cup Y) \subseteq f(X) \cup f(Y) \cup (X \cap Y)$

Proof $f(X \cup Y) \subseteq f(X) \cup Y$, $f(X \cup Y) \subseteq X \cup f(Y)$, so $f(X \cup Y) \subseteq (f(X) \cup Y) \cap (X \cup f(Y)) = f(X) \cup f(Y) \cup (X \cap Y)$. □

Proposition 2.3.1 *The following table "Logical and algebraic rules" (Table 2.4) is to be read as follows:*

Let a logic $\mid\!\sim$ *satisfy* (LLE) *and* (CCL), *and define a function* $f : \mathbf{D}_{\mathcal{L}} \to \mathbf{D}_{\mathcal{L}}$ *by* $f(M(T)) := M(\overline{\overline{T}})$. *Then f is well defined, satisfies* (μdp), *and* $\overline{\overline{T}} = Th(f(M(T)))$.

If $\mid\!\sim$ *satisfies a rule in the left hand side, then – provided the additional properties noted in the middle for* $\Rightarrow$ *hold too – f will satisfy the property in the right hand side.*

Conversely, if $f : \mathcal{Y} \to \mathcal{P}(M_{\mathcal{L}})$ *is a function, with* $\mathbf{D}_{\mathcal{L}} \subseteq \mathcal{Y}$, *and we define a logic* $\mid\!\sim$ by $\overline{\overline{T}} := Th(f(M(T)))$, *then* $\mid\!\sim$ *satisfies* (LLE) *and* (CCL). *If f satisfies* (μdp), *then* $f(M(T)) = M(\overline{\overline{T}})$.

If f satisfies a property in the right hand side, then – provided the additional properties noted in the middle for $\Leftarrow$ *hold too –* $\mid\!\sim$ *will satisfy the property in the left hand side.*

If "$T = \phi$*" is noted in the table, this means that, if one of the theories (the one named the same way in Definition 2.3* (p. 35)) *is equivalent to a formula, we do not need* (μdp).

Proof Set $f(T) := f(M(T))$, note that $f(T \cup T') := f(M(T \cup T')) = f(M(T) \cap M(T'))$.

Table 2.4 Logical and algebraic rules

		Basics	
(1.1)	(OR)	$\Rightarrow$	(μOR)
(1.2)		$\Leftarrow$	
(2.1)	$(disjOR)$	$\Rightarrow$	$(\mu disjOR)$
(2.2)		$\Leftarrow$	
(3.1)	(wOR)	$\Rightarrow$	(μwOR)
(3.2)		$\Leftarrow$	
(4.1)	(SC)	$\Rightarrow$	$(\mu \subseteq)$
(4.2)		$\Leftarrow$	
(5.1)	(CP)	$\Rightarrow$	$(\mu\emptyset)$
(5.2)		$\Leftarrow$	
(6.1)	(PR)	$\Rightarrow$	(μPR)
(6.2)		$\Leftarrow (\mu dp) + (\mu \subseteq)$	
(6.3)		$\not\Leftarrow -(\mu dp)$	
(6.4)		$\Leftarrow (\mu \subseteq)$ $T' = \phi$	
(6.5)	(PR)	$\Leftarrow$ $T' = \phi$	$(\mu PR')$
(7.1)	(CUT)	$\Rightarrow$	(μCUT)
(7.2)		$\Leftarrow$	
		Cumulativity	
(8.1)	(CM)	$\Rightarrow$	(μCM)
(8.2)		$\Leftarrow$	
(9.1)	$(ResM)$	$\Rightarrow$	$(\mu ResM)$
(9.2)		$\Leftarrow$	
(10.1)	$(\subseteq\supseteq)$	$\Rightarrow$	$(\mu \subseteq\supseteq)$
(10.2)		$\Leftarrow$	
(11.1)	(CUM)	$\Rightarrow$	(μCUM)
(11.2)		$\Leftarrow$	
		Rationality	
(12.1)	$(RatM)$	$\Rightarrow$	$(\mu RatM)$
(12.2)		$\Leftarrow (\mu dp)$	
(12.3)		$\not\Leftarrow -(\mu dp)$	
(12.4)		$\Leftarrow$ $T = \phi$	
(13.1)	$(RatM =)$	$\Rightarrow$	$(\mu =)$
(13.2)		$\Leftarrow (\mu dp)$	
(13.3)		$\not\Leftarrow -(\mu dp)$	
(13.4)		$\Leftarrow$ $T = \phi$	
(14.1)	$(Log =')$	$\Rightarrow$	$(\mu =')$
(14.2)		$\Leftarrow (\mu dp)$	
(14.3)		$\not\Leftarrow -(\mu dp)$	
(14.4)		$\Leftarrow T = \phi$	
(15.1)	$(Log \parallel)$	$\Rightarrow$	$(\mu \parallel)$
(15.2)		$\Leftarrow$	
(16.1)	$(Log\cup)$	$\Rightarrow (\mu \subseteq) + (\mu =)$	$(\mu\cup)$
(16.2)		$\Leftarrow (\mu dp)$	
(16.3)		$\not\Leftarrow -(\mu dp)$	
(17.1)	$(Log\cup')$	$\Rightarrow (\mu \subseteq) + (\mu =)$	$(\mu\cup')$
(17.2)		$\Leftarrow (\mu dp)$	
(17.3)		$\not\Leftarrow -(\mu dp)$	

We first show the general framework.

Let $\mid\!\sim$ satisfy (LLE) and (CCL). Let $f : \boldsymbol{D}_{\mathcal{L}} \rightarrow \boldsymbol{D}_{\mathcal{L}}$ be defined by $f(M(T)) := M(\overline{\overline{T}})$. If $M(T) = M(T')$, then $\overline{T} = \overline{T'}$, so by (LLE) $\overline{\overline{T}} = \overline{\overline{T'}}$, $f(M(T)) = f(M(T'))$, so f is well defined and satisfies (μdp). By (CCL) $Th(M(\overline{\overline{T}})) = \overline{\overline{T}}$.

Let f be given and $\mid\!\sim$ be defined by $\overline{\overline{T}} := Th(f(M(T)))$. Obviously, $\mid\!\sim$ satisfies (LLE) and (CCL) (and thus (RW)). If f satisfies (μdp), then $f(M(T)) = M(T')$ for some T', and $f(M(T)) = M(Th(f(M(T)))) = M(\overline{\overline{T}})$ by Fact 2.2.4 (p. 35). (We will use Fact 2.2.4 (p. 35) now without further mentioning.)

Next we show the following fact:

(a) If f satisfies (μdp), or T' is equivalent to a formula, then $Th(f(T) \cap M(T')) = \overline{\overline{\overline{T}} \cup T'}$.

Case 1: f satisfies (μdp). $Th(f(M(T)) \cap M(T')) = Th(M(\overline{\overline{T}}) \cap M(T') = \overline{\overline{\overline{T}} \cup T'}$ by Fact 2.2.3 (p. 35) (5).

Case 2: T' is equivalent to ϕ'. $Th(f(M(T)) \cap M(\phi')) = \overline{Th(f(M(T))) \cup \{\phi'\}} = \overline{\overline{\overline{T}} \cup \{\phi'\}}$ by Fact 2.2.3 (p. 35) (3).

We now prove the individual properties.

(1.1) $(OR) \Rightarrow (\mu OR)$

Let $X = M(T)$, $Y = M(T')$. $f(X \cup Y) = f(M(T) \cup M(T')) = f(M(T \vee T'))$ $:= M(\overline{\overline{T \vee T'}}) \subseteq_{(OR)} M(\overline{\overline{T}} \cap \overline{\overline{T'}}) =_{(CCL)} M(\overline{\overline{T}}) \cup M(\overline{\overline{T'}}) =: f(X) \cup f(Y)$.

(1.2) $(\mu OR) \Rightarrow (OR)$:

$\overline{\overline{T \vee T'}} := Th(f(M(T \vee T'))) = Th(f(M(T) \cup M(T'))) \supseteq_{(\mu OR)} Th(f(M(T)) \cup f(M(T'))) =$ (by Fact 2.2.2 (p. 34)) $Th(f(M(T))) \cap Th(f(M(T'))) =: \overline{\overline{T}} \cap \overline{\overline{T'}}$.

(2) By $\neg Con(T, T') \Leftrightarrow M(T) \cap M(T') = \emptyset$, we can use directly the proofs for 1.

(3.1) $(wOR) \Rightarrow (\mu wOR)$:

Let $X = M(T)$, $Y = M(T')$. $f(X \cup Y) = f(M(T) \cup M(T')) = f(M(T \vee T'))$ $:= M(\overline{\overline{T \vee T'}}) \subseteq_{(wOR)} M(\overline{\overline{T}} \cap \overline{T'}) =_{(CCL)} M(\overline{\overline{T}}) \cup M(\overline{T'}) =: f(X) \cup Y$.

(3.2) $(\mu wOR) \Rightarrow (wOR)$:

$\overline{\overline{T \vee T'}} := Th(f(M(T \vee T'))) = Th(f(M(T) \cup M(T'))) \supseteq_{(\mu wOR)} Th(f(M(T)) \cup M(T')) =$ (by Fact 2.2.2 (p. 34)) $Th(f(M(T))) \cap Th(M(T')) =: \overline{\overline{T}} \cap \overline{T'}$.

(4.1) $(SC) \Rightarrow (\mu \subseteq)$:

Trivial.

(4.2) $(\mu \subseteq) \Rightarrow (SC)$:

Trivial.

(5.1) $(CP) \Rightarrow (\mu\emptyset)$:

Trivial.

(5.2) $(\mu\emptyset) \Rightarrow (CP)$:

Trivial.

(6.1) $(PR) \Rightarrow (\mu PR)$:

Suppose $X := M(T)$, $Y := M(T')$, $X \subseteq Y$, we have to show $f(Y) \cap X \subseteq f(X)$. By prerequisite, $\overline{T'} \subseteq \overline{T}$, so $\overline{T \cup T'} = \overline{T}$, so $\overline{\overline{T \cup T'}} = \overline{\overline{T}}$ by (LLE). By (PR)

$\overline{\overline{T \cup T'}} \subseteq \overline{\overline{\overline{T'}} \cup T}$, so $f(Y) \cap X = f(T') \cap M(T) = M(\overline{\overline{T'}} \cup T) \subseteq M(\overline{\overline{T \cup T'}}) = M(\overline{\overline{T}}) = f(X)$.

(6.2) $(\mu PR) + (\mu dp) + (\mu \subseteq) \Rightarrow (PR)$:

$f(T) \cap M(T') =_{(\mu\subseteq)} f(T) \cap M(T) \cap M(T') = f(T) \cap M(T \cup T') \subseteq_{(\mu PR)} f(T \cup T')$, so $\overline{\overline{T \cup T'}} = Th(f(T \cup T')) \subseteq Th(f(T) \cap M(T')) = \overline{\overline{\overline{T}} \cup T'}$ by (a) above and (μdp).

(6.3) $(\mu PR) \not\Rightarrow (PR)$ without (μdp):

(μPR) holds in all preferential structures (see Definition 4.1.1 (p. 78)) by Fact 4.2.1 (p. 87). Example 4.2.1 (p. 87) shows that (DP) may fail in the resulting logic.

(6.4) $(\mu PR) + (\mu \subseteq) \Rightarrow (PR)$ if T' is classically equivalent to a formula:

It was shown in the proof of (6.2) that $f(T) \cap M(\phi') \subseteq f(T \cup \{\phi'\})$, so $\overline{\overline{T \cup \{\phi'\}}} = Th(f(T \cup \{\phi'\})) \subseteq Th(f(T) \cap M(\phi')) = \overline{\overline{\overline{T}} \cup \{\phi'\}}$ by (a) above.

(6.5) $(\mu PR') \Rightarrow (PR)$, if T' is classically equivalent to a formula:

$f(M(T)) \cap M(\phi') \subseteq_{(\mu PR')} f(M(T) \cap M(\phi')) = f(M(T \cup \{\phi'\}))$. So again $\overline{\overline{T \cup \{\phi'\}}} = Th(f(T \cup \{\phi'\})) \subseteq Th(f(T) \cap M(\phi')) = \overline{\overline{\overline{T}} \cup \{\phi'\}}$ by (a) above.

(7.1) $(CUT) \Rightarrow (\mu CUT)$:

So let $X = M(T)$, $Y = M(T')$, and $f(T) := M(\overline{\overline{T}}) \subseteq M(T') \subseteq M(T) \Rightarrow \overline{T} \subseteq \overline{T'} \subseteq \overline{\overline{T}} =_{(LLE)} \overline{\overline{(\overline{T})}} \Rightarrow$ (by (CUT)) $\overline{\overline{T}} = \overline{\overline{(\overline{T})}} \supseteq \overline{\overline{(\overline{T'})}} = \overline{\overline{T'}} \Rightarrow f(T) = M(\overline{\overline{T}}) \subseteq M(\overline{\overline{T'}}) = f(T')$, thus $f(X) \subseteq f(Y)$.

(7.2) $(\mu CUT) \Rightarrow (CUT)$:

Let $T \subseteq \overline{T'} \subseteq \overline{\overline{T}}$. Thus $f(T) \subseteq M(\overline{\overline{T}}) \subseteq M(T') \subseteq M(T)$, so by (μCUT) $f(T) \subseteq f(T')$, so $\overline{\overline{T}} = Th(f(T)) \supseteq Th(f(T')) = \overline{\overline{T'}}$.

(8.1) $(CM) \Rightarrow (\mu CM)$:

So let $X = M(T)$, $Y = M(T')$, and $f(T) := M(\overline{\overline{T}}) \subseteq M(T') \subseteq M(T) \Rightarrow \overline{T} \subseteq \overline{T'} \subseteq \overline{\overline{T}} =_{(LLE)} \overline{\overline{(\overline{T})}} \Rightarrow$ (by (LLE), (CM)) $\overline{\overline{T}} = \overline{\overline{(\overline{T})}} \subseteq \overline{\overline{(\overline{T'})}} = \overline{\overline{T'}} \Rightarrow f(T) = M(\overline{\overline{T}}) \supseteq M(\overline{\overline{T'}}) = f(T')$, thus $f(X) \supseteq f(Y)$.

(8.2) $(\mu CM) \Rightarrow (CM)$:

Let $T \subseteq \overline{T'} \subseteq \overline{\overline{T}}$. Thus by (μCM) and $f(T) \subseteq M(\overline{\overline{T}}) \subseteq M(T') \subseteq M(T)$, so $f(T) \supseteq f(T')$ by (μCM), so $\overline{\overline{T}} = Th(f(T)) \subseteq Th(f(T')) = \overline{\overline{T'}}$.

(9.1) $(ResM) \Rightarrow (\mu ResM)$:

Let $f(X) := M(\overline{\overline{\Delta}})$, $A := M(\alpha)$, $B := M(\beta)$. So $f(X) \subseteq A \cap B \Rightarrow \Delta \mid\!\sim \alpha, \beta \Rightarrow_{(ResM)} \Delta, \alpha \mid\!\sim \beta \Rightarrow M(\overline{\overline{\Delta, \alpha}}) \subseteq M(\beta) \Rightarrow f(X \cap A) \subseteq B$.

(9.2) $(\mu ResM) \Rightarrow (ResM)$:

Let $f(X) := M(\overline{\overline{\Delta}})$, $A := M(\alpha)$, $B := M(\beta)$. So $\Delta \mid\!\sim \alpha, \beta \Rightarrow f(X) \subseteq A \cap B \Rightarrow_{(\mu ResM)} f(X \cap A) \subseteq B \Rightarrow \Delta, \alpha \mid\!\sim \beta$.

(10.1) $(\subseteq\supseteq) \Rightarrow (\mu \subseteq\supseteq)$:

Let $f(T) \subseteq M(T')$, $f(T') \subseteq M(T)$. So $Th(M(T')) \subseteq Th(f(T))$, $Th(M(T)) \subseteq Th(f(T'))$, so $T' \subseteq \overline{T'} \subseteq \overline{\overline{T}}$, $T \subseteq \overline{T} \subseteq \overline{\overline{T'}}$, so by $(\subseteq\supseteq)$ $\overline{\overline{T}} = \overline{\overline{T'}}$, so $f(T) := M(\overline{\overline{T}}) = M(\overline{\overline{T'}}) =: f(T')$.

(10.2) $(\mu \subseteq\supseteq) \Rightarrow (\subseteq\supseteq)$:

Let $T \subseteq \overline{\overline{T'}}$ and $T' \subseteq \overline{\overline{T}}$. So by (CCL) $Th(M(T)) = \overline{T} \subseteq \overline{\overline{T'}} = Th(f(T'))$. But $Th(M(T)) \subseteq Th(X) \Rightarrow X \subseteq M(T) : X \subseteq M(Th(X)) \subseteq M(Th(M(T))) = M(T)$. So $f(T') \subseteq M(T)$, likewise $f(T) \subseteq M(T')$, so by $(\mu \subseteq\supseteq)$ $f(T) = f(T')$, so $\overline{\overline{T}} = \overline{\overline{T'}}$.

(11.1) $(CUM) \Rightarrow (\mu CUM)$:

So let $X = M(T)$, $Y = M(T')$, and $f(T) := M(\overline{\overline{T}}) \subseteq M(T') \subseteq M(T) \Rightarrow \overline{T} \subseteq \overline{T'} \subseteq \overline{\overline{T}} =_{(LLE)} \overline{\overline{(\overline{T})}} \Rightarrow \overline{\overline{T}} = \overline{\overline{(\overline{T})}} = \overline{\overline{(\overline{T'})}} = \overline{\overline{T'}} \Rightarrow f(T) = M(\overline{\overline{T}}) = M(\overline{\overline{T'}}) = f(T')$, thus $f(X) = f(Y)$.

(11.2) $(\mu CUM) \Rightarrow (CUM)$:

Let $T \subseteq \overline{T'} \subseteq \overline{\overline{T}}$. Thus by (μCUM) and $f(T) \subseteq M(\overline{\overline{T}}) \subseteq M(T') \subseteq M(T)$, $f(T) = f(T')$, so $\overline{\overline{T}} = Th(f(T)) = Th(f(T')) = \overline{\overline{T'}}$.

(12.1) $(RatM) \Rightarrow (\mu RatM)$:

Let $X = M(T)$, $Y = M(T')$, and $X \subseteq Y$, $X \cap f(Y) \neq \emptyset$, so $T \vdash T'$ and $M(T) \cap f(M(T')) \neq \emptyset$, so $Con(T, \overline{\overline{T'}})$, so $\overline{\overline{\overline{T'} \cup T}} \subseteq \overline{\overline{T}}$ by $(RatM)$, so $f(X) = f(M(T)) = M(\overline{\overline{T}}) \subseteq M(\overline{\overline{\overline{T'} \cup T}}) = M(\overline{\overline{T'}}) \cap M(T) = f(Y) \cap X$.

(12.2) $(\mu RatM) + (\mu dp) \Rightarrow (RatM)$:

Let $X = M(T)$, $Y = M(T')$, $T \vdash T'$, $Con(T, \overline{\overline{T'}})$, so $X \subseteq Y$ and by (μdp) $X \cap f(Y) \neq \emptyset$, so by $(\mu RatM)$ $f(X) \subseteq f(Y) \cap X$, so $\overline{\overline{T}} = \overline{\overline{T \cup T'}} = Th(f(T \cup T')) \supseteq Th(f(T') \cap M(T)) = \overline{\overline{\overline{T'} \cup T}}$ by (a) above and (μdp).

(12.3) $(\mu RatM) \not\Rightarrow (RatM)$ without (μdp):

$(\mu RatM)$ holds in all ranked preferential structures (see Definition 4.1.4 (p. 81)) by Fact 4.2.7 (p. 92). Example 2.3.6 (p. 50) (2) shows that $(RatM)$ may fail in the resulting logic.

(12.4) $(\mu RatM) \Rightarrow (RatM)$ if T is classically equivalent to a formula:

$\phi \vdash T' \Rightarrow M(\phi) \subseteq M(T')$. $Con(\phi, \overline{\overline{T'}}) \Leftrightarrow M(\overline{\overline{T'}}) \cap M(\phi) \neq \emptyset \Leftrightarrow f(T') \cap M(\phi) \neq \emptyset$ by Fact 2.2.3 (p. 35) (4). Thus, $f(M(\phi)) \subseteq f(M(T')) \cap M(\phi)$ by $(\mu RatM)$. Thus by (a) above $\overline{\overline{\overline{T'} \cup \{\phi\}}} \subseteq \overline{\overline{\phi}}$.

(13.1) $(RatM =) \Rightarrow (\mu =)$:

Let $X = M(T)$, $Y = M(T')$, and $X \subseteq Y$, $X \cap f(Y) \neq \emptyset$, so $T \vdash T'$ and $M(T) \cap f(M(T')) \neq \emptyset$, so $Con(T, \overline{\overline{T'}})$, so $\overline{\overline{\overline{T'} \cup T}} = \overline{\overline{T}}$ by $(RatM =)$, so $f(X) = f(M(T)) = M(\overline{\overline{T}}) = M(\overline{\overline{\overline{T'} \cup T}}) = M(\overline{\overline{T'}}) \cap M(T) = f(Y) \cap X$.

(13.2) $(\mu =) + (\mu dp) \Rightarrow (RatM =)$:

Let $X = M(T)$, $Y = M(T')$, $T \vdash T'$, $Con(T, \overline{\overline{T'}})$, so $X \subseteq Y$ and by (μdp) $X \cap f(Y) \neq \emptyset$, so by $(\mu =)$ $f(X) = f(Y) \cap X$. So $\overline{\overline{\overline{T'} \cup T}} = \overline{\overline{T}}$ (a) above and (μdp).

(13.3) $(\mu =) \not\Rightarrow (RatM =)$ without (μdp):

$(\mu =)$ holds in all ranked preferential structures (see Definition 4.1.4 (p. 81)) by Fact 4.2.7 (p. 92). Example 2.3.6 (p. 50) (1) shows that $(RatM =)$ may fail in the resulting logic.

(13.4) $(\mu =) \Rightarrow (RatM =)$ if T is classically equivalent to a formula:

The proof is almost identical to the one for (12.4). Again, the prerequisites of $(\mu =)$ are satisfied, so $f(M(\phi)) = f(M(T')) \cap M(\phi)$. Thus, $\overline{\overline{\overline{\overline{T'}} \cup \{\phi\}}} = \overline{\overline{\phi}}$ by (a) above.

Of the last four, we show (14), (15), (17), the proof for (16) is similar to the one for (17).

(14.1) $(Log =') \Rightarrow (\mu =')$:

$f(M(T')) \cap M(T) \neq \emptyset \Rightarrow Con(\overline{\overline{T'}} \cup T) \Rightarrow_{(Log=')} \overline{\overline{T \cup T'}} = \overline{\overline{\overline{\overline{T'}} \cup T}} \Rightarrow f(M(T \cup T')) = f(M(T')) \cap M(T)$.

(14.2) $(\mu =') + (\mu dp) \Rightarrow (Log =')$:

$Con(\overline{\overline{T'}} \cup T) \Rightarrow_{(\mu dp)} f(M(T')) \cap M(T) \neq \emptyset \Rightarrow f(M(T' \cup T)) = f(M(T') \cap M(T)) =_{(\mu=')} f(M(T')) \cap M(T)$, so $\overline{\overline{T' \cup T}} = \overline{\overline{\overline{\overline{T'}} \cup T}}$ by (a) above and (μdp).

(14.3) $(\mu =') \not\Rightarrow (Log =')$ without (μdp):

By Fact 4.2.7 (p. 92) $(\mu =')$ holds in ranked structures. Consider Example 2.3.6 (p. 50) (2). There, $Con(T, \overline{\overline{T'}})$, $T = T \cup T'$, and it was shown that $\overline{\overline{\overline{\overline{T'}} \cup T}} \not\subseteq \overline{\overline{T}} = \overline{\overline{T \cup T'}}$

(14.4) $(\mu =') \Rightarrow (Log =')$ if T is classically equivalent to a formula:

$Con(\overline{\overline{T'}} \cup \{\phi\}) \Rightarrow \emptyset \neq M(\overline{\overline{T'}}) \cap M(\phi) \Rightarrow f(T') \cap M(\phi) \neq \emptyset$ by Fact 2.2.3 (p. 35) (4). So $f(M(T' \cup \{\phi\})) = f(M(T') \cap M(\phi)) = f(M(T')) \cap M(\phi)$ by $(\mu =')$, so $\overline{\overline{T' \cup \{\phi\}}} = \overline{\overline{\overline{\overline{T'}} \cup \{\phi\}}}$ by (a) above.

(15.1) $(Log \parallel) \Rightarrow (\mu \parallel)$:

Trivial.

(15.2) $(\mu \parallel) \Rightarrow (Log \parallel)$:

Trivial.

(16) $(Log\cup) \Leftrightarrow (\mu\cup)$: Analogous to the proof of (17).

(17.1) $(Log\cup') + (\mu \subseteq) + (\mu =) \Rightarrow (\mu\cup')$:

$f(M(T')) \cap (M(T) - f(M(T))) \neq \emptyset \Rightarrow$ (by $(\mu \subseteq)$, $(\mu =)$, Fact 4.2.5 (p. 92)) $f(M(T')) \cap M(T) \neq \emptyset$, $f(M(T')) \cap f(M(T)) = \emptyset \Rightarrow Con(\overline{\overline{T'}}, T)$, $\neg Con(\overline{\overline{T'}}, \overline{\overline{T}})$ $\Rightarrow \overline{\overline{T \vee T'}} = \overline{\overline{T}} \Rightarrow f(M(T)) = f(M(T \vee T')) = f(M(T) \cup M(T'))$.

(17.2) $(\mu\cup') + (\mu dp) \Rightarrow (Log\cup')$:

$Con(\overline{\overline{T'}} \cup T)$, $\neg Con(\overline{\overline{T'}} \cup \overline{\overline{T}}) \Rightarrow_{(\mu dp)} f(T') \cap M(T) \neq \emptyset$, $f(T') \cap f(T) = \emptyset$ $\Rightarrow f(M(T')) \cap (M(T) - f(M(T))) \neq \emptyset \Rightarrow f(M(T)) = f(M(T) \cup M(T')) = f(M(T \vee T'))$. So $\overline{\overline{T}} = \overline{\overline{T \vee T'}}$.

(17.3) and (16.3) are solved by Example 2.3.6 (p. 50) (3). □

Example 2.3.6 (1) $(\mu =)$ without (μdp) does not imply $(RatM =)$:

Take $\{p_i : i \in \omega\}$ and put $m := m_{\bigwedge p_i}$, the model which makes all p_i true, in the top layer, all the other in the bottom layer. Let $m' \neq m$, $T' := \emptyset$, $T := Th(m, m')$. Then $\overline{\overline{T'}} = T'$, so $Con(\overline{\overline{T'}}, T)$, $\overline{\overline{T}} = Th(m')$, $\overline{\overline{\overline{\overline{T'}} \cup T}} = T$.

So $(RatM =)$ fails, but $(\mu =)$ holds in all ranked structures.

(2) $(\mu RatM)$ without (μdp) does not imply $(RatM)$:

Take $\{p_i : i \in \omega\}$ and let $m := m_{\bigwedge p_i}$, the model which makes all p_i true.

Let $X := M(\neg p_0) \cup \{m\}$ be the top layer, put the rest of $M_{\mathcal{L}}$ in the bottom layer. Let $Y := M_{\mathcal{L}}$. The structure is ranked, as shown in Fact 4.2.7 (p. 92), $(\mu RatM)$ holds.

Let $T' := \emptyset$, $T := Th(X)$. We have to show that $Con(T, \overline{\overline{T'}})$, $T \vdash T'$, but $\overline{\overline{\overline{T'}} \cup T} \not\subseteq \overline{\overline{T}}$. $\overline{\overline{T'}} = Th(M(p_0) - \{m\}) = \overline{p_0}$. $T = \overline{\{\neg p_0\} \vee Th(m)}$, $\overline{\overline{T}} = T$. So $Con(T, \overline{\overline{T'}})$. $M(\overline{\overline{T'}}) = M(p_0)$, $M(T) = X$, $M(\overline{\overline{T'}} \cup T) = M(\overline{\overline{T'}}) \cap M(T) = \{m\}$, $m \models p_1$, so $p_1 \in \overline{\overline{\overline{T'}} \cup T}$, but $X \not\models p_1$.

(3) This example shows that we need (μdp) to go from $(\mu\cup)$ to $(Log\cup)$ and from $(\mu\cup')$ to $(Log\cup')$.

Let $v(\mathcal{L}) := \{p, q\} \cup \{p_i : i < \omega\}$. Let m make all variables true.

Put all models of $\neg p$, and m, in the upper layer, all other models in the lower layer. This is ranked, so by Fact 4.2.7 (p. 92) $(\mu\cup)$ and $(\mu\cup')$ hold. Set $X := M(\neg q) \cup \{m\}$, $X' := M(q) - \{m\}$, $T := Th(X) = \neg q \vee Th(m)$, $T' := Th(X') = \overline{q}$. Then $\overline{\overline{T}} = \overline{p \wedge \neg q}$, $\overline{\overline{T'}} = \overline{p \wedge q}$. We have $Con(\overline{\overline{T'}}, T)$, $\neg Con(\overline{\overline{T'}}, \overline{\overline{T}})$. But $\overline{\overline{T \vee T'}} = \overline{p} \neq \overline{\overline{T}} = \overline{p \wedge \neg q}$ and $Con(\overline{\overline{T \vee T'}}, T')$, so $(Log\cup)$ and $(Log\cup')$ fail.

Fact 2.3.3 $(CUT) \not\Rightarrow (PR)$

Proof We give two proofs:

(1) If $(CUT) \Rightarrow (PR)$, then by $(\mu PR) \Rightarrow$ (by Fact 2.3.1 (p. 40) (3)) (μCUT) $\Rightarrow$ (by Proposition 2.3.1 (p. 45) (7.2) $(CUT) \Rightarrow (PR)$ we would have a proof of $(\mu PR) \Rightarrow (PR)$ without (μdp), which is impossible, as shown by Example 4.2.1 (p. 87).

(2) Reconsider Example 2.3.2 (p. 43), and say $a \models p \wedge q$, $b \models p \wedge \neg q$, $c \models \neg p \wedge q$. It is shown there that (μCUM) holds, so (μCUT) holds, so by Proposition 2.3.1 (p. 45) (7.2) (CUT) holds, if we define $\overline{\overline{T}} := Th(f(M(T))$. Set $T := \{p \vee (\neg p \wedge q)\}$, $T' := \{p\}$, then $\overline{\overline{T \cup T'}} = \overline{\overline{T'}} = \overline{\{p \wedge \neg q\}}$, $\overline{\overline{T}} = \overline{T}$, $\overline{T \cup T'} = \overline{T'} = \overline{\{p\}}$, so (PR) fails. □

Chapter 3
Abstract Semantics by Size

3.1 The First-Order Setting

We first introduce a generalized quantifier in a first-order setting, as this is very natural, and prepares the more abstract discussion to come.

Definition 3.1.1 Augment the language of first-order logic by the new quantifier: If ϕ and ψ are formulas, then so are $\nabla x\phi(x)$, $\nabla x\phi(x) : \psi(x)$, for any variable x. The :-versions are the restricted variants. We call any formula of $\mathcal{L}$, possibly containing ∇ a $\nabla - \mathcal{L}$-formula.

Definition 3.1.2 Let $\mathcal{L}$ be a first-order language and M be a $\mathcal{L}$-structure. Let $\mathcal{N}(M)$ be a weak filter, or $\mathcal{N}$-system – $\mathcal{N}$ for normal – over M. Define $\langle M, \mathcal{N}(M)\rangle \models \phi$ for any $\nabla - \mathcal{L}$-formula inductively as usual with one additional induction step:

$\langle M, \mathcal{N}(M)\rangle \models \nabla x\phi(x)$ iff there is $A \in \mathcal{N}(M)$ s.t. $\forall a \in A\ (\langle M, \mathcal{N}(M)\rangle \models \phi[a])$.

Definition 3.1.3 Let any axiomatization of predicate calculus be given. Augment this with the axiom schemata

(1) $\nabla x\phi(x) \wedge \forall x(\phi(x) \rightarrow \psi(x)) \Rightarrow \nabla x\psi(x)$,
(2) $\nabla x\phi(x) \Rightarrow \neg\nabla x\neg\phi(x)$,
(3) $\forall x\phi(x) \Rightarrow \nabla x\phi(x)$ and $\nabla x\phi(x) \Rightarrow \exists x\phi(x)$, and
(4) $\nabla x\phi(x) \leftrightarrow \nabla y\phi(y)$ if x does not occur free in $\phi(y)$ and y does not occur free in $\phi(x)$.

(for all ϕ, ψ).

Proposition 3.1.1 *The axioms given in Definition 3.1.3 (p.* 53*) are sound and complete for the semantics of Definition 3.1.2 (p.* 53*) See* [Sch95-1] *or* [Sch04].

Definition 3.1.4 Call $\mathcal{N}^+(M) = \langle \mathcal{N}(N) : N \subseteq M\rangle$ a $\mathcal{N}^+$-system or system of weak filters over M iff for each $N \subseteq M$, $\mathcal{N}(N)$ is a weak filter or $\mathcal{N}$-system over N. (It suffices to consider the definable subsets of M.)

D.M. Gabbay, K. Schlechta, *Logical Tools for Handling Change in Agent-Based Systems*, Cognitive Technologies, DOI 10.1007/978-3-642-04407-6_3,

Definition 3.1.5 Let $\mathcal{L}$ be a first-order language, and M a $\mathcal{L}$-structure. Let $\mathcal{N}^+(M)$ be a $\mathcal{N}^+$-system over M.

Define $\langle M, \mathcal{N}^+(M)\rangle \models \phi$ for any formula inductively as usual with the additional induction steps:

1. $\langle M, \mathcal{N}^+(M)\rangle \models \nabla x \phi(x)$ iff there is $A \in \mathcal{N}(M)$ s.t. $\forall a \in A$ $(\langle M, \mathcal{N}^+(M)\rangle \models \phi[a])$ and
2. $\langle M, \mathcal{N}^+(M)\rangle \models \nabla x \phi(x) : \psi(x)$ iff there is $A \in \mathcal{N}(\{x : \langle M, \mathcal{N}^+(M)\rangle \models \phi(x)\})$ s.t. $\forall a \in A$ $(\langle M, \mathcal{N}^+(M)\rangle \models \psi[a])$.

Definition 3.1.6 Extend the logic of first-order predicate calculus by adding the axiom schemata

(1) a. $\nabla x \phi(x) \Leftrightarrow \nabla x(x = x) : \phi(x)$, b. $\forall x(\sigma(x) \leftrightarrow \tau(x)) \wedge \nabla x \sigma(x) : \phi(x) \Rightarrow \nabla x \tau(x) : \phi(x)$,
(2) $\nabla x \phi(x) : \psi(x) \wedge \forall x(\phi(x) \wedge \psi(x) \rightarrow \vartheta(x)) \Rightarrow \nabla x \phi(x) : \vartheta(x)$,
(3) $\exists x \phi(x) \wedge \nabla x \phi(x) : \psi(x) \Rightarrow \neg \nabla x \phi(x) : \neg \psi(x)$,
(4) $\forall x(\phi(x) \rightarrow \psi(x)) \Rightarrow \nabla x \phi(x) : \psi(x)$ and $\nabla x \phi(x) : \psi(x) \rightarrow [\exists x \phi(x) \rightarrow \exists x(\phi(x) \wedge \psi(x))]$, and
(5) $\nabla x \phi(x) : \psi(x) \leftrightarrow \nabla y \phi(y) : \psi(y)$ (under the usual caveat for substitution).
(for all ϕ, ψ, ϑ, σ, τ).

Proposition 3.1.2 *The axioms of Definition 3.1.6 (p. 54) are sound and complete for the $\mathcal{N}^+$-semantics of ∇ as defined in Definition* 3.1.5 *(p. 54). See* [Sch95-1] *or* [Sch04].

3.2 General Size Semantics

3.2.1 Introduction

3.2.1.1 Context

We show how one can develop a multitude of rules for nonmonotonic logics from a very small set of principles about reasoning with size. The notion of size gives an algebraic semantics to nonmonotonic logics, in the sense that α implies β iff the set of cases where $\alpha \wedge \neg\beta$ holds is a small subset of all α-cases. In a similar way e.g. Heyting algebras are an algebraic semantics for intuitionistic logic.

In our understanding, algebraic semantics describe the abstract properties that corresponding model sets have. Structural semantics, on the other hand, give intuitive concepts like accessibility or preference from which properties of model sets, and thus algebraic semantics, originate.

Varying properties of structural semantics (e.g. transitivity) result in varying properties of algebraic semantics, and thus of logical rules. We consider operations directly on the algebraic semantics and their logical consequences, and we see that simple manipulations of the size concept result in most rules of nonmonotonic logics. Even more, we show how to generate new rules from those manipulations.

The result is one big table, which, in a much more modest scale, can be seen as a "periodic table" of the "elements" of nonmonotonic logic. Some simple underlying principles allow to generate them all.

Historical remarks: The first time that abstract size was related to nonmonotonic logics was, to our knowledge, in the second author's [Sch90] and [Sch95-1] and independently in [BB94]. More detailed remarks can, for example, be found in [GS08c]. But, again to our knowledge, connections are elaborated systematically and in fine detail here for the first time.

3.2.1.2 Overview

The main part of this section is the big table in Sect. 3.2.2.6 (p. 58). It shows connections and how to develop a multitude of logical rules known from nonmonotonic logics by combining a small number of principles about size. We use them as building blocks to construct the rules from.

These principles are some basic and very natural postulates, (Opt), (iM), $(eM\mathcal{I})$, $(eM\mathcal{F})$, and a continuum of power of the notion of "small" or, dually, "big", from $(1 * s)$ to $(< \omega * s)$. From these, we can develop the rest except, essentially, rational monotony, and thus an infinity of different rules.

This is a conceptual section and it does not contain any more difficult formal results. The interest lies, in our opinion, in the simplicity, paucity, and naturalness of the basic building blocks. We hope that this schema brings more and deeper order into the rich fauna of nonmonotonic and related logics.

3.2.2 Main Table

3.2.2.1 Notation

(1) $\mathcal{I}(X) \subseteq \mathcal{P}(X)$ and $\mathcal{F}(X) \subseteq \mathcal{P}(X)$ are dual abstract notions of size, $\mathcal{I}(X)$ is the set of "small" subsets of X, $\mathcal{F}(X)$ the set of "big" subsets of X. They are dual in the sense that $A \in \mathcal{I}(X) \Leftrightarrow X - A \in \mathcal{F}(X)$. "$\mathcal{I}$" evokes "ideal", "$\mathcal{F}$" evokes "filter" though the full strength of both is reached only in $(< \omega * s)$. "s" evokes "small" and "$(x * s)$" stands for "x small sets together are still not everything".

(2) If $A \subseteq X$ is neither in $\mathcal{I}(X)$ nor in $\mathcal{F}(X)$, we say it has medium size, and we define $\mathcal{M}(X) := \mathcal{P}(X) - (\mathcal{I}(X) \cup \mathcal{F}(X))$. $\mathcal{M}^+(X) := \mathcal{P}(X) - \mathcal{I}(X)$ is the set of subsets which are not small.

(3) $\nabla x \phi$ is a generalized first-order quantifier; it is read as "almost all x have property ϕ". $\nabla x(\phi : \psi)$ is the relativized version, read: "almost all x with property ϕ have also property ψ". To keep the table simple, we write mostly only the non-relativized versions. Formally, we have $\nabla x \phi :\Leftrightarrow \{x : \phi(x)\} \in \mathcal{F}(U)$, where U is the universe and $\nabla x(\phi : \psi) :\Leftrightarrow \{x : (\phi \wedge \psi)(x)\} \in \mathcal{F}(\{x : \phi(x)\})$. Soundness and completeness results on ∇ can be found in [Sch95-1].

(4) Analogously, for propositional logic, we define

$$\alpha \mid\sim \beta :\Leftrightarrow M(\alpha \wedge \beta) \in \mathcal{F}(M(\alpha)),$$

where $M(\phi)$ is the set of models of ϕ.

(5) In preferential structures, $\mu(X) \subseteq X$ is the set of minimal elements of X. This generates a principal filter by $\mathcal{F}(X) := \{A \subseteq X : \mu(X) \subseteq A\}$. Corresponding properties about μ are not listed systematically.

(6) The usual rules (AND), etc. are named here (AND_ω), as they are in a natural ascending line of similar rules, based on strengthening of the filter/ideal properties.

3.2.2.2 The Group of Rules

The rules concern properties of $\mathcal{I}(X)$ or $\mathcal{F}(X)$, or dependencies between such properties for different X and Y. All X, Y, etc. will be subsets of some universe, say V. Intuitively, V is the set of all models of some fixed propositional language. It is not necessary to consider all subsets of V, the intention is to consider subsets of V, which are definable by a formula or a theory. So we assume all X, Y, etc. taken from some $\mathcal{Y} \subseteq \mathcal{P}(V)$, which we call the domain. In the former case, $\mathcal{Y}$ is closed under set difference, while in the latter case not necessarily so. (We will mention it when we need some particular closure property.)

The rules are divided into five groups:

(1) (Opt), which says that "All" is optimal – i.e. when there are no exceptions, then a soft rule $\mid\sim$ holds.

(2) Three monotony rules are as follows:

(2.1) (iM) is inner monotony: a subset of a small set is small,

(2.2) $(eM\mathcal{I})$ external monotony for ideals: enlarging the base set keeps small sets small, and

(2.3) $(eM\mathcal{F})$ external monotony for filters: a big subset stays big when the base set shrinks.

These three rules are very natural if "size" is anything coherent over change of base sets. In particular, they can be seen as weakening.

(3) $(\approx)$ keeps proportions, it is here mainly to point the possibility out.

(4) A group of rules $x * s$, which say how many small sets will not yet add to the base set. The notation "$(< \omega * s)$" is an allusion to the full filter property that filters are closed under *finite* intersections.

(5) Rational monotony, which can best be understood as robustness of $\mathcal{M}^+$, see $(\mathcal{M}^{++})$ (3).

We will assume all base sets to be nonempty in order to avoid pathologies and in particular, clashes between (Opt) and $(1 * s)$.

Note that the full strength of the usual definitions of a filter and an ideal are reached only in line $(< \omega * s)$.

Regularities

(1) The group of rules $(x * s)$ use ascending strength of $\mathcal{I}/\mathcal{F}$.
(2) The column $(\mathcal{M}^+)$ contains interesting algebraic properties. In particular, they show a strengthening from $(3 * s)$ up to rationality. They are not necessarily equivalent to the corresponding (I_x) rules, not even in the presence of the basic rules. The examples show that care has to be taken when considering the different variants.
(3) Adding the somewhat superfluous (CM_2), we have increasing cautious monotony from (wCM) to full (CM_ω).
(4) We have increasing "or" from (wOR) to full (OR_ω).
(5) The line $(2 * s)$ is only there because there seems to be no $(\mathcal{M}_2^+)$, otherwise we could begin $(n * s)$ at $n = 2$.

3.2.2.3 Direct Correspondences

Several correspondences are trivial and are mentioned now. Somewhat less obvious (independencies are given in Sect. 3.2.3 (p. 64). Finally, the connections with the μ-rules are given in Sect. 3.2.4 (p. 69). In those rules, (I_ω) is implicit, as they are about principal filters. Still, the μ-rules are written in the table "Rules on size" in their intuitively adequate place.

(1) The columns "Ideal" and "Filter" are mutually dual, when both entries are defined.
(2) The correspondence between the ideal/filter column and the ∇-column is obvious, the latter is added only for completeness' sake and to point out the trivial translation to (augmented) first-order logic. Note that the rules in the ∇-column are object-level axioms.
(3) The ideal/filter and the AND-column correspond directly.
(4) We can construct logical rules from the $\mathcal{M}^+$-column by direct correspondence, e.g. for $\left(\mathcal{M}_\omega^+\right)$, (1):
Set $Y := M(\gamma)$, $X := M(\gamma \wedge \beta)$, $A := M(\gamma \wedge \beta \wedge \alpha)$.

- $X \in \mathcal{M}^+(Y)$ will become $\gamma \not\mid\!\sim \neg\beta$,
- $A \in \mathcal{F}(X)$ will become $\gamma \wedge \beta \mid\!\sim \alpha$,
- $A \in \mathcal{M}^+(Y)$ will become $\gamma \not\mid\!\sim \neg(\alpha \wedge \beta)$.

So we obtain $\gamma \not\mid\!\sim \neg\beta,\ \gamma \wedge \beta \mid\!\sim \alpha \Rightarrow \gamma \not\mid\!\sim \neg(\alpha \wedge \beta)$.
We did not want to make the table too complicated, so such rules are not listed in the table.
(5) Various direct correspondences:

- In the line (Opt), the filter/ideal entry corresponds to (SC),
- in the line (iM), the filter/ideal entry corresponds to (RW),
- in the line $(eM\mathcal{I})$, the ideal entry corresponds to (PR') and (wOR),
- in the line $(eM\mathcal{F})$, the filter entry corresponds to (wCM),
- in the line $(\approx)$, the filter/ideal entry corresponds to $(disjOR)$ and (NR),

- in the line $(1 * s)$, the filter/ideal entry corresponds to (CP),
- in the line $(2 * s)$, the filter/ideal entry corresponds to $(CM_2) = (OR_2)$.

(6) Note that one can, e.g. write (AND_2) in two flavours:

- $\alpha \mid\sim \beta, \alpha \mid\sim \beta' \Rightarrow \alpha \not\mid\sim \neg\beta \vee \neg\beta'$ or
- $\alpha \mid\sim \beta \Rightarrow \alpha \not\mid\sim \neg\beta$

– this is $(CM_2) = (OR_2)$.
For reasons of simplicity, we mention only one.

3.2.2.4 Rational Monotony

$(RatM)$ does not fit into adding small sets. We have exhausted the combination of small sets by $(< \omega * s)$, unless we go to languages with infinitary formulas.

The next idea would be to add medium size sets. But, by definition, $2 * medium$ can be all. Adding small and medium sets would not help either: Suppose we have a rule $medium + n * small \neq all$. Taking the complement of the first medium set, which is again medium, we have the rule $2 * n * small \neq all$. So we do not see any meaningful new internal rule, i.e. without changing the base set.

(For a similar reason, rules like $A, B \in \mathcal{I}(X) \Rightarrow A \cup B \notin \mathcal{F}(X)$ will give us nothing new, as then $A \cup B \in \mathcal{I}(X)$ is the full filter/ideal property, and $A \cup B \in \mathcal{M}(X)$ means just that four subsets of X may add up to the full X.)

Probably, $(RatM)$ has more to do with independence: by default, all "normalities" are independent, and intersecting with another formula preserves normality as much as possible. One should not forget here either the double use of "small" sets in our context: Nonmonotonicity works with "small" exception sets and many rules concern modifying the base set by a "small" subset. See also Remark 3.2.1 (p. 58) below.

Still, rational monotony has its natural place in an ascending chain of conditions, which can be seen by looking at the $\mathcal{M}^+$ conditions, especially at $\left(\mathcal{M}^+_\omega\right)$ (1) – (4) and $(\mathcal{M}^{++})$ (3). Further research might tell whether there are still deeper connections behind this formal series.

More remarks on rules beyond rationality can be found in Sect. 3.2.2.8 (p. 63).

3.2.2.5 Summary

We can obtain all rules except $(RatM)$ and $(\approx)$ from (Opt), the monotony rules – (iM), $(eM\mathcal{I})$, $(eM\mathcal{F})$ – and $(x * s)$ with increasing x.

3.2.2.6 Main Table

The following table is split into two (Tables 3.1 and 3.2), as it is too big for printing in one page.

Remark 3.2.1 There is, however, an important conceptual distinction to make here. Filters express "size" in an abstract way, in the context of nonmonotonic logics,

Table 3.1 Rules on size – Part I

	"Ideal"	"Filter"	$\mathcal{M}^+$	∇
		Optimal proportion		
(Opt)	$\emptyset \in \mathcal{I}(X)$	$X \in \mathcal{F}(X)$		$\forall x\alpha \to \nabla x\alpha$
	Monotony (improving proportions). (iM): internal monotony, $(eM\mathcal{I})$: external monotony for ideals, $(eM\mathcal{F})$: external monotony for filters			
(iM)	$A \subseteq B \in \mathcal{I}(X)$ $\Rightarrow$ $A \in \mathcal{I}(X)$	$A \in \mathcal{F}(X)$, $A \subseteq B \subseteq X$ $\Rightarrow B \in \mathcal{F}(X)$		$\nabla x\alpha \wedge \forall x(\alpha \to \alpha')$ $\to \nabla x\alpha'$
$(eM\mathcal{I})$	$X \subseteq Y \Rightarrow$ $\mathcal{I}(X) \subseteq \mathcal{I}(Y)$			$\nabla x(\alpha : \beta) \wedge$ $\forall x(\alpha' \to \beta) \to$ $\nabla x(\alpha \vee \alpha' : \beta)$
$(eM\mathcal{F})$		$X \subseteq Y \Rightarrow$ $\mathcal{F}(Y) \cap \mathcal{P}(X) \subseteq$ $\mathcal{F}(X)$		$\nabla x(\alpha : \beta) \wedge$ $\forall x(\beta \wedge \alpha \to \alpha') \to$ $\nabla x(\alpha \wedge \alpha' : \beta)$
		Keeping proportions		
$(\approx)$	$(\mathcal{I} \cup disj)$ $A \in \mathcal{I}(X)$, $B \in \mathcal{I}(Y)$, $X \cap Y = \emptyset \Rightarrow$ $A \cup B \in \mathcal{I}(X \cup Y)$	$(\mathcal{F} \cup disj)$ $A \in \mathcal{F}(X)$, $B \in \mathcal{F}(Y)$, $X \cap Y = \emptyset \Rightarrow$ $A \cup B \in \mathcal{F}(X \cup Y)$	$(\mathcal{M}^+ \cup disj)$ $A \in \mathcal{M}^+(X)$, $B \in \mathcal{M}^+(Y)$, $X \cap Y = \emptyset \Rightarrow$ $A \cup B \in \mathcal{M}^+(X \cup Y)$	$\nabla x(\alpha : \beta) \wedge$ $\nabla x(\alpha' : \beta) \wedge$ $\neg\exists x(\alpha \wedge \alpha') \to$ $\nabla x(\alpha \vee \alpha' : \beta)$
		Robustness of proportions: $n * small \neq All$		
$(1 * s)$	$(\mathcal{I}_1)$ $X \notin \mathcal{I}(X)$	$(\mathcal{F}_1)$ $\emptyset \notin \mathcal{F}(X)$		(∇_1) $\nabla x\alpha \to \exists x\alpha$
$(2 * s)$	$(\mathcal{I}_2)$ $A, B \in \mathcal{I}(X) \Rightarrow$ $A \cup B \neq X$	$(\mathcal{F}_2)$ $A, B \in \mathcal{F}(X) \Rightarrow$ $A \cap B \neq \emptyset$		(∇_2) $\nabla x\alpha \wedge \nabla x\beta$ $\to \exists x(\alpha \wedge \beta)$
$(n * s)$ $(n \geq 3)$	$(\mathcal{I}_n)$ $A_1, \ldots, A_n \in \mathcal{I}(X)$ $\Rightarrow$ $A_1 \cup \cdot \cup A_n \neq X$	$(\mathcal{F}_n)$ $A_1, \ldots, A_n \in \mathcal{I}(X)$ $\Rightarrow$ $A_1 \cap \cdot \cap A_n \neq \emptyset$	$(\mathcal{M}^+_n)$ $X_1 \in \mathcal{F}(X_2), \ldots,$ $X_{n-1} \in \mathcal{F}(X_n) \Rightarrow$ $X_1 \in \mathcal{M}^+(X_n)$	(∇_n) $\nabla x\alpha_1 \wedge \cdot \wedge \nabla x\alpha_n$ $\to$ $\exists x(\alpha_1 \wedge \cdot \wedge \alpha_n)$
$(< \omega * s)$	$(\mathcal{I}_\omega)$ $A, B \in \mathcal{I}(X) \Rightarrow$ $A \cup B \in \mathcal{I}(X)$	$(\mathcal{F}_\omega)$ $A, B \in \mathcal{F}(X) \Rightarrow$ $A \cap B \in \mathcal{F}(X)$	$(\mathcal{M}^+_\omega)$ (1) $A \in \mathcal{F}(X)$, $X \in \mathcal{M}^+(Y)$ $\Rightarrow A \in \mathcal{M}^+(Y)$ (2) $A \in \mathcal{M}^+(X)$, $X \in \mathcal{F}(Y)$ $\Rightarrow A \in \mathcal{M}^+(Y)$ (3) $A \in \mathcal{F}(X)$, $X \in \mathcal{F}(Y)$ $\Rightarrow A \in \mathcal{F}(Y)$ (4) $A, B \in \mathcal{I}(X) \Rightarrow$ $A - B \in \mathcal{I}(X{-}B)$	(∇_ω) $\nabla x\alpha \wedge \nabla x\beta \to$ $\nabla x(\alpha \wedge \beta)$

Table 3.1 (continued)

	"Ideal" . "Filter"	$\mathcal{M}^+$	∇
	Robustness of $\mathcal{M}^+$		
$(\mathcal{M}^{++})$	. . . .	$(\mathcal{M}^{++})$ (1) $A \in \mathcal{I}(X)$, $B \notin \mathcal{F}(X)$ $\Rightarrow A - B \in \mathcal{I}(X - B)$ (2) $A \in \mathcal{F}(X)$, $B \notin \mathcal{F}(X)$ $\Rightarrow A - B \in \mathcal{F}(X - B)$ (3) $A \in \mathcal{M}^+(X)$, $X \in \mathcal{M}^+(Y)$ $\Rightarrow A \in \mathcal{M}^+(Y)$	

$\alpha \mid\!\sim \beta$ iff the set of $\alpha \wedge \neg\beta$ is small in α. But here, we were interested in "small" changes in the reference set X (or α in our example). So we have two quite different uses of "size", one for nonmonotonic logics, abstractly expressed by a filter, the other for coherence conditions. It is possible, but not necessary, to consider both essentially the same notions. But we should not forget that we have two conceptually different uses of size here.

3.2.2.7 The Filter and the Partial Order View

N. Friedman and J. Halpern have presented an alternative abstract view on non-monotonic logics, perhaps best accessible in [FH96]. Their approach is via a partial order, essentially re-writing $A \in \mathcal{I}(X)$ as $A < X$. Thus, ideals and filters are replaced by a binary relation, expressing size. The (almost perfect) equivalence to the approach of [BB94] was shown in [Sch97-1].

Makinson (in personal communication) has suggested the following approach in the spirit of [FH96]:

> Define a 2-place function f into a set with a linear order $\leq$, and with threshold elements s and l, $s \leq l$, and rewrite $A \in \mathcal{I}(X)$ as $f(A, X) \leq s$, etc.
> To speak about medium size elements, we need another element m between s and l.
> In this way, we can translate
>
> - (Opt) to: $f(\emptyset, X) = s$, $f(X, X) = l$.
> - (im) to: if $A \subseteq X \subseteq Y$, then $f(X, Y) = s \Rightarrow f(A, Y) = s$ and $f(A, Y) = l \Rightarrow f(X, Y) = l$.
> - (eMI) to: if $A \subseteq X \subseteq Y$, then $f(A, X) = s \Rightarrow f(A, Y) = s$.
> - (eMF) to: if $A \subseteq X \subseteq Y$, then $f(A, Y) = l \Rightarrow f(A, X) = l$.
>
> In this translation, (eMI) and (eMF) are special cases of the principle:
> $(Order2)$ if $A \subseteq X \subseteq Y$, then $f(A, Y) \leq f(A, X)$,
> and (im) can be written as
> $(Order1)$ if $A \subseteq X \subseteq Y$, then $f(A, Y) \leq f(X, Y)$,

Table 3.2 Rules on size – Part II

	Various rules	AND	OR	Caut./rat.mon.
Optimal proportion				
(Opt)	(SC) $\alpha \vdash \beta \Rightarrow \alpha \mid\sim \beta$			
Monotony (improving proportions)				
(iM)	(RW) $\alpha \mid\sim \beta, \beta \vdash \beta' \Rightarrow$ $\alpha \mid\sim \beta'$			
$(eM\mathcal{I})$	(PR') $\alpha \mid\sim \beta, \alpha \vdash \alpha',$ $\alpha' \wedge \neg\alpha \vdash \beta \Rightarrow$ $\alpha' \mid\sim \beta$ (μPR) $X \subseteq Y \Rightarrow$ $\mu(Y) \cap X \subseteq \mu(X)$		(wOR) $\alpha \mid\sim \beta, \alpha' \vdash \beta \Rightarrow$ $\alpha \vee \alpha' \mid\sim \beta$ (μwOR) $\mu(X \cup Y) \subseteq \mu(X) \cup Y$	
$(eM\mathcal{F})$				(wCM) $\alpha \mid\sim \beta, \alpha' \vdash \alpha,$ $\alpha \wedge \beta \vdash \alpha' \Rightarrow$ $\alpha' \mid\sim \beta$
Keeping proportions				
$(\approx)$	(NR) $\alpha \mid\sim \beta \Rightarrow$ $\alpha \wedge \gamma \mid\sim \beta$ or $\alpha \wedge \neg\gamma \mid\sim \beta$		$(disjOR)$ $\alpha \mid\sim \beta, \alpha' \mid\sim \beta'$ $\alpha \vdash \neg\alpha', \Rightarrow$ $\alpha \vee \alpha' \mid\sim \beta \vee \beta'$ $(\mu disjOR)$ $X \cap Y = \emptyset \Rightarrow$ $\mu(X \cup Y) \subseteq \mu(X) \cup \mu(Y)$	

Table 3.2 (continued)

	Various rules	AND	OR	Caut./rat.mon.
		Robustness of proportions: $n * small \neq All$		
$(1 * s)$	(CP) $\alpha \mid\sim \bot \Rightarrow \alpha \vdash \bot$	(AND_1) $\alpha \mid\sim \beta \Rightarrow \alpha \not\vdash \neg\beta$		
$(2 * s)$		(AND_2) $\alpha \mid\sim \beta, \alpha \mid\sim \beta' \Rightarrow$ $\alpha \not\vdash \neg\beta \vee \neg\beta'$	(OR_2) $\alpha \mid\sim \beta \Rightarrow \alpha \not\mid\sim \neg\beta$	(CM_2) $\alpha \mid\sim \beta \Rightarrow \alpha \not\mid\sim \neg\beta$
$(n * s)$ $(n \geq 3)$		(AND_n) $\alpha \mid\sim \beta_1, \ldots, \alpha \mid\sim \beta_n$ $\Rightarrow$ $\alpha \not\vdash \neg\beta_1 \vee \cdot \vee \neg\beta_n$	(OR_n) $\alpha_1 \mid\sim \beta, \ldots, \alpha_{n-1} \mid\sim \beta$ $\Rightarrow$ $\alpha_1 \vee \cdot \vee \alpha_{n-1} \not\mid\sim \neg\beta$	(CM_n) $\alpha \mid\sim \beta_1, \ldots, \alpha \mid\sim \beta_{n-1}$ $\Rightarrow$ $\alpha \wedge \beta_1 \wedge \cdot \wedge \beta_{n-2} \not\mid\sim$ $\neg\beta_{n-1}$
$(< \omega * s)$		(AND_ω) $\alpha \mid\sim \beta, \alpha \mid\sim \beta' \Rightarrow$ $\alpha \mid\sim \beta \wedge \beta'$	(OR_ω) $\alpha \mid\sim \beta, \alpha' \mid\sim \beta \Rightarrow$ $\alpha \vee \alpha' \mid\sim \beta$ (μOR) $\mu(X \cup Y) \subseteq \mu(X) \cup \mu(Y)$	(CM_ω) $\alpha \mid\sim \beta, \alpha \mid\sim \beta' \Rightarrow$ $\alpha \wedge \beta \mid\sim \beta'$ (μCM) $\mu(X) \subseteq Y \subseteq X \Rightarrow$ $\mu(Y) \subseteq \mu(X)$
		Robustness of $\mathcal{M}^+$		
$(\mathcal{M}^{++})$				$(RatM)$ $\alpha \mid\sim \beta, \alpha \not\mid\sim \neg\beta' \Rightarrow$ $\alpha \wedge \beta' \mid\sim \beta$ $(\mu RatM)$ $X \subseteq Y,$ $X \cap \mu(Y) \neq \emptyset \Rightarrow$ $\mu(X) \subseteq \mu(Y) \cap X$

both can be put together to obtain

$(Order)$ if $A \subseteq X \subseteq Y$, then $f(A, Y) \leq \min\{f(X, Y), f(A, X)\}$,

and also (Opt) be added to obtain finally

$(Order^+)$ if $A \subseteq X \subseteq Y$, then $s = f(\emptyset, X) \leq f(A, Y) \leq \min\{f(X, Y), f(A, X)\} \leq f(X, X) = l$.

It might be a question of taste which language one prefers. Above principle $(Order2)$ has the advantage to unite the intuitively close (eMI) and (eMF), whereas our separate notation underlines that our intuition might be false, as there are examples which separate them.

3.2.2.8 Discussion of Other, Related, Rules

We discuss here briefly more rules found in the literature – and express our gratitude to the referee who pointed them out to us. One of them, (NR), fits into our "keeping proportions" line and is integrated there; another one, (DR), is the unitary version of our $(Log \parallel)$, and put in the table "Logical rules".

From [BMP97], we cite the following, see Table 2.1:

- Disjunctive rationality
  $(DR)\ \alpha \vee \beta \mid\sim \gamma \Rightarrow \alpha \mid\sim \gamma$ or $\beta \mid\sim \gamma$
  holds in ranked structures and corresponds to our $(Log \parallel)$.
- Negation rationality
  $(NR)\ \alpha \mid\sim \beta \Rightarrow \alpha \wedge \gamma \mid\sim \beta$ or $\alpha \wedge \neg\gamma \mid\sim \beta$
  corresponds to the $(\mathcal{M}^+)$ variant of $(\approx)$, $(\mathcal{M}^+ \cup disj)$.

The following four rules, also taken from [BMP97], all fall in a class we may call $(ULTRA)$. Looking at them from a size perspective, they say that some filters must be ultrafilters (i.e. there are no medium size sets, everything is either small or big). From the preferential perspective, it means that there is at most one minimal model.

- Determinacy preservation
  $(DP)\ \alpha \mid\sim \beta \Rightarrow \alpha \wedge \gamma \mid\sim \beta$ or $\alpha \wedge \gamma \mid\sim \neg\beta$.
- Rational transitivity
  $(RT)\ \alpha \mid\sim \beta, \beta \mid\sim \gamma \Rightarrow \alpha \mid\sim \gamma$ or $\alpha \mid\sim \neg\gamma$.
- Rational contraposition
  $(RC)\ \alpha \mid\sim \beta \Rightarrow \neg\beta \mid\sim \alpha$ or $\neg\beta \mid\sim \neg\alpha$.
- Weak determinacy
  $(WD)\ True \mid\sim \neg\alpha \Rightarrow \alpha \mid\sim \beta$ or $\alpha \mid\sim \neg\beta$.

We think that these rules may well be adapted to certain situations, but will not be able to solve more general problems like the blonde Swedes problem. (Swedes are normally blonde and tall, but even not-blonde Swedes are tall. This is the problem of subideal situations, where still as much as possible of the ideal is preserved.)

We cite from [HM07], see also [Haw96], [Haw07], the following rules satisfied by probabilistic consequences relation (defined as validity above a fixed value threshold):

- Very cautious monotony
  $(VCM)\ \alpha \mid\sim \beta \wedge \gamma \Rightarrow \alpha \wedge \beta \mid\sim \gamma$,
  which we can deduce from (I_ω) the same way as (CM_ω), using other standard rules.
- Weak OR
  $(WOR)\ \alpha \wedge \beta \mid\sim \gamma, \alpha \wedge \neg\beta \mid\sim \gamma \Rightarrow \alpha \mid\sim \gamma$
  corresponds to our $(disjOR)$.
- Weak AND
  $(WAND)\ \alpha \mid\sim \gamma, \alpha \wedge \neg\beta \mid\sim \beta \Rightarrow \alpha \mid\sim \beta \wedge \gamma$
  will not be discussed here, as the prerequisite $\alpha \wedge \neg\beta \mid\sim \beta$ entails that $\emptyset$ can be a big subset.

3.2.3 Coherent Systems

3.2.3.1 Definition and Basic Facts

Note that whenever we work with model sets, the rule (LLE), left logical equivalence, $\vdash \alpha \leftrightarrow \alpha' \Rightarrow (\alpha \mid\sim \beta \Leftrightarrow \alpha' \mid\sim \beta)$ will hold. We will use this without further mentioning.

Definition 3.2.1 A coherent system of sizes, $\mathcal{CS}$, consists of a universe U, $\emptyset \notin \mathcal{Y} \subseteq \mathcal{P}(U)$, and for all $X \in \mathcal{Y}$ a system $\mathcal{I}(X) \subseteq \mathcal{P}(X)$ (dually $\mathcal{F}(X)$, i.e. $A \in \mathcal{F}(X) \Leftrightarrow X - A \in \mathcal{I}(X)$). $\mathcal{Y}$ may satisfy certain closure properties like closure under $\cup$, $\cap$, complementation, etc. We will mention this when needed, and not obvious.

> We say that $\mathcal{CS}$ satisfies a certain property iff all $X, Y \in \mathcal{Y}$ satisfy this property. $\mathcal{CS}$ is called basic or level 1 iff it satisfies (Opt), (iM), $(eM\mathcal{I})$, $(eM\mathcal{F})$, $(1{*}s)$. $\mathcal{CS}$ is level x iff it satisfies (Opt), (iM), $(eM\mathcal{I})$, $(eM\mathcal{F})$, $(x * s)$.

Fact 3.2.1 Note that if for any Y, $\mathcal{I}(Y)$ consists only of subsets of at most one element, then $(eM\mathcal{F})$ is trivially satisfied for Y and its subsets by (Opt).

Fact 3.2.2 Let a $\mathcal{CS}$ be given s.t. $\mathcal{Y} = \mathcal{P}(U)$. If $X \in \mathcal{Y}$ satisfies $(\mathcal{M}^{++})$, but not $(< \omega * s)$, then there is $Y \in \mathcal{Y}$ which does not satisfy $(2 * s)$.

Proof We work with version (1) of $(\mathcal{M}^{++})$; we will see in Fact 3.2.8 (p. 68) that all three versions are equivalent.
As X does not satisfy $(< \omega * s)$, there are $A, B \in \mathcal{I}(X)$ s.t. $A \cup B \in \mathcal{M}^+(X)$. $A \in \mathcal{I}(X)$, $A \cup B \in \mathcal{M}^+(X) \Rightarrow X - (A \cup B) \notin \mathcal{F}(X)$, so by $(\mathcal{M}^{++})(1)$ $A = A-(X-(A\cup B)) \in \mathcal{I}(X-(X-(A\cup B))) = \mathcal{I}(A\cup B)$. Likewise $B \in \mathcal{I}(A\cup B)$, so $(2 * s)$ does not hold for $A \cup B$. □

Fact 3.2.3 $(eM\mathcal{I})$ and $(eM\mathcal{F})$ are formally independent, though intuitively equivalent.

Proof Let $U := \{x, y, z\}$, $X := \{x, z\}$, $\mathcal{Y} := \mathcal{P}(U) - \{\emptyset\}$

(1) Let $\mathcal{F}(U) := \{A \subseteq U : z \in A\}$, $\mathcal{F}(Y) = \{Y\}$ for all $Y \subset U$. (Opt), (iM) hold, $(eM\mathcal{I})$ holds trivially, so does $(< \omega * s)$, but $(eM\mathcal{F})$ fails for U and X.
(2) Let $\mathcal{F}(X) := \{\{z\}, X\}$, $\mathcal{F}(Y) := \{Y\}$ for all $Y \subseteq U$, $Y \neq X$. (Opt), (iM), $(< \omega * s)$ hold trivially, $(eM\mathcal{F})$ holds by Fact 3.2.1 (p. 64). $(eM\mathcal{I})$ fails, as $\{x\} \in \mathcal{I}(X)$, but $\{x\} \notin \mathcal{I}(U)$. □

Fact 3.2.4 A level n system is strictly weaker than a level $n + 1$ system.

Proof Consider $U := \{1, \ldots, n + 1\}$, $\mathcal{Y} := \mathcal{P}(U) - \{\emptyset\}$. Let $\mathcal{I}(U) := \{\emptyset\} \cup \{\{x\} : x \in U\}$, $\mathcal{I}(X) := \{\emptyset\}$ for $X \neq U$. (iM), $(eM\mathcal{I})$, $(eM\mathcal{F})$ hold trivially. $(n * s)$ holds trivially for $X \neq U$, but also for U. $((n + 1) * s)$ does not hold for U. □

Remark 3.2.2 Note that our schemata allow us to generate infinitely many new rules, here is an example:

Start with A, add $s_{1,1}$, $s_{1,2}$ two sets small in $A \cup s_{1,1}$ ($A \cup s_{1,2}$, respectively). Consider now $A \cup s_{1,1} \cup s_{1,2}$ and s_2 s.t. s_2 is small in $A \cup s_{1,1} \cup s_{1,2} \cup s_2$. Continue with $s_{3,1}$, $s_{3,2}$ small in $A \cup s_{1,1} \cup s_{1,2} \cup s_2 \cup s_{3,1}$, etc. Without additional properties this system creates a new rule, which is not equivalent to any usual rules.

3.2.3.2 Implications Between the Finite Versions

Fact 3.2.5

(1) $(I_n) + (eM\mathcal{I}) \Rightarrow (\mathcal{M}_n^+)$,
(2) $(I_n) + (eM\mathcal{I}) \Rightarrow (CM_n)$, and
(3) $(I_n) + (eM\mathcal{I}) \Rightarrow (OR_n)$.

Proof (1) Let $X_1 \subseteq \ldots \subseteq X_n$, so $X_n = X_1 \cup (X_2 - X_1) \cup \ldots \cup (X_n - X_{n-1})$. Let $X_i \in \mathcal{F}(X_{i+1})$, so $X_{i+1} - X_i \in \mathcal{I}(X_{i+1}) \subseteq \mathcal{I}(X_n)$ by $(eM\mathcal{I})$ for $1 \leq i \leq n-1$, so by (I_n) $X_1 \in \mathcal{M}^+(X_n)$.
(2) Suppose $\alpha \mid\sim \beta_1, \ldots, \alpha \mid\sim \beta_{n-1}$, but $\alpha \wedge \beta_1 \wedge \ldots \wedge \beta_{n-2} \mid\sim \neg\beta_{n-1}$. Then $M(\alpha \wedge \neg\beta_1), \ldots, M(\alpha \wedge \neg\beta_{n-1}) \in \mathcal{I}(M(\alpha))$ and $M(\alpha \wedge \beta_1 \wedge \ldots \wedge \beta_{n-2} \wedge \beta_{n-1}) \in \mathcal{I}(M(\alpha \wedge \beta_1 \wedge \ldots \wedge \beta_{n-2})) \subseteq \mathcal{I}(M(\alpha))$ by $(eM\mathcal{I})$. But $M(\alpha) = M(\alpha \wedge \neg\beta_1) \cup \ldots \cup M(\alpha \wedge \neg\beta_{n-1}) \cup M(\alpha \wedge \beta_1 \wedge \ldots \wedge \beta_{n-2} \wedge \beta_{n-1})$ is now the union of n small subsets, *contradiction*.
(3) Let $\alpha_1 \mid\sim \beta, \ldots, \alpha_{n-1} \mid\sim \beta$, so $M(\alpha_i \wedge \neg\beta) \in \mathcal{I}(M(\alpha_i))$ for $1 \leq i \leq n-1$, so $M(\alpha_i \wedge \neg\beta) \in \mathcal{I}(M(\alpha_1 \vee \ldots \vee \alpha_{n-1}))$ for $1 \leq i \leq n-1$ by $(eM\mathcal{I})$, so $M((\alpha_1 \vee \ldots \vee \alpha_{n-1}) \wedge \beta) = M(\alpha_1 \vee \ldots \vee \alpha_{n-1}) - \bigcup\{M(\alpha_i \wedge \neg\beta) : 1 \leq i \leq n-1\}$ $\notin \mathcal{I}(M(\alpha_1 \vee \ldots \vee \alpha_{n-1}))$ by (I_n), so $\alpha_1 \vee \ldots \vee \alpha_{n-1} \not\mid\sim \neg\beta$. □

In the following example, (OR_n), $(\mathcal{M}_n^+)$, (CM_n) hold, but $(\mathcal{I}_n)$ fails, so by Fact 3.2.5 (p. 65) $(\mathcal{I}_n)$ is strictly stronger than (OR_n), $(\mathcal{M}_n^+)$, (CM_n).

Example 3.2.1 Let $n \geq 3$. Consider $X := \{1, \ldots, n\}$, $\mathcal{Y} := \mathcal{P}(X) - \{\emptyset\}$, $\mathcal{I}(X) := \{\emptyset\} \cup \{\{i\} : 1 \leq i \leq n\}$, and for all $Y \subset X$ $\mathcal{I}(Y) := \{\emptyset\}$.

(Opt), (iM), $(eM\mathcal{I})$, $(eM\mathcal{F})$ (by Fact 3.2.1 (p. 64)), $(1 * s)$, $(2 * s)$ hold, (I_n) fails, of course.

(1) (OR_n) holds: Suppose $\alpha_1 \mid\sim \beta, \ldots, \alpha_{n-1} \mid\sim \beta, \alpha_1 \vee \ldots \vee \alpha_{n-1} \mid\sim \neg\beta$.
Case 1: $\alpha_1 \vee \ldots \vee \alpha_{n-1} \vdash \neg\beta$, then for all i $\alpha_i \vdash \neg\beta$, so for no i $\alpha_i \mid\sim \beta$ by $(1 * s)$ and thus (AND_1), *contradiction*.
Case 2: $\alpha_1 \vee \ldots \vee \alpha_{n-1} \not\vdash \neg\beta$, then $M(\alpha_1 \vee \ldots \vee \alpha_{n-1}) = X$, and there is exactly 1 $k \in X$ s.t. $k \models \beta$. Fix this k. By prerequisite, $\alpha_i \mid\sim \beta$. If $M(\alpha_i) = X$, $\alpha_i \vdash \beta$ cannot be, so there must be exactly 1 k' s.t. $k' \models \neg\beta$, but $card(X) \geq 3$, *contradiction*. So $M(\alpha_i) \subset X$ and $\alpha_i \vdash \beta$, so $M(\alpha_i) = \emptyset$ or $M(\alpha_i) = \{k\}$ for all i, so $M(\alpha_1 \vee \ldots \vee \alpha_{n-1}) \neq X$, *contradiction*.
(2) $(\mathcal{M}_n^+)$ holds: $(\mathcal{M}_n^+)$ is a consequence of $(\mathcal{M}_\omega^+)$ (3), so it suffices to show that the latter holds. Let $X_1 \in \mathcal{F}(X_2)$, $X_2 \in \mathcal{F}(X_3)$. Then $X_1 = X_2$ or $X_2 = X_3$, so the result is trivial.
(3) (CM_n) holds: Suppose $\alpha \mid\sim \beta_1, \ldots, \alpha \mid\sim \beta_{n-1}, \alpha \wedge \beta_1 \wedge \ldots \wedge \beta_{n-2} \mid\sim \neg\beta_{n-1}$.
Case 1: For all i, $1 \leq i \leq n-2$, $\alpha \vdash \beta_i$, then $M(\alpha \wedge \beta_1 \wedge \ldots \wedge \beta_{n-2}) = M(\alpha)$, so $\alpha \mid\sim \beta_{n-1}$ and $\alpha \mid\sim \neg\beta_{n-1}$, *contradiction*.
Case 2: There is i, $1 \leq i \leq n-2$, $\alpha \not\vdash \beta_i$, then $M(\alpha) = X$, $M(\alpha \wedge \beta_1 \wedge \ldots \wedge \beta_{n-2}) \subset M(\alpha)$, so $\alpha \wedge \beta_1 \wedge \ldots \wedge \beta_{n-2} \vdash \neg\beta_{n-1}$. $Card(M(\alpha \wedge \beta_1 \wedge \ldots \wedge \beta_{n-2})) \geq n-(n-2) = 2$, so $card(M(\neg\beta_{n-1})) \geq 2$, so $\alpha \not\mid\sim \beta_{n-1}$, *contradiction*.

3.2.3.3 Implications Between the ω Versions

Fact 3.2.6 $(CM_\omega) \Leftrightarrow (\mathcal{M}_\omega^+)$ (4)

Proof "$\Rightarrow$" Suppose all sets are definable.

Let $A, B \in \mathcal{I}(X)$, $X = M(\alpha)$, $A = M(\alpha \wedge \neg\beta)$, $B = M(\alpha \wedge \neg\beta')$, so $\alpha \mid\sim \beta$, $\alpha \mid\sim \beta'$, so by (CM_ω) $\alpha \wedge \beta' \mid\sim \beta$, so $A - B = M(\alpha \wedge \beta' \wedge \neg\beta) \in \mathcal{I}(M(\alpha \wedge \beta')) = \mathcal{I}(X - B)$.

"$\Leftarrow$"

Let $\alpha \mid\sim \beta$, $\alpha \mid\sim \beta'$, so $M(\alpha \wedge \neg\beta) \in \mathcal{I}(M(\alpha))$, $M(\alpha \wedge \neg\beta') \in \mathcal{I}(M(\alpha))$, so by prerequisite $M(\alpha \wedge \neg\beta') - M(\alpha \wedge \neg\beta) = M(\alpha \wedge \beta \wedge \neg\beta') \in \mathcal{I}(M(\alpha) - M(\alpha \wedge \neg\beta)) = \mathcal{I}(M(\alpha \wedge \beta))$, so $\alpha \wedge \beta \mid\sim \beta'$. □

Fact 3.2.7

(1) $(I_\omega) + (eM\mathcal{I}) \Rightarrow (OR_\omega)$,
(2) $(I_\omega) + (eM\mathcal{I}) \Rightarrow (\mathcal{M}_\omega^+)$ (1),
(3) $(I_\omega) + (eM\mathcal{F}) \Rightarrow (\mathcal{M}_\omega^+)$ (2),
(4) $(I_\omega) + (eM\mathcal{I}) \Rightarrow (\mathcal{M}_\omega^+)$ (3),
(5) $(I_\omega) + (eM\mathcal{F}) \Rightarrow (\mathcal{M}_\omega^+)$ (4) (and thus, by Fact 3.2.6 (p. 66), (CM_ω)).

Proof (1) Let $\alpha \mid\sim \beta$, $\alpha' \mid\sim \beta \Rightarrow M(\alpha \wedge \neg\beta) \in \mathcal{I}(M(\alpha))$, $M(\alpha' \wedge \neg\beta) \in \mathcal{I}(M(\alpha'))$, so by $(eM\mathcal{I})$ $M(\alpha \wedge \neg\beta) \in \mathcal{I}(M(\alpha \vee \alpha'))$, $M(\alpha' \wedge \neg\beta) \in \mathcal{I}(M(\alpha \vee \alpha'))$, so $M((\alpha \vee \alpha') \wedge \neg\beta) \in \mathcal{I}(M(\alpha \vee \alpha'))$ by (I_ω), so $\alpha \vee \alpha' \mid\sim \beta$.

(2) Let $A \subseteq X \subseteq Y$, $A \in \mathcal{I}(Y)$, $X - A \in \mathcal{I}(X) \subseteq_{(eM\mathcal{I})} \mathcal{I}(Y) \Rightarrow X = (X-A) \cup A \in \mathcal{I}(Y)$ by (I_ω).

(3) Let $A \subseteq X \subseteq Y$, let $A \in \mathcal{I}(Y)$, $Y - X \in \mathcal{I}(Y) \Rightarrow A \cup (Y - X) \in \mathcal{I}(Y)$ by (I_ω) $\Rightarrow X - A = Y - (A \cup (Y - X)) \in \mathcal{F}(Y) \Rightarrow X - A \in \mathcal{F}(X)$ by $(eM\mathcal{F})$.

(4) Let $A \subseteq X \subseteq Y$, $A \in \mathcal{F}(X)$, $X \in \mathcal{F}(Y)$, so $Y - X \in \mathcal{I}(Y)$, $X - A \in \mathcal{I}(X) \subseteq_{(eM\mathcal{I})} \mathcal{I}(Y) \Rightarrow Y - A = (Y - X) \cup (X - A) \in \mathcal{I}(Y)$ by $(\mathcal{I}_\omega) \Rightarrow A \in \mathcal{F}(Y)$.

(5) Let $A, B \subseteq X$, $A, B \in \mathcal{I}(X) \Rightarrow_{(I_\omega)} A \cup B \in \mathcal{I}(X) \Rightarrow X - (A \cup B) \in \mathcal{F}(X)$, but $X - (A \cup B) \subseteq X - B$, so $X - (A \cup B) \in \mathcal{F}(X - B)$ by $(eM\mathcal{F})$, so $A - B = (X - B) - (X - (A \cup B)) \in \mathcal{I}(X - B)$. □

We give three examples of independence of the various versions of $(\mathcal{M}^+_\omega)$.

Example 3.2.2 All numbers refer to the versions of $(\mathcal{M}^+_\omega)$.

For easier reading, we rewrite for $A \subseteq X \subseteq Y$

$$(\mathcal{M}^+_\omega)\ (1)\text{: } A \in \mathcal{F}(X),\ A \in \mathcal{I}(Y) \Rightarrow X \in \mathcal{I}(Y),$$

$$(\mathcal{M}^+_\omega)\ (2)\text{: } X \in \mathcal{F}(Y),\ A \in \mathcal{I}(Y) \Rightarrow A \in \mathcal{I}(X).$$

We give three examples. Investigating all possibilities exhaustively seems quite tedious, and might best be done with the help of a computer. Fact 3.2.1 (p. 64) will be used repeatedly.

- (1), (2), (4) fail, (3) holds: Let $Y := \{a, b, c\}$, $\mathcal{Y} := \mathcal{P}(Y) - \{\emptyset\}$, $\mathcal{F}(Y) := \{\{a, c\}, \{b, c\}, Y\}$ Let $X := \{a, b\}$, $\mathcal{F}(X) := \{\{a\}, X\}$, $A := \{a\}$, and $\mathcal{F}(Z) := \{Z\}$ for all $Z \neq X, Y$. (Opt), (iM), $(eM\mathcal{I})$, $(eM\mathcal{F})$ hold, (I_ω) fails, of course.

  (1) fails: $A \in \mathcal{F}(X)$, $A \in \mathcal{I}(Y)$, $X \notin \mathcal{I}(Y)$.
  (2) fails: $\{a, c\} \in \mathcal{F}(Y)$, $\{a\} \in \mathcal{I}(Y)$, but $\{a\} \notin \mathcal{I}(\{a, c\})$.
  (3) holds: If $X_1 \in \mathcal{F}(X_2)$, $X_2 \in \mathcal{F}(X_3)$, then $X_1 = X_2$ or $X_2 = X_3$, so (3) holds trivially (note that $X \notin \mathcal{F}(Y)$).
  (4) fails: $\{a\}, \{b\} \in \mathcal{I}(Y)$, $\{a\} \notin \mathcal{I}(Y - \{b\}) = \mathcal{I}(\{a, c\}) = \{\emptyset\}$.

- (2), (3), (4) fail, (1) holds:
  Let $Y := \{a, b, c\}$, $\mathcal{Y} := \mathcal{P}(Y) - \{\emptyset\}$, $\mathcal{F}(Y) := \{\{a, b\}, \{a, c\}, Y\}$
  Let $X := \{a, b\}$, $\mathcal{F}(X) := \{\{a\}, X\}$, and $\mathcal{F}(Z) := \{Z\}$ for all $Z \neq X, Y$.
  (Opt), (iM), $(eM\mathcal{I})$, $(eM\mathcal{F})$ hold, (I_ω) fails, of course.

  (1) holds: Let $X_1 \in \mathcal{F}(X_2)$, $X_1 \in \mathcal{I}(X_3)$, we have to show $X_2 \in \mathcal{I}(X_3)$. If $X_1 = X_2$, then this is trivial. Consider $X_1 \in \mathcal{F}(X_2)$. If $X_1 \neq X_2$, then X_1 has to be $\{a\}$ or $\{a, b\}$ or $\{a, c\}$. But none of these are in $\mathcal{I}(X_3)$ for any X_3, so the implication is trivially true.
  (2) fails: $\{a, c\} \in \mathcal{F}(Y)$, $\{c\} \in \mathcal{I}(Y)$, $\{c\} \notin \mathcal{I}(\{a, c\})$.

(3) fails: $\{a\} \in \mathcal{F}(X)$, $X \in \mathcal{F}(Y)$, $\{a\} \notin \mathcal{F}(Y)$.
(4) fails: $\{b\}, \{c\} \in \mathcal{I}(Y)$, $\{c\} \notin \mathcal{I}(Y - \{b\}) = \mathcal{I}(\{a, c\}) = \{\emptyset\}$.

- (1), (2), (4) hold, (3) fails:
  Let $Y := \{a, b, c\}$, $\mathcal{Y} := \mathcal{P}(Y) - \{\emptyset\}$, $\mathcal{F}(Y) := \{\{a, b\}, \{a, c\}, Y\}$
  Let $\mathcal{F}(\{a, b\}) := \{\{a\}, \{a, b\}\}$, $\mathcal{F}(\{a, c\}) := \{\{a\}, \{a, c\}\}$, and $\mathcal{F}(Z) := \{Z\}$ for all other Z.
  (Opt), (iM), $(eM\mathcal{I})$, $(eM\mathcal{F})$ hold, (I_ω) fails, of course.

  (1) holds: Let $X_1 \in \mathcal{F}(X_2)$, $X_1 \in \mathcal{I}(X_3)$, we have to show $X_2 \in \mathcal{I}(X_3)$. Consider $X_1 \in \mathcal{I}(X_3)$. If $X_1 = X_2$, this is trivial. If $\emptyset \neq X_1 \in \mathcal{I}(X_3)$, then $X_1 = \{b\}$ or $X_1 = \{c\}$, but then by $X_1 \in \mathcal{F}(X_2)$, X_2 has to be $\{b\}$ or $\{c\}$, so $X_1 = X_2$.
  (2) holds: Let $X_1 \subseteq X_2 \subseteq X_3$, let $X_2 \in \mathcal{F}(X_3)$, $X_1 \in \mathcal{I}(X_3)$, we have to show $X_1 \in \mathcal{I}(X_2)$. If $X_1 = \emptyset$, this is trivial, likewise if $X_2 = X_3$. Otherwise $X_1 = \{b\}$ or $X_1 = \{c\}$, and $X_3 = Y$. If $X_1 = \{b\}$, then $X_2 = \{a, b\}$, and the condition holds, likewise if $X_1 = \{c\}$, then $X_2 = \{a, c\}$, and it holds again.
  (3) fails: $\{a\} \in \mathcal{F}(\{a, c\})$, $\{a, c\} \in \mathcal{F}(Y)$, $\{a\} \notin \mathcal{F}(Y)$.
  (4) holds: If $A, B \in \mathcal{I}(X)$ and $A \neq B$, $A, B \neq \emptyset$, then $X = Y$ and, e.g. $A = \{c\}$, $B = \{b\}$, and $\{c\} \in \mathcal{I}(Y - \{b\}) = \mathcal{I}(\{a, c\})$.

3.2.3.4 Rational Monotony

Fact 3.2.8
The three versions of $(\mathcal{M}^{++})$ are equivalent.
(We assume closure of the domain under set difference. For the third version of $(\mathcal{M}^{++})$, we use (iM).)

Proof For (1) and (2), we have $A, B \subseteq X$, for (3) we have $A \subseteq X \subseteq Y$. For $A, B \subseteq X$, $(X - B) - ((X - A) - B) = A - B$ holds.

(1) $\Rightarrow$ (2): Let $A \in \mathcal{F}(X)$, $B \notin \mathcal{F}(X)$, so $X - A \in \mathcal{I}(X)$, so by prerequisite $(X - A) - B \in \mathcal{I}(X - B)$, so $A - B = (X - B) - ((X - A) - B) \in \mathcal{F}(X - B)$.

(2) $\Rightarrow$ (1): Let $A \in \mathcal{I}(X)$, $B \notin \mathcal{F}(X)$, so $X - A \in \mathcal{F}(X)$, so by prerequisite $(X - A) - B \in \mathcal{F}(X - B)$, so $A - B = (X - B) - ((X - A) - B) \in \mathcal{I}(X - B)$.

(1) $\Rightarrow$ (3): Suppose $A \notin \mathcal{M}^+(Y)$, but $X \in \mathcal{M}^+(Y)$, we show $A \notin \mathcal{M}^+(X)$. So $A \in \mathcal{I}(Y)$, $Y - X \notin \mathcal{F}(Y)$, so by (1) $A = A - (Y - X) \in \mathcal{I}(Y - (Y - X)) = \mathcal{I}(X)$.

(3) $\Rightarrow$ (1): Suppose $A - B \notin \mathcal{I}(X - B)$, $B \notin \mathcal{F}(X)$, we show $A \notin \mathcal{I}(X)$. By prerequisite $A - B \in \mathcal{M}^+(X - B)$, $X - B \in \mathcal{M}^+(X)$, so by (3) $A - B \in \mathcal{M}^+(X)$, so by (iM) $A \in \mathcal{M}^+(X)$, so $A \notin \mathcal{I}(X)$. □

Fact 3.2.9 We assume that all sets are definable by a formula.
$(RatM) \Leftrightarrow (\mathcal{M}^{++})$.

Proof We show equivalence of $(RatM)$ with version (1) of $(\mathcal{M}^{++})$.
"$\Rightarrow$"

We have $A, B \subseteq X$, so we can write $X = M(\phi)$, $A = M(\phi \wedge \neg\psi)$, $B = M(\phi \wedge \neg\psi')$. $A \in \mathcal{I}(X)$, $B \notin \mathcal{F}(X)$, so $\phi \mid\sim \psi$, $\phi \not\mid\sim \neg\psi'$, so by $(RatM)$ $\phi \wedge \psi' \mid\sim \psi$, so $A - B = M(\phi \wedge \neg\psi) - M(\phi \wedge \neg\psi') = M(\phi \wedge \psi' \wedge \neg\psi) \in \mathcal{I}(M(\phi \wedge \psi')) = \mathcal{I}(X - B)$.

"$\Leftarrow$"

Let $\phi \mid\sim \psi$, $\phi \not\mid\sim \neg\psi'$, so $M(\phi \wedge \neg\psi) \in \mathcal{I}(M(\phi))$, $M(\phi \wedge \neg\psi') \notin \mathcal{F}(M(\phi))$, so by $(\mathcal{M}^{++})$ (1) $M(\phi \wedge \psi' \wedge \neg\psi) = M(\phi \wedge \neg\psi) - M(\phi \wedge \neg\psi') \in \mathcal{I}(M(\phi \wedge \psi'))$, so $\phi \wedge \psi' \mid\sim \psi$. □

3.2.4 Size and Principal Filter Logic

The connection with logical rules was shown in Tables 2.1 and 2.2 of Section 2.3 (p. 35).

(1)–(7) of the following proposition (in different notation, as the more systematic connections were found only afterwards) was already published in [GS08c], we give it here in totality to complete the picture.

Proposition 3.2.1 *If $f(X)$ is the smallest A s.t. $A \in \mathcal{F}(X)$, then, given the property on the left, the one on the right follows.*

Conversely, when we define $\mathcal{F}(X) := \{X' : f(X) \subseteq X' \subseteq X\}$, given the property on the right, the one on the left follows. For this direction, we assume that we can use the full powerset of some base set U – as is the case for the model sets of a finite language. This is perhaps not too restrictive, as we mainly want to stress here the intuitive connections, without putting any weight on definability questions.

We assume (iM) to hold.

Note that there is no (μwCM), as the conditions $(\mu \ldots)$ imply that the filter is principal, and thus that (I_ω) holds – we cannot "see" (wCM) alone with principal filters, as (I_ω) will hold automatically in all filters, so a fortiori in all principal filters (Table 3.3).

Table 3.3 Size and principal filter rules

(1)	$(eM\mathcal{I})$	$\Leftrightarrow$	(μwOR)
(2)	$(eM\mathcal{I}) + (I_\omega)$	$\Leftrightarrow$	(μOR)
(3)	$(eM\mathcal{I}) + (I_\omega)$	$\Leftrightarrow$	(μPR)
(4)	$(I \cup disj)$	$\Leftrightarrow$	$(\mu disjOR)$
(5)	$(\mathcal{M}^+_\omega)(4)$	$\Leftrightarrow$	(μCM)
(6)	$(\mathcal{M}^{++})$	$\Leftrightarrow$	$(\mu RatM)$
(7)	(I_ω)	$\Leftrightarrow$	(μAND)
(8.1)	$(eM\mathcal{I}) + (I_\omega)$	$\Rightarrow$	(μCUT)
(8.2)		$\not\Leftarrow$	
(9.1)	$(eM\mathcal{I}) + (I_\omega) + (\mathcal{M}^+_\omega)(4)$	$\Rightarrow$	(μCUM)
(9.2)		$\not\Leftarrow$	
(10.1)	$(eM\mathcal{I}) + (I_\omega) + (eM\mathcal{F})$	$\Rightarrow$	$(\mu \subseteq\supseteq)$
(10.2)		$\not\Leftarrow$	

Proof

(1) $(eM\mathcal{I}) \Rightarrow (\mu wOR)$:
$X - f(X)$ is small in X, so it is small in $X \cup Y$ by $(eM\mathcal{I})$, so $A := X \cup Y - (X - f(X)) \in \mathcal{F}(X \cup Y)$, but $A \subseteq f(X) \cup Y$, and $f(X \cup Y)$ is the smallest element of $\mathcal{F}(X \cup Y)$, so $f(X \cup Y) \subseteq A \subseteq f(X) \cup Y$.
$(\mu wOR) \Rightarrow (eM\mathcal{I})$:
Let $X \subseteq Y$, $X' := Y - X$. Let $A \in \mathcal{I}(X)$, so $X - A \in \mathcal{F}(X)$, so $f(X) \subseteq X - A$, so $f(X \cup X') \subseteq f(X) \cup X' \subseteq (X - A) \cup X'$ by prerequisite, so $(X \cup X') - ((X - A) \cup X') = A \in \mathcal{I}(X \cup X')$.

(2) $(eM\mathcal{I}) + (I_\omega) \Rightarrow (\mu OR)$:
$X - f(X)$ is small in X, $Y - f(Y)$ is small in Y, so both are small in $X \cup Y$ by $(eM\mathcal{I})$, so $A := (X - f(X)) \cup (Y - f(Y))$ is small in $X \cup Y$ by (I_ω), but $X \cup Y - (f(X) \cup f(Y)) \subseteq A$, so $f(X) \cup f(Y) \in \mathcal{F}(X \cup Y)$; so, as $f(X \cup Y)$ is the smallest element of $\mathcal{F}(X \cup Y)$, $f(X \cup Y) \subseteq f(X) \cup f(Y)$.
$(\mu OR) \Rightarrow (eM\mathcal{I}) + (I_\omega)$:
Let again $X \subseteq Y$, $X' := Y - X$. Let $A \in \mathcal{I}(X)$, so $X - A \in \mathcal{F}(X)$, so $f(X) \subseteq X - A$. $f(X') \subseteq X'$, so $f(X \cup X') \subseteq f(X) \cup f(X') \subseteq (X - A) \cup X'$ by prerequisite, so $(X \cup X') - ((X - A) \cup X') = A \in \mathcal{I}(X \cup X')$.
(I_ω) holds by definition.

(3) $(eM\mathcal{I}) + (I_\omega) \Rightarrow (\mu PR)$:
Let $X \subseteq Y$. $Y - f(Y)$ is the largest element of $\mathcal{I}(Y)$, $X - f(X) \in \mathcal{I}(X) \subseteq \mathcal{I}(Y)$ by $(eM\mathcal{I})$, so $(X - f(X)) \cup (Y - f(Y)) \in \mathcal{I}(Y)$ by (I_ω), so by "largest" $X - f(X) \subseteq Y - f(Y)$, so $f(Y) \cap X \subseteq f(X)$.
$(\mu PR) \Rightarrow (eM\mathcal{I}) + (I_\omega)$:
Let again $X \subseteq Y$, $X' := Y - X$. Let $A \in \mathcal{I}(X)$, so $X - A \in \mathcal{F}(X)$, so $f(X) \subseteq X - A$, so by prerequisite $f(Y) \cap X \subseteq X - A$, so $f(Y) \subseteq X' \cup (X - A)$, so $(X \cup X') - (X' \cup (X - A)) = A \in \mathcal{I}(Y)$.
Again, (I_ω) holds by definition.

(4) $(I \cup disj) \Rightarrow (\mu disjOR)$:
If $X \cap Y = \emptyset$, then (1) $A \in \mathcal{I}(X)$, $B \in \mathcal{I}(Y) \Rightarrow A \cup B \in \mathcal{I}(X \cup Y)$ and (2) $A \in \mathcal{F}(X)$, $B \in \mathcal{F}(Y) \Rightarrow A \cup B \in \mathcal{F}(X \cup Y)$ are equivalent. (By $X \cap Y = \emptyset$, $(X - A) \cup (Y - B) = (X \cup Y) - (A \cup B)$.) So $f(X) \in \mathcal{F}(X)$, $f(Y) \in \mathcal{F}(Y)$ $\Rightarrow$ (by prerequisite) $f(X) \cup f(Y) \in \mathcal{F}(X \cup Y)$. $f(X \cup Y)$ is the smallest element of $\mathcal{F}(X \cup Y)$, so $f(X \cup Y) \subseteq f(X) \cup f(Y)$.
$(\mu disjOR) \Rightarrow (I \cup disj)$:
Let $X \subseteq Y$, $X' := Y - X$. Let $A \in \mathcal{I}(X)$, $A' \in \mathcal{I}(X')$, so $X - A \in \mathcal{F}(X)$, $X' - A' \in \mathcal{F}(X')$, so $f(X) \subseteq X - A$, $f(X') \subseteq X' - A'$, so $f(X \cup X') \subseteq f(X) \cup f(X') \subseteq (X - A) \cup (X' - A')$ by prerequisite, so $(X \cup X') - ((X - A) \cup (X' - A')) = A \cup A' \in \mathcal{I}(X \cup X')$.

(5) $(\mathcal{M}^+_\omega) \Rightarrow (\mu CM)$:
$f(X) \subseteq Y \subseteq X \Rightarrow X - Y \in \mathcal{I}(X)$, $X - f(X) \in \mathcal{I}(X) \Rightarrow$ (by $(\mathcal{M}^+_\omega)$, (4)) $A := (X - f(X)) - (X - Y) \in \mathcal{I}(Y) \Rightarrow Y - A = f(X) - (X - Y) \in \mathcal{F}(Y)$ $\Rightarrow f(Y) \subseteq f(X) - (X - Y) \subseteq f(X)$.
$(\mu CM) \Rightarrow (\mathcal{M}^+_\omega)$:

Let $X - A \in \mathcal{I}(X)$, so $A \in \mathcal{F}(X)$, let $B \in \mathcal{I}(X)$, so $f(X) \subseteq X - B \subseteq X$, so by prerequisite $f(X - B) \subseteq f(X)$. As $A \in \mathcal{F}(X)$, $f(X) \subseteq A$, so $f(X - B) \subseteq f(X) \subseteq A \cap (X - B) = A{-}B$, and $A - B \in \mathcal{F}(X{-}B)$, so $(X - A) - B = X - (A \cup B) = (X - B) - (A - B) \in \mathcal{I}(X{-}B)$, so $(\mathcal{M}^{+}_{\omega})$, (4) holds.

(6) $(\mathcal{M}^{++}) \Rightarrow (\mu RatM)$:
Let $X \subseteq Y$, $X \cap f(Y) \neq \emptyset$. If $Y - X \in \mathcal{F}(Y)$, then $A := (Y - X) \cap f(Y) \in \mathcal{F}(Y)$, but by $X \cap f(Y) \neq \emptyset$ $A \subset f(Y)$, contradicting "smallest" of $f(Y)$. So $Y - X \notin \mathcal{F}(Y)$, and by $(\mathcal{M}^{++})$ $X - f(Y) = (Y - f(Y)) - (Y - X) \in \mathcal{I}(X)$, so $X \cap f(Y) \in \mathcal{F}(X)$, so $f(X) \subseteq f(Y) \cap X$.
$(\mu RatM) \Rightarrow (\mathcal{M}^{++})$:
Let $A \in \mathcal{F}(Y)$, $B \notin \mathcal{F}(Y)$. $B \notin \mathcal{F}(Y) \Rightarrow Y - B \notin \mathcal{I}(Y) \Rightarrow (Y - B) \cap f(Y) \neq \emptyset$. Set $X := Y{-}B$, so $X \cap f(Y) \neq \emptyset$, $X \subseteq Y$, so $f(X) \subseteq f(Y) \cap X$ by prerequisite. $f(Y) \subseteq A \Rightarrow f(X) \subseteq f(Y) \cap X = f(Y) - B \subseteq A{-}B$.

(7) Trivial in both directions.

(8.1) Let $f(X) \subseteq Y \subseteq X$. $Y - f(Y) \in \mathcal{I}(Y) \subseteq \mathcal{I}(X)$ by $(eM\mathcal{I})$. $f(X) \subseteq Y \Rightarrow X - Y \subseteq X - f(X) \in \mathcal{I}(X)$, so by (iM) $X - Y \in \mathcal{I}(X)$. Thus by (I_ω) $X - f(Y) = (X - Y) \cup (Y - f(Y)) \in \mathcal{I}(X)$, so $f(Y) \in \mathcal{F}(X)$, so $f(X) \subseteq f(Y)$ by definition.

(8.2) (μCUT) is too special to allow to deduce $(eM\mathcal{I})$. Consider $U := \{a, b, c\}$, $X := \{a, b\}$, $\mathcal{F}(X) = \{X, \{a\}\}$, $\mathcal{F}(Z) = \{Z\}$ for all other $X \neq Z \subseteq U$. Then $(eM\mathcal{I})$ fails, as $\{b\} \in \mathcal{I}(X)$, but $\{b\} \notin \mathcal{I}(U)$. (iM) and $(eM\mathcal{F})$ hold. We have to check $f(A) \subseteq B \subseteq A \Rightarrow f(A) \subseteq f(B)$. The only case where it might fail is $A = X$, $B = \{a\}$, but it holds there too.

(9.1) Obviously (μCM)+$(\mu CUT) \Leftrightarrow (\mu CUM)$, so the result follows from (5.1) and (8.1).

(9.2) Consider the same example as in (8.2). $f(A) \subseteq B \subseteq A \Rightarrow f(A) = f(B)$ holds there, too, by the same argument as above.

(10.1) Let $f(X) \subseteq Y$, $f(Y) \subseteq X$. So $f(X), f(Y) \subseteq X \cap Y$, and $X - (X \cap Y) \in \mathcal{I}(X)$, $Y - (X \cap Y) \in \mathcal{I}(Y)$ by (iM). Thus $f(X), f(Y) \in \mathcal{F}(X \cap Y)$ by $(eM\mathcal{F})$ and $f(X) \cap f(Y) \in \mathcal{F}(X \cap Y)$ by (I_ω). So $X \cap Y - (f(X) \cap f(Y)) \in \mathcal{I}(X \cap Y)$, so $X \cap Y - (f(X) \cap f(Y)) \in \mathcal{I}(X), \mathcal{I}(Y)$ by $(eM\mathcal{I})$, so $(X - (X \cap Y)) \cup (X \cap Y - f(X) \cap f(Y)) = X - f(X) \cap f(Y) \in \mathcal{I}(X)$ by (I_ω), so $f(X) \cap f(Y) \in \mathcal{F}(X)$, likewise $f(X) \cap f(Y) \in \mathcal{F}(Y)$, so $f(X) \subseteq f(X) \cap f(Y)$, $f(Y) \subseteq f(X) \cap f(Y)$, and $f(X) = f(Y)$.

(10.2) Consider again the same example as in (8.2), we have to show that $f(A) \subseteq B$, $f(B) \subseteq A \Rightarrow f(A) = f(B)$. The only interesting case is when one of A, B is X, but not both. Let, e.g. $A = X$. We then have $f(X) = \{a\}$, $f(B) = B \subseteq X$, and $f(X) = \{a\} \subseteq B$, so $B = \{a\}$, and the condition holds. □

The product size defined by principal filters is discussed in Sect. 7.1.2.1 (p. 179).

Chapter 4
Preferential Structures – Part I

This chapter, Preferential structures – Part I, is dedicated to the basic case without conditions for the domain. The following chapter, Part II, see Chap. 5 (p. 119), will treat the case with supplementary conditions for the domain, as well as applications and special cases. Higher preferential structures will be treated in the next but one chapter, see Chap. 6 (p. 157).

4.1 Introduction

After the present section, we will treat in Sect. 4.2 (p. 87) the case without conditions on the domain, and in Sect. 5.1 (p. 119) the case with the usual conditions on the domain, in particular closure under finite unions and finite intersections. But, first some general remarks.

4.1.1 Remarks on Nonmonotonic Logics and Preferential Semantics

Nonmonotonic logics were, historically, studied from two different points of view: The syntactic side, where rules like (AND), (CUM) (see Table 2.1) were postulated for their naturalness in reasoning, and from the semantic side, by the introduction of preferential structures (see Definition 4.1.1 (p. 78) and Definition 4.1.2 (p. 79) below). This work was done on the one hand by Gabbay [Gab85], Makinson [Mak94], and others, and for the second approach by Shoham and others, see [Sho87b], [BS85]. Both approaches were brought together by Kraus, Lehmann, Magidor and others, see [KLM90], [LM92], in their completeness results.

A preferential structure $\mathcal{M}$ defines a logic $\mid\!\sim$ by $T \mid\!\sim \phi$ iff ϕ holds in all $\mathcal{M}$-minimal models of T. This is made precise in Definition 4.1.1 (p. 78) and Definition 4.1.2 (p. 79) below. At the same time, $\mathcal{M}$ defines also a model set function by assigning to the set of models of T the set of its minimal models. As logics can speak only about definable model sets (here the model set defined by T), $\mathcal{M}$ defines a function from the definable sets of models to arbitrary model sets: $\mu_{\mathcal{M}} : \boldsymbol{D}(\mathcal{L}) \rightarrow$

D.M. Gabbay, K. Schlechta, *Logical Tools for Handling Change in Agent-Based Systems*, Cognitive Technologies, DOI 10.1007/978-3-642-04407-6_4,

$\mathcal{P}(M(\mathcal{L}))$. This is the general framework within which we will work most of the time. Different logics and situations (see, e.g. plausibility logic, Sect. 5.3 (p. 131), but also update situations, etc., Chap. 8 (p. 219)) will force us to generalize, we then consider functions $f : \mathcal{Y} \to \mathcal{P}(W)$, where W is an arbitrary set and $\mathcal{Y} \subseteq \mathcal{P}(W)$.

This chapter is about representation proofs in the realm of preferential and related structures and concerns mainly the following points:

(1) the importance of @T closure properties @t of the domain, in particular under finite unions,
(2) the conditions affected by lack of definability preservation,
(3) the limit version of preferential structures,
(4) the problems and solutions for "hidden dimensions", i.e. dimensions not directly observable.

Concerning (1), the main new result is probably the fact that, in the absence of closure under finite unions, cumulativity fans out to an infinity of different conditions (see Example 4.2.4 (p. 100)). We also separate here clearly the main proof technique from simplifications possible in the case of closure under finite unions.

Concerning (2), we examine in a systematic way conditions affected by absence of definability preservation, and use now more carefully crafted proofs which can be used in the cases with and those without definability preservation, achieving thus a simplification and a conceptually clearer approach.

Concerning (3), we introduce the concept of an algebraic limit to separate logical problems (i.e. due to lack of definability preservation) from algebraic ones of the limit variant. Again, this results in a better insight into problems and solutions of this variant.

Concerning (4), we describe a problem common to several representation questions, where we cannot directly observe all dimensions of a result.

Conceptually, one can subsume (2) and (4) under a more general notion of "blurred observation", but the solutions appear to be sufficiently different to merit a separate treatment – at least at the present stage of investigation.

Throughout the text, we emphasize those points in proofs where arbitrary choices are made, where we go beyond that which we know, where we "loose ignorance", as we think that an ideal proof is also in this aspect as general as possible, and avoids such losses, or, at least, reveals them clearly. For example, ignorance is lost when we complete in arbitrary manner a partial order to a complete one or when we choose arbitrarily some copy to be smaller than suitable other ones. The authors think that this, perhaps seemingly pedantic, attention will reveal itself as fruitful in future work.

The present text is a continuation of the second author's [Sch04]. Many results presented here were already shown in [Sch04], but in a less systematic and more ad hoc way. The emphasis of this book is on describing the problems and machinery "behind the scene" on a clear separation of general problems and possible simplifications. In particular, we take a systematic look at closure of the domain (mainly under finite unions), conditions affected by lack of definability preservation, and an analysis of the limit variant, leading to the notion of an algebraic limit.

The systematic investigation of $H(U, u)$ and $H(U)$, see Definition 4.2.4 (p. 100), is also new. Thus, many proofs have been systematized, prerequisites have been better isolated, and the results are now more general, so they can be reused more easily. In particular, we often first look at the case without closure under finite union and obtain in the second step the case with this closure as a simplification of the former.

The cases of "hidden dimensions" are now looked at in a more systematic way, and infinite variants are discussed and solved for the first time – to the authors' knowledge. In particular, such hidden dimensions are present in update situations, where we only observe the outcome, and can conclude about starting and intermediate steps only indirectly. We will see that this question leads to a (it seems) non trivial problem, and how to circumvent it in a natural way, using the limit approach.

The separation of the limit variant into structural limit, algebraic limit, and logical limit allows us to better identify problems, and also see that how a number of problems here are just those arising also from lack of definability preservation.

Finally, we solve some (to the authors' knowledge) open representation problems: Problems around Aqvist's deontic logic and Booth's revision approach – see Sect. 7.2 (p. 198) and Sect. 8.2.2 (p. 236).

The core of the completeness proofs consists in a general proof strategy, which is, essentially, a mathematical reformulation of the things we have to do. For instance, in general preferential structures, if $x \in X - \mu(X)$, we need to "minimize" x by some other element. Yet, we do not know by which one. This leads in a natural way to consider copies of x, one for each $x' \in X$, and to minimize such $\langle x, x' \rangle$ by x' – somehow. Essentially, this is all there is to do in the most general case. As x may be in several $X - \mu(X)$, we have to do the construction for all such X simultaneously, leading us to consider the product and choice functions. This is the basic proof construction – the mathematical counterpart of the representation idea. Of course, the same has to be done for above x', so we will have $\langle x, f \rangle$, $\langle x', f' \rangle$, etc. Now, there is a problem: Do we make all such $\langle x', f' \rangle$ smaller than the $\langle x, f \rangle$, only one, or some of them? We simply do not know, they all give the same result in the basic construction, as we are basically interested only in the first coordinate – and will see in the outcome only this. Choosing one possibility is completely arbitrary, and such a choice is a loss of ignorance. An ideal proof should not do this, it should not commit beyond the necessary, it should "preserve ignorance". There is an easy way out: We just construct for all possibilities one structure – just as a classical theory may have more than one model. This inflation of structures is the honest solution. Of course, once we are aware of the problem, we can construct just one structure, but should remember the arbitrariness of our decision. This attention is not purely academic or pedantic, as we will see immediately when we try to make the construction transitive. If we make all copies of x' smaller than $\langle x, f \rangle$, then the structure cannot be made transitive by simple closure: We did not pay enough attention to what can happen "in future". The solution is not to consider simple functions for copies, but trees of elements and functions, giving us complete control.

So far we did not care about prerequisites, but we just looked at the necessary construction. This will be done in the second step, where we will see that sufficiently strong prerequisites, especially about closure of the domain of μ, in particular whether for X, X' in the domain $X \cup X'$ is also in the domain, can simplify the construction considerably.

Thus, we have a core (the basic construction) and an initial part – eventual simplifications due to closure properties – but are not yet finished: So far, we looked only at model sets (or, more generally, just arbitrary sets) and the algebraic properties of the μ-functions to represent, and we still have to translate our results to logic. As long as our μ-functions preserve definability, i.e. $\mu(M(T))$, $M(T)$ the set of models of some theory T, is again $M(T')$ for some, usually other, theory T', this is simple. But, of course, this is a very strong assumption in the general infinite case. Usually, $\mu(M(T))$ will only generate a theory, but it will not be all of the models of this theory – logic is too poor to see that something is missing. So our observation of the result is blurred.

Thus, there is still a final part to consider. We have shown in [Sch04] that the general case is impossible to characterize by traditional means and we need other means – we have to work with "small" model sets: Our observation can be a bit off the true situation, but not too much. This, again, can be seen as a problem of ignorance: We do not know the exact result, but only that it is not too far off from what we see. Thus, characterization will be exactly this: It gives a sort of upper limit of how far we can be off, and anything within those limits is possible, but it does not give an exact result.

This last part will be considered in a more systematic way too. In particular, we will look at conditions which might be affected and at those which will not be affected. At the same time, we take care to make our basic construction sufficiently general, so we do not need that $\mu(X)$ is an element of the domain of μ.

To summarize, we clearly distinguish here three parts: A core part with the essential construction, an initial part with possible simplifications when domain closure conditions are sufficient, and a final part concerning the sharpness of our observations.

A problem which is conceptually similar to the definability preservation question is that of "hidden dimensions". For example, in update situations (but also in situations of a not necessarily symmetric distance-based revision), we may see only the result, the last dimension, and initial and intermediate steps are hidden from direct observation. More precisely, we may know that the endpoints of preferred developments are all in some set X, but have no direct information where intermediate points are. Again, our tools of observation are not sufficiently sharp to see the precise picture. This may generate non trivial problems, especially when we may have infinite descending chains. A natural solution is to consider here (partly or totally) the limit approach, where those things hold, which hold in the limit – i.e. the further we "approach" the (nonexisting) minimum, these properties will finally hold.

This is another subject of the present text:

We will introduce the concepts of the structural, the algebraic, and the logical limit, and will see that this allows us to separate problems in this usually quite difficult case. Some problems are simply due to the fact that a seemingly nice structural limit does not have nice algebraic properties anymore, so it should not be considered. So, to have a "good" limit, the limit should not only capture the idea of a structural limit, but its algebraic counterpart should also capture the essential algebraic properties of the minimal choice functions. Other problems are due to the fact that the nice algebraic limit does not translate to nice logical properties, and we will see that this is often due to the same problems we saw in the absence of definability preservation.

Thus, in a way, we come back in a cycle to the same problems again. This is one of the reasons the book form seems adequate for our results and problems: They are often interconnected, and a unified presentation seems the best.

It might be useful to emphasize the parallel investigation in the minimal and the limit variant:

For minimal preferential structures, we have

- logical laws or descriptions like $\alpha \mid\sim \alpha$ – they are the (imperfect – by definability preservation problems) reflection of the abstract description,
- abstract or algebraic semantics, like $\mu(A) \subseteq A$ – they are the abstract description of the foundation,
- structural semantics – they are the intuitive foundation.

Likewise, for the limit situation, we have

- structural limits – they are again the foundation,
- resulting abstract behaviour, which, again, has to be an abstract or algebraic limit, resulting from the structural limit,
- a logical limit, which reflects the abstract limit, and may be plagued by definability preservation problems, etc. when going from the model to the logics side.

Note that these clear distinctions have some philosophical importance, too. The structures need an intuitive or philosophical justification, why do we describe preference by transitive relations, why do we admit copies, etc.? The resulting algebraic choice functions are void of such questions.

4.1.2 Basic Definitions

The following two definitions make preferential structures precise. We first give the algebraic definition and then the definition of the consequence relation generated by a preferential structure. In the algebraic definition, the set U is an arbitrary set, while in the application to logic, this will be the set of classical models of the underlying propositional language.

In both cases, we first present the simpler variant without copies and then the one with copies. (Note that e.g. [KLM90], [LM92] use labelling functions instead, the version without copies corresponds to injective labelling functions, the one with copies to the general case. These are just different ways of speaking.) We will discuss the difference between the version without and the version with copies below, where we show that the version with copies is strictly more expressive than the version without copies, and that transitivity of the relation adds new properties in the case without copies. When we summarize our own results below (see Sect. 4.2.2.2 (p. 95)), we will mention that, in the general case with copies, transitivity can be added without changing properties.

We give here the "minimal version" and the much more complicated "limit version" is presented and discussed in Sect. 5.5 (p. 145). Recall the intuition that the relation $\prec$ expresses "normality" or "importance" – the $\prec$-smaller an element is, the more normal or important it is. The smallest elements are those which count.

Definition 4.1.1 Fix $U \neq \emptyset$ and consider arbitrary X. Note that this X has not necessarily anything to do with U, or $\mathcal{U}$ below. Thus, the functions $\mu_{\mathcal{M}}$ below are in principle functions from V to V – where V is the set-theoretical universe we work in.

Note that we work here often with copies of elements (or models). In other areas of logic, most authors work with valuation functions. Both definitions – copies or valuation functions – are equivalent, a copy $\langle x, i\rangle$ can be seen as a state $\langle x, i\rangle$ with valuation x. In the beginning of research on preferential structures, the notion of copies was widely used, whereas, e.g. [KLM90] used that of valuation functions. There is perhaps a weak justification of the former terminology. In modal logic, even if two states have the same valid classical formulas, they might still be distinguishable by their valid modal formulas. But this depends on the fact that modality is in the object language. In most work on preferential stuctures, the consequence relation is outside the object language, so different states with same valuation are in a stronger sense copies of each other.

(1.1) *Preferential models* or *structures*:

(1.1) The version without copies:
A pair $\mathcal{M} := \langle U, \prec\rangle$ with U an arbitrary set and $\prec$ an arbitrary binary relation on U is called a *preferential model* or *structure*.

(1.2) The version with copies:
A pair $\mathcal{M} := \langle \mathcal{U}, \prec\rangle$ with $\mathcal{U}$ an arbitrary set of pairs and $\prec$ an arbitrary binary relation on $\mathcal{U}$ is called a *preferential model* or *structure*.
If $\langle x, i\rangle \in \mathcal{U}$, then x is intended to be an element of U and i the index of the copy.
We sometimes also need copies of the relation $\prec$. We will then replace $\prec$ by one or several arrows α attacking non minimal elements, e.g. $x \prec y$ will be written as $\alpha : x \to y$, $\langle x, i\rangle \prec \langle y, i\rangle$ will be written as $\alpha : \langle x, i\rangle \to \langle y, i\rangle$, and finally we might have $\langle \alpha, k\rangle : x \to y$ and $\langle \alpha, k\rangle : \langle x, i\rangle \to \langle y, i\rangle$, etc.

(2) *Minimal elements*, the functions $\mu_{\mathcal{M}}$:

(2.1) The version without copies:
Let $\mathcal{M} := \langle U, \prec \rangle$, and define
$\mu_{\mathcal{M}}(X) := \{x \in X : x \in U \wedge \neg\exists x' \in X \cap U.x' \prec x\}$.
$\mu_{\mathcal{M}}(X)$ is called the set of *minimal elements* of X (in $\mathcal{M}$).
Thus, $\mu_{\mathcal{M}}(X)$ is the set of elements such that there is no smaller one in X.

(2.2) The version with copies:
Let $\mathcal{M} := \langle \mathcal{U}, \prec \rangle$ be as above. Define
$\mu_{\mathcal{M}}(X) := \{x \in X : \exists \langle x, i \rangle \in \mathcal{U}.\neg\exists \langle x', i' \rangle \in \mathcal{U}(x' \in X \wedge \langle x', i' \rangle' \prec \langle x, i \rangle)\}$.
Thus, $\mu_{\mathcal{M}}(X)$ is the projection on the first coordinate of the set of elements such that there is no smaller one in X.
Again, by abuse of language, we say that $\mu_{\mathcal{M}}(X)$ is the set of *minimal elements* of X in the structure. If the context is clear, we will also write just μ.
We sometimes say that $\langle x, i \rangle$ "*kills*" or "*minimizes*" $\langle y, j \rangle$ if $\langle x, i \rangle \prec \langle y, j \rangle$. By abuse of language we also say a set X *kills* or *minimizes* a set Y if for all $\langle y, j \rangle \in \mathcal{U}$, $y \in Y$ there is $\langle x, i \rangle \in \mathcal{U}$, $x \in X$ s.t. $\langle x, i \rangle \prec \langle y, j \rangle$.
$\mathcal{M}$ is also called *injective* or 1-copy, iff there is always at most one copy $\langle x, i \rangle$ for each x. Note that the existence of copies corresponds to a non-injective labelling function – as is often used in nonclassical logic, e.g. modal logic.

We say that $\mathcal{M}$ is *transitive*, *irreflexive*, etc., iff $\prec$ is. Note that $\mu(X)$ might well be empty, even if X is not.

Definition 4.1.2 We define the consequence relation of a preferential structure for a given propositional language $\mathcal{L}$.

(1.1) If m is a classical model of a language $\mathcal{L}$, we say by abuse of language
$\langle m, i \rangle \models \phi$ iff $m \models \phi$,
and if X is a set of such pairs, then
$X \models \phi$ iff for all $\langle m, i \rangle \in X$ $m \models \phi$.

(1.2) If $\mathcal{M}$ is a preferential structure and X is a set of $\mathcal{L}$-models for a classical propositional language $\mathcal{L}$, or a set of pairs $\langle m, i \rangle$, where the m are such models, we call $\mathcal{M}$ a *classical preferential structure* or *model*.

(2) *Validity* in a preferential structure, or the *semantical consequence relation* defined by such a structure:
Let $\mathcal{M}$ be as above.
We define:
$T \models_{\mathcal{M}} \phi$ iff $\mu_{\mathcal{M}}(M(T)) \models \phi$, i.e. $\mu_{\mathcal{M}}(M(T)) \subseteq M(\phi)$.

(3) $\mathcal{M}$ will be called *definability preserving* iff for all $X \in \boldsymbol{D}_{\mathcal{L}}$, $\mu_{\mathcal{M}}(X) \in \boldsymbol{D}_{\mathcal{L}}$.

As $\mu_{\mathcal{M}}$ is defined on $\boldsymbol{D}_{\mathcal{L}}$, but need by no means always result in some new definable set, this is (and reveals itself as a quite strong) additional property.

Example 4.1.1 This simple example illustrates the importance of copies. Such examples seem to have appeared for the first time in print in [KLM90], but can probably be attributed to folklore.
Consider the propositional language $\mathcal{L}$ of two propositional variables p, q, and the classical preferential model $\mathcal{M}$ defined by

$m \models p \wedge q, m' \models p \wedge q, m_2 \models \neg p \wedge q, m_3 \models \neg p \wedge \neg q,$ with $m_2 \prec m, m_3 \prec m'$, and let $\models_{\mathcal{M}}$ be its consequence relation.(m and m' are logically identical.)

Obviously, $Th(m) \vee \{\neg p\} \models_{\mathcal{M}} \neg p$, but there is no complete theory T' s.t. $Th(m) \vee T' \models_{\mathcal{M}} \neg p$. (If there were one, T' would correspond to m, m_2, m_3, or the missing $m_4 \models p \wedge \neg q$, but we need two models to kill all copies of m.) On the other hand, if there were just one copy of m, then one other model, i.e. a complete theory would suffice. More formally, if we admit at most one copy of each model in a structure $\mathcal{M}$, $m \not\models T$, and $Th(m) \vee T \models_{\mathcal{M}} \phi$ for some ϕ s.t. $m \models \neg\phi$ – i.e. m is not minimal in the models of $Th(m) \vee T$ – then there is a complete T' with $T' \vdash T$ and $Th(m) \vee T' \models_{\mathcal{M}} \phi$, i.e. there is m'' with $m'' \models T'$ and $m'' \prec m$.

We define now two additional properties of the relation, smoothness and rankedness.

Definition 4.1.3 Let $\mathcal{Y} \subseteq \mathcal{P}(U)$. (In applications to logic, $\mathcal{Y}$ will be $\boldsymbol{D}_{\mathcal{L}}$.)
A preferential structure $\mathcal{M}$ is called $\mathcal{Y}$-smooth iff for every $X \in \mathcal{Y}$ every element $x \in X$ is either minimal in X or above an element, which is minimal in X. More precisely:

(1) The version without copies:
If $x \in X \in \mathcal{Y}$, then either $x \in \mu(X)$ or there is $x' \in \mu(X)$ s.t. $x' \prec x$.
(2) The version with copies:
If $x \in X \in \mathcal{Y}$ and $\langle x, i\rangle \in \mathcal{U}$, then either there is no $\langle x', i'\rangle \in \mathcal{U}$, $x' \in X$, $\langle x', i'\rangle \prec \langle x, i\rangle$ or there is $\langle x', i'\rangle \in \mathcal{U}$, $\langle x', i'\rangle \prec \langle x, i\rangle$, $x' \in X$, s.t. there is no $\langle x'', i''\rangle \in \mathcal{U}$, $x'' \in X$, with $\langle x'', i''\rangle \prec \langle x', i'\rangle$.
(Writing down all details here again might make it easier to read applications of the definition later on.)

When considering the models of a language $\mathcal{L}$, $\mathcal{M}$ will be called *smooth* iff it is $\boldsymbol{D}_{\mathcal{L}}$-smooth ; $\boldsymbol{D}_{\mathcal{L}}$ is the default. Obviously, the richer the set $\mathcal{Y}$ is, the stronger the condition $\mathcal{Y}$-smoothness will be.

Fact 4.1.1 Let $\prec$ be an irreflexive, binary relation on X, then the following two conditions are equivalent:

(1) There is Ω and an irreflexive, total, binary relation $\prec'$ on Ω and a function $f : X \to \Omega$ s.t. $x \prec y \Leftrightarrow f(x) \prec' f(y)$ for all $x, y \in X$.
(2) Let $x, y, z \in X$ and $x \perp y$ w.r.t. $\prec$ (i.e. neither $x \prec y$ nor $y \prec x$), then $z \prec x \Rightarrow z \prec y$ and $x \prec z \Rightarrow y \prec z$.

Proof

(1) $\Rightarrow$ (2): Let $x \perp y$, thus neither $fx \prec' fy$ nor $fy \prec' fx$, but then $fx = fy$. Let now $z \prec x$, so $fz \prec' fx = fy$, so $z \prec y$. $x \prec z \Rightarrow y \prec z$ is similar.

(2) $\Rightarrow$ (1): For $x \in X$, let $[x] := \{x' \in X : x \perp x'\}$ and $\Omega := \{[x] : x \in X\}$. For $[x], [y] \in \Omega$, let $[x] \prec' [y] :\Leftrightarrow x \prec y$. This is well defined: Let $x \perp x'$, $y \perp y'$, and $x \prec y$, then $x \prec y'$ and $x' \prec y'$. Obviously, $\prec'$ is an irreflexive, total binary relation. Define $f : X \to \Omega$ by $fx := [x]$, then $x \prec y \Leftrightarrow [x] \prec' [y] \Leftrightarrow fx \prec' fy$. □

Definition 4.1.4 We call an irreflexive, binary relation $\prec$ on X, which satisfies (1) (equivalently (2)) of Fact 4.1.1 (p. 80) as ranked. By abuse of language, we also call a preferential structure $\langle X, \prec \rangle$ ranked, iff $\prec$ is.

The smoothness condition says that if $x \in X$ is not a minimal element of X, then there is $x' \in \mu(X)$ $x' \prec x$. In the finite case without copies, smoothness is a trivial consequence of transitivity and lack of cycles. But note that in the other cases infinite descending chains might still exist, even if the smoothness condition holds, they are just "short-circuited": We might have such chains, but below every element in the chain is a minimal element. In the authors' opinion, smoothness is difficult to justify as a structural property (or, in a more philosophical spirit, as a property of the world): Why should we always have such minimal elements below non minimal ones? Smoothness has, however, a justification from its consequences. Its attractiveness comes from two sides:

First, it generates a very valuable logical property, cumulativity (CUM): If $\mathcal{M}$ is smooth and $\overline{\overline{T}}$ is the set of $\models_{\mathcal{M}}$-consequences, then for $T \subseteq \overline{T'} \subseteq \overline{\overline{T}} \Rightarrow \overline{\overline{T}} = \overline{\overline{T'}}$. Second, for certain approaches, it facilitates completeness proofs, as we can look directly at "ideal" elements, without having to bother about intermediate stages. See in particular the work by Lehmann and his co-authors, [KLM90],[LM92].

"Smoothness" or, as it is also called, "stopperedness" seems – in the authors' opinion – a misnamer. We think it should better be called something like "weak transitivity": Consider the case where $a \succ b \succ c$, but $c \not\prec a$, with $c \in \mu(X)$. It is then not necessarily the case that $a \succ c$, but there is c' "sufficiently close to c", i.e. in $\mu(X)$, such that $a \succ c'$. Results and proof techniques underline this idea. First, in the general case with copies and in the smooth case (in the presence of $(\cup)$!), transitivity does not add new properties, it is "already present"; Second, the construction of smoothness by sequences σ (see below in Sect. 4.2.2.3 (p. 99)) is very close in spirit to a transitive construction.

The second condition, rankedness, seems easier to justify already as a property of the structure. It says that, essentially, the elements are ordered in layers: If a and b are not comparable, then they are in the same layer. So, if c is above (below) a, it will also be above (below) b – like pancakes or geological strata. Apart from the triangle inequality (and leaving aside cardinality questions), this is then just a distance from some imaginary, ideal point. Again, this property has important consequences on the resulting model choice functions and consequence relations, making proof techniques for the non ranked and ranked case very different.

$\mathcal{Y}$ can have certain properties; in classical propositional logic for instance, if $\mathcal{Y}$ is the set of formula-defined model sets, then $\mathcal{Y}$ is closed under complements, finite unions, and finite intersection. If $\mathcal{Y}$ is the set of theory-defined model sets, $\mathcal{Y}$ is closed under finite unions, arbitrary intersections, but not complements anymore.

A careful consideration of closure conditions of the domain was motivated by Lehmann's plausibility logic, see [Leh92a], and remotivated by the work of Arieli and Avron, see [AA00]. In both cases, the language does not have a built-in "or" – resulting in absence $(\cup)$ of the domain.

When trying to show completeness of Lehmann's system, the second author noted the importance of the closure of the domain under $(\cup)$, see [Sch96-3]. The work of Arieli and Avron incited him to look at this property in a more systematic way, which lead to the discovery of Example 4.2.4 (p. 100), and thus of the enormous strength of closure of the domain under finite unions, and, more generally, of the importance of domain closure conditions.

In the resulting completeness proofs again, a strategy of "divide and conquer" is useful. This helps us to unify (or extend) our past completeness proofs for the smooth case in the following way: We will identify more clearly than in the past a more or less simple algebraic property – (HU), (HU, u), etc. – which allows us to split the proofs into two parts: The first part (see Sect. 4.2.2 (p. 94)) shows validity of the property, and this demonstration depends on closure properties; the second part shows how to construct a representing structure using the algebraic property. This second part will be totally independent from closure properties, and is essentially an "administrative" way to use the property for a construction. This split approach allows us thus to isolate the demonstration of the used property from the construction itself, bringing both parts clearer to light, and simplifying the proofs by using common parts.

The readers will see that the successively more complicated conditions (HU), (HU, u) reflect well the successively more complicated situations of representation:

(HU): smooth (and transitive) structures in the presence of $(\cup)$,
(HU, u): smooth structures in the absence of $(\cup)$,

This comparison becomes clearer when we see that in the final, most complicated case, we will have to carry around all the history of minimization, $\langle Y_0, \ldots, Y_n \rangle$, necessary for transitivity, which could be summarized in the first case with finite unions. Thus, from an abstract point of view, it is a very natural development.

In the rest of this Sect. 4.1 (p. 73), we will only describe the problems to solve without giving a solution. This will be done in the next sections. Moreover, we will assume that we have precise knowledge of f, i.e. what we see as $f(X)$ for $X \in \mathcal{Y}$ is really the result and not some approximation – as we will permit later in Sect. 5.4 (p. 134).

So this part is a leisurely description of problems and things to do. We start with the most general case, arbitrary preferential structures, turn to transitive such structures, then to smooth, then to smooth transitive ones, and conclude by ranked and ranked smooth structures.

Throughout, we will try to preserve ignorance, i.e. not to assume anything we are not forced to assume. This will become clearer in a moment. Once we have understood the problem, we will sometimes just gloss over it by choosing one solution, but we should always be conscious that there is a problem.

We will consider here choice functions $f : \mathcal{Y} \to \mathcal{P}(W)$, where $\mathcal{Y} \subseteq \mathcal{P}(W)$, and the problems to represent them by various preferential structures.

We will see the following basic representation problems and the constructions to solve them, i.e. to find representing structures for

(1) General preferential structures;
(2) general transitive preferential structures;
(3) smooth preferential structures;
(4) smooth transitive preferential structures;
(5) ranked preferential structures; and
(6) Smooth-ranked preferential structures;

The problems of and solutions to the ranked case are quite different from the first four cases. In particular, the situation when $\emptyset \neq U \in \mathcal{Y}$, but $f(U) = \emptyset$ does not present major difficulties in cases (1)–(4), but is quite nasty in the last case.

4.1.2.1 The Situation

We work in some universe W, there is a function $f : \mathcal{Y} \to \mathcal{P}(W)$, where $\mathcal{Y} \subseteq \mathcal{P}(W)$, f will have certain properties, and perhaps $\mathcal{Y}$ too, and we will try to represent f by a preferential structure $\mathcal{Z}$ of a certain type, i.e. we want $f = \mu_{\mathcal{Z}}$, with $\mu_{\mathcal{Z}}$ the μ-function or choice function of a preferential structure $\mathcal{Z}$. Note that the codomain of f is not necessarily a subset of $\mathcal{Y}$ – so we have to pay attention not to apply f twice.

Before we go into details, we now give an overview of the results.

The following table "preferential representation", see Table 8.4 (p. 234), summarizes representation by preferential structures. The positive implications on the right are shown in Proposition 2.3.1 (p. 45) (going via the μ-functions), those on the left are shown in the respective representation theorems.

"Singletons" means that the domain must contain all singletons, "1 copy" or "$\geq$ 1 copy" means that the structure may contain only one copy for each point or several, and "$(\mu\emptyset)$", etc. for the preferential structure mean that the μ-function of the structure has to satisfy this property.

We call a characterization "normal" iff it is a universally quantified Boolean combination (of any fixed, but perhaps infinite, length) of rules of the usual form. We do not go into details here.

In the second column from the left "$\Rightarrow$" means, for instance for the smooth case, that for any $\mathcal{Y}$ closed under finite unions and any choice function f which satisfies the conditions in the left hand column, there is a (here $\mathcal{Y}$-smooth) preferential structure $\mathcal{X}$ which represents it, i.e. for all $Y \in \mathcal{Y}$, $f(Y) = \mu_{\mathcal{X}}(Y)$, where $\mu_{\mathcal{X}}$ is the model choice function of the structure $\mathcal{X}$. The inverse arrow $\Leftarrow$ means that

Table 4.1 Preferential representation

μ-function		Pref. Structure		Logic
$(\mu \subseteq)$	$\Leftrightarrow$ Proposition 6.2.1 (p. 163)	Reactive	$\Leftrightarrow$ Proposition 6.5.1 (p. 173)	$(LLE) + (CCL) + (SC)$
$(\mu \subseteq) + (\mu CUM)$	$\Rightarrow (\cap)$ Proposition 6.4.1 (p. 171)	Reactive + essentially smooth		
$(\mu \subseteq) + (\mu \subseteq \supseteq)$	$\Rightarrow$ Proposition 6.4.1 (p. 171)	Reactive + essentially smooth	$\Leftrightarrow$ Proposition 6.5.1 (p. 173)	$(LLE) + (CCL) + (SC) + (\subseteq\supseteq)$
$(\mu \subseteq) + (\mu CUM) + (\mu \subseteq \supseteq)$	$\Leftarrow$ Fact 6.1.2 (p. 161)	Reactive + essentially smooth		
$(\mu \subseteq) + (\mu PR)$	$\Leftarrow$ Fact 4.2.1 (p. 87)	General	$\Rightarrow (\mu dp)$	$(LLE) + (RW) + (SC) + (PR)$
	$\Rightarrow$ Proposition 4.2.1 (p. 94)		$\Leftarrow$	
			$\not\Rightarrow$ without (μdp) Example 4.2.1 (p. 87)	
			$\not\Leftrightarrow$ without (μdp) Proposition 5.2.15 in [Sch04]	Any "normal" characterization of any size
$(\mu \subseteq) + (\mu PR)$	$\Leftarrow$ Fact 4.2.1 (p. 87)	Transitive	$\Rightarrow (\mu dp)$	$(LLE) + (RW) + (SC) + (PR)$
	$\Rightarrow$ Proposition 4.2.2 (p. 95)		$\Leftarrow$	
			$\not\Rightarrow$ without (μdp) Example 4.2.1 (p. 87)	

Table 4.1 (continued)

μ-function		Pref. Structure		Logic
			$\Leftrightarrow$ without (μdp) Proposition 5.2.5, 5.2.11 in [Sch04]	Using "small" exception sets
$(\mu \subseteq) + (\mu PR) + (\mu CUM)$	$\Leftarrow$ Fact 4.2.2 (p. 87)	Smooth	$\Rightarrow (\mu dp)$ Proposition 5.1.5 (p. 124)	$(LLE) + (RW) +$ $(SC) + (PR) +$ (CUM)
	$\Rightarrow (\cup)$ Proposition 3.3.4 in [Sch04], Proposition 5.1.1 (p. 120)		$\Leftarrow (\cup)$ Proposition 5.1.5 (p. 124)	
	$\not\Rightarrow$ without $(\cup)$ See [Sch04]		$\not\Leftrightarrow$ without (μdp) Example 4.2.1 (p. 87)	
$(\mu \subseteq) + (\mu PR) + (\mu CUM)$	$\Leftarrow$ Fact 4.2.2 (p. 87)	Smooth+transitive	$\Rightarrow (\mu dp)$ Proposition 5.1.5 (p. 124)	$(LLE) + (RW) +$ $(SC) + (PR) +$ (CUM)
	$\Rightarrow (\cup)$ Proposition 3.3.8 in [Sch04], Proposition 5.1.1 (p. 120)		$\Leftarrow (\cup)$ Proposition 5.1.5 (p. 124)	
			$\not\Leftrightarrow$ without (μdp) Example 4.2.1 (p. 87)	
			$\Leftrightarrow$ without (μdp) Proposition 5.2.9, 5.2.11 in [Sch04]	Using "small" exception sets

Table 4.1 (continued)

μ-function		Pref. Structure		Logic
$(\mu\subseteq)+(\mu=)+(\mu PR)+$ $(\mu=')+(\mu\parallel)+(\mu\cup)+$ $(\mu\cup')+(\mu\in)+(\mu RatM)$	$\Leftarrow$ Fact 4.2.7 (p. 92)	Ranked, ≥ 1 copy		
$(\mu\subseteq)+(\mu=)+(\mu PR)+$ $(\mu\cup)+(\mu\in)$	$\not\Rightarrow$ Example 4.2.2 (p. 94)	Ranked		
$(\mu\subseteq)+(\mu=)+(\mu\emptyset)$	$\Leftrightarrow$, $(\cup)$ Proposition 3.10.11 in [Sch04]	Ranked, 1 copy + $(\mu\emptyset)$		
$(\mu\subseteq)+(\mu=)+(\mu\emptyset)$	$\Leftrightarrow$, $(\cup)$ Proposition 3.10.11 in [Sch04]	Ranked, smooth, 1 copy + $(\mu\emptyset)$		
$(\mu\subseteq)+(\mu=)+(\mu\emptyset fin)+$ $(\mu\in)$	$\Leftrightarrow$, $(\cup)$, singletons Proposition 3.10.12 in [Sch04]	Ranked, smooth, ≥ 1 copy + $(\mu\emptyset fin)$		
$(\mu\subseteq)+(\mu PR)+(\mu\parallel)+$ $(\mu\cup)+(\mu\in)$	$\Leftrightarrow$, $(\cup)$, singletons Proposition 3.10.14 in [Sch04]	Ranked ≥ 1 copy	$\not\Rightarrow$ without (μdp) Example 2.3.6 (p. 50)	$(RatM)$, $(RatM=)$, $(Log\cup)$, $(Log\cup')$
			$\not\Leftrightarrow$ without (μdp) Proposition 5.2.16 in [Sch04]	Any "normal" characterization of any size

the model choice function for any smooth $\mathcal{X}$ defined on such $\mathcal{Y}$ will satisfy the conditions on the left.

4.2 Preferential Structures Without Domain Conditions

4.2.1 General Discussion

We treat in this section the general case without conditions on the domain. We will see that it is more difficult than when we can impose the usual conditions (closure under finite intersections and finite unions). The latter case will be dealt with briefly (as most of it was already done in [Sch04]) in Sect. 5.1 (p. 119).

4.2.1.1 General Preferential Structures

We give now just three simple facts to put the reader in the right mood for what follows.

Fact 4.2.1 $(\mu \subseteq)$ and (μPR) hold in all preferential structures.

Proof Trivial. The central argument is: If $x, y \in X \subseteq Y$ and $x \prec y$ in X, then also $x \prec y$ in Y. □

Fact 4.2.2 $(\mu \subseteq)$, (μPR), and (μCUM) hold in all smooth preferential structures.

Proof By Fact 4.2.1 (p. 87), we only have to show (μCUM). By Fact 2.3.1 (p. 40), (μCUT) follows from (μPR), so it remains to show (μCM). So suppose $\mu(X) \subseteq Y \subseteq X$, we have to show $\mu(Y) \subseteq \mu(X)$. Let $x \in X - \mu(X)$, so there is $x' \in X$, $x' \prec x$, by smoothness, there must be $x'' \in \mu(X)$, $x'' \prec x$, so $x'' \in Y$ and $x \notin \mu(Y)$. The proof for the case with copies is analogous. □

Example 4.2.1 This example was first given in [Sch92]. It shows that condition (PR) may fail in preferential structures which are not definability preserving.

Let $v(\mathcal{L}) := \{p_i : i \in \omega\}$, $n, n' \in M_{\mathcal{L}}$ be defined by $n \models \{p_i : i \in \omega\}$, $n' \models \{\neg p_0\} \cup \{p_i : 0 < i < \omega\}$.

Let $\mathcal{M} := \langle M_{\mathcal{L}}, \prec \rangle$ where only $n \prec n'$, i.e. just two models are comparable. Note that the structure is transitive and smooth. Thus, by Fact 4.2.2 (p. 87) $(\mu \subseteq)$, (μPR), and (μCUM) hold.

Let $\mu := \mu_{\mathcal{M}}$, and $\mid\!\sim$ be defined as usual by μ.

Set $T := \emptyset$, $T' := \{p_i : 0 < i < \omega\}$. We have $M_T = M_{\mathcal{L}}$, $f(M_T) = M_{\mathcal{L}} - \{n'\}$, $M_{T'} = \{n, n'\}$, $f(M_{T'}) = \{n\}$. So by the result of Example 2.2.1 (p. 35), f is not definability preserving, and, furthermore, $\overline{\overline{\overline{T}}} = \overline{T}$, $\overline{\overline{\overline{T'}}} = \overline{\{p_i : i < \omega\}}$, *so* $p_0 \in \overline{\overline{\overline{T \cup T'}}}$, *but* $\overline{\overline{\overline{\overline{T}}} \cup T'} = \overline{\overline{\overline{T} \cup T'}} = \overline{T'}$, *so* $p_0 \notin \overline{\overline{\overline{\overline{T}}} \cup T'}$, contradicting (PR), which holds in all definability preserving preferential structures.

We know from Fact 4.2.1 (p. 87) that f has to satisfy $(\mu \subseteq)$ and (μPR). Let then $u \in U \in \mathcal{Y}$.

If $u \in f(U)$, then, for this u, this U there is nothing to do, we just have to take care that at least one copy of u will be minimal in U.

If $u \notin f(U)$, then u must be minimized by some $u' \in U$ – more precisely: it might well be that in all smaller U' is not minimized: $U' \subset U$, $U' \in \mathcal{Y}$, $u \in U' \Rightarrow u \in f(U')$. If for example, $U = \{u, u', u''\}$, $U' = \{u, u'\}$, and $U'' = \{u, u''\}$ with $U', U'' \in \mathcal{Y}$, then there cannot be $u \prec u$, nor $u' \prec u$, nor $u'' \prec u$, and we have to make copies of u, so that only in U, but neither in U' nor in U'' all copies are minimized. Thus, what we have to do is to create $\langle u, u\rangle$, $\langle u, u'\rangle$, $\langle u, u''\rangle$ and to make $u \prec \langle u, u\rangle$, $u' \prec \langle u, u'\rangle$, $u'' \prec \langle u, u''\rangle$ (or something similar). Thus, in the presence of full U, all copies will be minimized, but in all $U' \subset U$ at least one copy of u is not minimized. When we look now at our construction, we note the following: (a) u is minimized in U; (b) we took no commitment for other $U' \subseteq U$. Thus, we might not know anything about such other U', and leave this question totally open – we preserved our ignorance; (c) the construction is independent of all other U' – except that any U' with $U \subseteq U'$ will also minimize u, but this was an inevitable fact. Note that there might also be $U' \subseteq U$ with $u \in U' - f(U')$, but no minimal one (with respect to $\subseteq$).

We can see the problem of copies also as preservation of ignorance, and can also solve it with many structures – as is done in [SGMRT00]. We have to do this construction now also for other u and U, so we will perhaps introduce copies for other elements, too. Suppose we have copies $\langle u', y\rangle$ and $\langle u', y'\rangle$ for the above u'. As μ is insensitive to the particular index and it wants only at least one copy of u' to be smaller than $\langle u, u'\rangle$, we have a problem we cannot decide: Shall we make $\langle u', y\rangle \prec \langle u, u'\rangle$, or $\langle u', y'\rangle \prec \langle u, u'\rangle$, or both? Deciding for one solution would go beyond our knowledge (though it would do no harm, representation would be preserved), and we would not preserve our ignorance. The only honest solution to the problem is to admit that we do not know, and branch into all possible cases, i.e. for any nonempty subset X' of the copies of u', we make all copies $\langle u', y\rangle \in X'$ smaller than $\langle u, u'\rangle$. Thus, we construct many structures instead of one, and say: The real one is one of them, but we do not know.

Note that all these structures will be different, as points which are logically indiscernible will be different from an order theoretic point of view. We should also note the parallel here to Kripke models for modal logic, where the standard construction works with complete consistent theories in the full $\Box$-language, with nested $\Box$'s etc., where we might see the differences between two points only when following arrows to some depths. Here, the situation is similar: $\langle u, u'\rangle \succ \langle u', y\rangle$ and $\langle u, u'\rangle \succ \langle u', y'\rangle$ are the same on level 0; it is in both cases u. On level 1, they are the same too, as we see in both cases u', and only in level 2, they may begin to look differently: y and y' may choose different successors.

Once we are aware of the problem – i.e. we do not know enough for a decision – we can, of course, choose one sufficient for our representation purpose. But it is important to see the arbitrariness of the decision we take. The natural solution will then be to decide for making ALL copies of u' smaller than $\langle u, u'\rangle$.

Perhaps U is not the only set such that $u \in U - f(U)$, but there is also U' with $u \in U' - f(U')$. In this case, we repeat the construction for U', and choose now for each copy of u (at least) one element in U, and one in U', which minimizes this copy. Then, in the presence of all elements of U, or all elements of U', all copies will be minimized. The solution is thus to consider all copies $\langle u, g \rangle$, where $g \in \Pi\{X \in \mathcal{Y} : u \in X - f(X)\}$.

We will define $\langle u, g \rangle \succ \langle y, h \rangle$ iff $y \in ran(g)$ – this is the adequate condition – and forget about h, but keep in mind that we took here an arbitrary decision, and should, to preserve ignorance, branch into all possibilities. We will see its importance immediately now.

4.2.1.2 Transitive Preferential Structures

The Example 4.2.3 (p. 97) shows that we cannot just make the above construction transitive and preserve representation. This is an illustration of the fact that we have to be careful about excessive relations.

The new construction avoids this, as it "looks ahead":

Seen from one fixed point, an arbitrary relation is a graph where we identify $\succ$ with $\rightarrow$, i.e. $u \succ x$ will be written $u \rightarrow x$. The picture is perhaps easier to read when we write this graph as a tree, repeating nodes when necessary. So, from the starting point u, we can go to x and x', from x to y and y', from x' to w and w' and w'', etc. So we can write the tree of all direct and indirect successors of u, $t(u)$, and if x is a direct successor of u, then the tree for x, $t(x)$ will be a subtree of $t(u)$, beginning at the successor x of the root in $t(u)$.

This gives us now a method to control all direct and indirect successors of an element. We write as index above tree, and define $\langle u, t(u) \rangle \succ \langle x, t(x) \rangle$ iff $t(x)$ is the subtree of $t(u)$ which begins at the direct successor x of u. In the next step, we make the relation transitive, of course, we now have to see that this can be done without destroying representation, and we will use in our construction special choice functions, which always choose for $u \in Y - f(u)$ u itself – this is allowed, and they will do what we want. The details are given in the formal construction below.

4.2.1.3 Smooth Structures

In analogy to Case (1), and with the same argument, we will consider choice functions $g \in \Pi\{f(X) : x \in X - f(X)\}$.

(In the final construction, we will construct simultaneously for all u, U such that $u \in f(U)$ a U-minimal copy; so in the following intuitive discussion, it will suffice to find minimal u, x, etc. with the required properties. This remark is destined for readers who wonder how this will all fit together. We should also note that we will again be in the dilemma which copy to make smaller, and will do so for all candidates – violating our principle of preserving ignorance. Yet, as before, as long as we are aware of it, it will do no harm.)

To see the new problem arising now, we start with U, and suppose that $u \in f(U)$. Let now $u \in X - f(X)$, then we have to find $x \in f(X)$ below u. First, x must not be in U, as we would have destroyed minimality of u in U, this is analogous to

Case (1), so we need $f(X) - U \neq \emptyset$. But let now $u \in f(Y)$, $x \in Y$. In Case (1), it was sufficient to find another copy of u, which is minimal in Y. Now, we have to do more: To find a $y \in f(Y)$, y below u, so smoothness will hold. We will call the following process the "repairing process for u, x, and Y". Suppose then that $u \in f(Y)$ and $x \in f(Y)$ for $Y \in \mathcal{Y}$. Then we have destroyed minimality of u in Y, but have repaired smoothness immediately again by finding the minimal x. The situation is different if $x \in Y - f(Y)$ (and there was no $x' \prec u$, $x' \in f(Y)$ chosen at the same time). Then we have destroyed minimality of u in Y, without repairing smoothness, and we have to repair it now by finding suitable $y \prec u$, $y \in f(Y)$. Of course, y must not be in U, as this would destroy minimality of u in U.

Thus, we have to find for all Y with $u \in f(Y)$, $x \in Y - f(Y)$ some $y \in f(Y)$, $y \prec u$, $y \notin U$. Note that this repair process is individual, i.e. we do not have to find one universal y which repairs lost minimality for ALL such Y at the same time, but it suffices to do it one by one, individually for every single such Y.

But now, the solutions y for such Y may have introduced new problems: Not only x is below u, but also y is below u. If there is now $Z \in \mathcal{Y}$ such that $u \in f(Z)$ and $y \in Z - f(Z)$, then we have to do the same repairing process for u, y, Z: Find suitable $z \in f(Z)$ below u, $z \notin U$, etc. So we will have an infinite repairing process, where each step may introduce new problems, which will be repaired in the next step.

To illustrate that the problem is still a bit more complicated, we make a definition, and see that we have to avoid in above situation not only U, but $H(U, u)$ to be defined now.

$H(U, u)_0 := U$,
$H(U, u)_{\alpha+1} := H(U, u)_\alpha \cup \bigcup\{X : u \in X \wedge \mu(X) \subseteq H(U, u)_\alpha\}$,
$H(U, u)_\lambda := \bigcup\{H(U, u)_\alpha : \alpha < \lambda\}$ for $limit(\lambda)$,
$H(U, u) := \bigcup\{H(U, u)_\alpha : \alpha < \kappa\}$ for κ sufficiently big
($card(Z)$ suffices, as the procedure trivializes, when we cannot add any new elements).
(HU, u) is the property: (HU, u)
$u \in \mu(U)$, $u \in Y - \mu(Y) \Rightarrow \mu(Y) \not\subseteq H(U, u)$ – of course for all u and U.
$(U, Y \in \mathcal{Y})$.

Fact 4.2.3 (HU, u) holds in smooth structures.

The proof is given in Fact 4.2.12 (p. 108) (2).

We note now that we have to consider our principle of preserving ignorance again: We can choose first arbitrary $y \in f(Y)$ to repair for u, x, Y. So which one we choose is – a priori – an arbitrary choice. Yet, this choice might have repercussions later, as different y and y' chosen to repair for u, x, Y might force different repairs for u, y, Z or u, y', Z', as Z might be such that $u \in f(Z)$, $y \in Z - f(Z)$, and Z' such that $u \in f(Z)$, $y' \in Z - f(Z')$, and it might be possible to find suitable $z \in f(Z)$, $z \notin H(U, u)$, but no suitable z', etc. So, at first sight, the arbitrary choice might reveal an impasse later on. We will see that we can easily solve the problem in the not necessarily transitive case, but we do not see at the time of writing any easy solution in the transitive case, if the domain is not necessarily closed under finite unions.

4.2.1.4 Transitive Smooth Structures

The basic, now more complicated, situation to consider is the following:

Let again $u \in f(U)$, $u \in X - f(X)$, we have to find $x \in f(X)$ – outside $H(U, u)$ as in Case (3). Thus, we need again $f(X) - H(U, u) \neq \emptyset$. Again, we have to repair all damage done, i.e. for all u, x, Y as discussed in Case (3), the infinite repair process discussed there.

Suppose now that $x \in Y - f(Y)$, so we have to find $y \in f(Y)$, outside $H(U, u)$ by transitivity of the relation, as $y \prec x \prec u$, and, in addition outside $H(X, x)$, as in Case (3), now for X and x. Thus, we need $f(Y) - (H(U, u) \cup H(X, x)) \neq \emptyset$. Moreover, we have to do the same for all elements y introduced by the above repairing process. Again, we have to do repairing: $y \prec u$ and $y \prec x$, so for all Y' such that $u \in f(Y')$, $y \in Y' - f(Y')$ we have to repair for u, y, Y', and if $x \in f(Y')$, $y \in Y' - f(Y')$ we have to repair for x, y, Y', creating new smaller elements, etc.

If $y \in Z - f(Z)$, we have to find $z \in f(Z)$, outside $H(U, u)$, $H(X, x)$, $H(Y, y)$, etc., so the further we go down, the longer the condition will be. Thus, we need $f(Z) - (H(U, u) \cup H(X, x) \cup H(Y, y)) \neq \emptyset$. And, again we have to repair for u, z, x, z, and y, z. And so on.

Note again the arbitrariness of choice, when there is not a unique solution, i.e. no unique x, y, z, etc. This has to be considered when we want to respect preservation of ignorance, but also an early wrong choice might lead to an impasse, leading to backtracking to this early wrong choice.

We will see that the closure of the domain under $(\cup)$ makes all this easily possible, but the authors do not see an easy solution in the absence of $(\cup)$ at the time of writing – the problem is an initial potentially wrong choice, which we do not see how to avoid other than by trying.

So we will give here only a formal negative result by an example, see Example 4.2.5 (p. 112), and essentially repeat the result given in [Sch04] using $(\cup)$, see Proposition 5.1.1 (p. 120), presented in Sect. 5.1 (p. 119).

4.2.1.5 Ranked Structures

We give here some definitions and show elementary facts about ranked structures. We also prove a general abstract nonsense fact about extending relations, to be used here and also later on.

The crucial fact will be Lemma 4.2.1 (p. 93). It shows that we can do with either one or infinitely many copies of each model. The reason behind it is the following: Suppose we have exactly two copies of one model, m, m', where m and m' have the same logical properties. If, e.g. $m \prec m'$, then, as we consider only minimal elements, m' will be "invisible". If m and m' are incomparable, then, by rankedness (modularity), they will have the same elements above (and below) themselves; they have the same behaviour in the preferential structure. An immediate consequence is the "singleton property" of Lemma 4.2.1 (p. 93): One element suffices to destroy minimality, and it suffices to look at pairs (and singletons).

We first note the following trivial.

Fact 4.2.4 In a ranked structure, smoothness and the condition

$$(\mu\emptyset)\ X \neq \emptyset \Rightarrow \mu(X) \neq \emptyset$$

are (almost) equivalent.

Proof Suppose $(\mu\emptyset)$ holds, and let $x \in X - \mu(X)$, $x' \in \mu(X)$. Then $x' \prec x$ by rankedness. Conversely, if the structure is smooth and there is an element $x \in X$ in the structure (recall that structures may have "gaps", but this condition is a minor point, which we shall neglect here – this is the precise meaning of "almost"), then either $x \in \mu(X)$ or there is $x' \prec x$, $x' \in \mu(X)$, so $\mu(X) \neq \emptyset$. □

Fact 4.2.5 In the presence of $(\mu =)$ and $(\mu \subseteq)$, $f(Y) \cap (X - f(X)) \neq \emptyset$ is equivalent to $f(Y) \cap X \neq \emptyset$ and $f(Y) \cap f(X) = \emptyset$.

Proof $f(Y) \cap (X - f(X)) = (f(Y) \cap X) - (f(Y) \cap f(X))$.

"$\Leftarrow$": Let $f(Y) \cap X \neq \emptyset$, $f(Y) \cap f(X) = \emptyset$, so $f(Y) \cap (X - f(X)) \neq \emptyset$.

"$\Rightarrow$": Suppose $f(Y) \cap (X - f(X)) \neq \emptyset$, so $f(Y) \cap X \neq \emptyset$. Suppose $f(Y) \cap f(X) \neq \emptyset$, so by $(\mu \subseteq)$ $f(Y) \cap X \cap Y \neq \emptyset$, so by $(\mu =)$ $f(Y) \cap X \cap Y = f(X \cap Y)$ and $f(X) \cap X \cap Y \neq \emptyset$, so by $(\mu =)$ $f(X) \cap X \cap Y = f(X \cap Y)$, so $f(X) \cap Y = f(Y) \cap X$ and $f(Y) \cap (X - f(X)) = \emptyset$. □

Fact 4.2.6 If $\prec$ on X is ranked, and free of cycles, then $\prec$ is transitive.

Proof Let $x \prec y \prec z$. If $x \bot z$, then $y \succ z$, resulting in a cycle of length 2. If $z \prec x$, then we have a cycle of length 3. So $x \prec z$. □

The following fact is essentially Fact 3.10.8 of [Sch04].

Fact 4.2.7 In all ranked structures, $(\mu \subseteq)$, $(\mu =)$, (μPR), $(\mu =')$, $(\mu \parallel)$, $(\mu\cup)$, $(\mu\cup')$, $(\mu \in)$, $(\mu RatM)$ will hold, if the corresponding closure conditions are satisfied.

Proof $(\mu \subseteq)$ and (μPR) hold in all preferential structures.

$(\mu =)$ and $(\mu =')$ are trivial.

$(\mu\cup)$ and $(\mu\cup')$: All minimal copies of elements in $f(Y)$ have the same rank. If some $y \in f(Y)$ has all its minimal copies killed by an element $x \in X$, by rankedness, x kills the rest too.

$(\mu \in)$: If $f(\{a\}) = \emptyset$, we are done. Take the minimal copies of a in $\{a\}$, they are all killed by one element in X.

$(\mu \parallel)$: Case $f(X) = \emptyset$: If below every copy of $y \in Y$ there is a copy of some $x \in X$, then $f(X \cup Y) = \emptyset$. Otherwise, $f(X \cup Y) = f(Y)$. Suppose now $f(X) \neq \emptyset$,

$f(Y) \neq \emptyset$, then the minimal ranks decide: if they are equal, $f(X \cup Y) = f(X) \cup f(Y)$, etc.

$(\mu RatM)$: Let $X \subseteq Y$, $y \in X \cap f(Y) \neq \emptyset$, $x \in f(X)$. By rankedness, $y \prec x$, or $y \perp x$, $y \prec x$ is impossible, as $y \in X$, so $y \perp x$, and $x \in f(Y)$. □

Definition 4.2.1 Let $\mathcal{Z} = \langle \mathcal{X}, \prec \rangle$ be a preferential structure. Call $\mathcal{Z}$, $1 - \infty$ over Z, iff for all $x \in Z$ there are exactly one or infinitely many copies of x, i.e. for all $x \in Z$ $\{u \in \mathcal{X} : u = \langle x, i \rangle$ for some $i\}$ has cardinality 1 or $\geq \omega$.

The following lemma is Lemma 3.10.4 of [Sch04].

Lemma 4.2.1 *Let $\mathcal{Z} = \langle \mathcal{X}, \prec \rangle$ be a preferential structure and $f : \mathcal{Y} \to \mathcal{P}(Z)$ with $\mathcal{Y} \subseteq \mathcal{P}(Z)$ be represented by $\mathcal{Z}$, i.e. for $X \in \mathcal{Y}$, $f(X) = \mu_{\mathcal{Z}}(X)$, and $\mathcal{Z}$ be ranked and free of cycles. Then there is a structure $\mathcal{Z}'$, $1 - \infty$ over Z, ranked and free of cycles, which also represents f.*

Proof We construct $\mathcal{Z}' = \langle \mathcal{X}', \prec' \rangle$.

Let $A := \{x \in Z$: there is some $\langle x, i \rangle \in \mathcal{X}$, but for all $\langle x, i \rangle \in \mathcal{X}$ there is $\langle x, j \rangle \in \mathcal{X}$ with $\langle x, j \rangle \prec \langle x, i \rangle\}$,
let $B := \{x \in Z$: there is some $\langle x, i \rangle \in \mathcal{X}$, s.t. for no $\langle x, j \rangle \in \mathcal{X}$ $\langle x, j \rangle \prec \langle x, i \rangle\}$,
let $C := \{x \in Z$: there is no $\langle x, i \rangle \in \mathcal{X}\}$.

Let $c_i : i < \kappa$ be an enumeration of C. We introduce for each such c_i ω many copies $\langle c_i, n \rangle : n < \omega$ into $\mathcal{X}'$, put all $\langle c_i, n \rangle$ above all elements in $\mathcal{X}$, and order the $\langle c_i, n \rangle$ by $\langle c_i, n \rangle \prec' \langle c_{i'}, n' \rangle :\Leftrightarrow (i = i'$ and $n > n')$ or $i > i'$. Thus, all $\langle c_i, n \rangle$ are comparable.

If $a \in A$, then there are infinitely many copies of a in $\mathcal{X}$, as $\mathcal{X}$ was cycle-free, we put them all into $\mathcal{X}'$. If $b \in B$, we choose exactly one such minimal element $\langle b, m \rangle$ (i.e. there is no $\langle b, n \rangle \prec \langle b, m \rangle$) into $\mathcal{X}'$ and omit all other elements. (For definiteness, assume in all applications $m = 0$.) For all elements from A and B, we take the restriction of the order $\prec$ of $\mathcal{X}$. This is the new structure $\mathcal{Z}'$.

Obviously, adding the $\langle c_i, n \rangle$ does not introduce cycles, irreflexivity and rankedness are preserved. Moreover, any substructure of a cycle-free, irreflexive, ranked structure also has these properties, so $\mathcal{Z}'$ is $1 - \infty$ over Z, ranked, and free of cycles.

We show that $\mathcal{Z}$ and $\mathcal{Z}'$ are equivalent. Let then $X \subseteq Z$, we have to prove $\mu(X) = \mu'(X)$ $(\mu := \mu_{\mathcal{Z}}, \mu' := \mu_{\mathcal{Z}'})$.

Let $z \in X - \mu(X)$. If $z \in C$ or $z \in A$, then $z \notin \mu'(X)$. If $z \in B$, let $\langle z, m \rangle$ be the chosen element. As $z \notin \mu(X)$, there is $x \in X$ s.t. some $\langle x, j \rangle \prec \langle z, m \rangle$. x cannot be in C. If $x \in A$, then also $\langle x, j \rangle \prec' \langle z, m \rangle$. If $x \in B$, then there is some $\langle x, k \rangle$ also in $\mathcal{X}'$. $\langle x, j \rangle \prec \langle x, k \rangle$ is impossible. If $\langle x, k \rangle \prec \langle x, j \rangle$, then $\langle z, m \rangle \succ \langle x, k \rangle$ by transitivity. If $\langle x, k \rangle \perp \langle x, j \rangle$, then also $\langle z, m \rangle \succ \langle x, k \rangle$ by rankedness. In any case, $\langle z, m \rangle \succ' \langle x, k \rangle$, and thus $z \notin \mu'(X)$.

Let $z \in X - \mu'(X)$. If $z \in C$ or $z \in A$, then $z \notin \mu(X)$. Let $z \in B$ and some $\langle x, j \rangle \prec' \langle z, m \rangle$. x cannot be in C, as they were sorted on top, so $\langle x, j \rangle$ exists in $\mathcal{X}$ too and $\langle x, j \rangle \prec \langle z, m \rangle$. But if any other $\langle z, i \rangle$ is also minimal in $\mathcal{Z}$ among the $\langle z, k \rangle$, then by rankedness also $\langle x, j \rangle \prec \langle z, i \rangle$, as $\langle z, i \rangle \perp \langle z, m \rangle$, so $z \notin \mu(X)$. □

Notation 4.2.1 We fix the following notation: $A := \{x \in Z : f(x) = \emptyset\}$ and $B := Z - A$ (here and in future we sometimes write $f(x)$ for $f(\{x\})$, likewise $f(x, x') = x$ for $f(\{x, x'\}) = \{x\}$, etc., when the meaning is obvious).

Corollary 4.2.1 *If f can be represented by a ranked $\mathcal{Z}$ free of cycles, then there is $\mathcal{Z}'$, which is also ranked and cycle-free, all $b \in B$ occur in 1 copy, all $a \in A$ ∞ often.*

The following example was presented in Fact 3.10.13 of [Sch04].

Example 4.2.2 This example shows that the conditions $(\mu \subseteq) + (\mu PR) + (\mu =) + (\mu \cup) + (\mu \in)$ can be satisfied, and still representation by a ranked structure is impossible.
Consider $\mu(\{a, b\}) = \emptyset$, $\mu(\{a\}) = \{a\}$, $\mu(\{b\}) = \{b\}$. The conditions $(\mu \subseteq) + (\mu PR) + (\mu =) + (\mu \cup) + (\mu \in)$ hold trivially. This is representable, e.g. by $a_1 \succeq b_1 \succeq a_2 \succeq b_2 \ldots$ without transitivity. (Note that rankedness implies transitivity, $a \preceq b \preceq c$, but not for $a = c$.) But this cannot be represented by a ranked structure: As $\mu(\{a\}) \neq \emptyset$, there must be a copy a_i of minimal rank, likewise for b and some b_i. If they have the same rank, $\mu(\{a, b\}) = \{a, b\}$, otherwise it will be $\{a\}$ or $\{b\}$.

In the general situation we have possibly $U \neq \emptyset$, but $f(U) = \emptyset$. In this case, we only know that below each $u \in U$, there must be infinitely many $u' \in U$ or infinitely many copies of such $u' \in U$. (It is only in such cases that we need copies for representation in ranked structures, see Lemma 4.2.1 (p. 93).) Thus, the amount of information we have is very small. It is not surprising that representation problems are now difficult, as we will see below (see Sect. 4.2.2.5 (p. 113)), and we will not go into more details here.

4.2.2 Detailed Discussion

4.2.2.1 General Preferential Structures

The material in this section is taken from [Sch04], Sect. 3.2.1 there, the result was already shown in [Sch92] with the same methods.

Proposition 4.2.1 *Let $\mu : \mathcal{Y} \to \mathcal{P}(U)$ satisfy $(\mu \subseteq)$ and (μPR). Then there is a preferential structure $\mathcal{X}$ s.t. $\mu = \mu_{\mathcal{X}}$. See e.g.* [Sch04].

Proof The preferential structure is defined in Construction 4.2.1 (p. 95) and Claim 4.2.2 (p. 95) shows the representation. The construction is basic for much of the rest of the material on nonranked structures. □

Definition 4.2.2 For $x \in Z$, let $\mathcal{Y}_x := \{Y \in \mathcal{Y}: x \in Y - \mu(Y)\}$, $\Pi_x := \Pi\mathcal{Y}_x$.

Note that $\emptyset \notin \mathcal{Y}_x$, $\Pi_x \neq \emptyset$, and that $\Pi_x = \{\emptyset\}$ iff $\mathcal{Y}_x = \emptyset$.

Claim 4.2.1 Let $\mu : \mathcal{Y} \to \mathcal{P}(Z)$ satisfy $(\mu \subseteq)$ and (μPR), and let $U \in \mathcal{Y}$. Then $x \in \mu(U) \Leftrightarrow x \in U \wedge \exists f \in \Pi_x . ran(f) \cap U = \emptyset$.

Proof
Case 1: $\mathcal{Y}_x = \emptyset$, thus $\Pi_x = \{\emptyset\}$. "$\Rightarrow$": Take $f := \emptyset$. "$\Leftarrow$": $x \in U \in \mathcal{Y}$, $\mathcal{Y}_x = \emptyset$ $\Rightarrow x \in \mu(U)$ by definition of $\mathcal{Y}_x$.

Case 2: $\mathcal{Y}_x \neq \emptyset$. "$\Rightarrow$": Let $x \in \mu(U) \subseteq U$. It suffices to show $Y \in \mathcal{Y}_x \Rightarrow Y - U \neq \emptyset$. But if $Y \subseteq U$ and $Y \in \mathcal{Y}_x$, then $x \in Y - \mu(Y)$, contradicting (μPR). "$\Leftarrow$": If $x \in U - \mu(U)$, then $U \in \mathcal{Y}_x$, so $\forall f \in \Pi_x . ran(f) \cap U \neq \emptyset$. □

Construction 4.2.1 Let $\mathcal{X} := \{\langle x, f \rangle : x \in Z \wedge f \in \Pi_x\}$, $\langle x', f' \rangle \prec \langle x, f \rangle :\Leftrightarrow x' \in ran(f)$, and $\mathcal{Z} := \langle \mathcal{X}, \prec \rangle$.

Claim 4.2.2 For $U \in \mathcal{Y}$, $\mu(U) = \mu_{\mathcal{Z}}(U)$.

Proof By Claim 4.2.1 (p. 95), it suffices to show that for all $U \in \mathcal{Y}$, $x \in \mu_{\mathcal{Z}}(U) \Leftrightarrow x \in U$ and $\exists f \in \Pi_x . ran(f) \cap U = \emptyset$. So let $U \in \mathcal{Y}$. "$\Rightarrow$": If $x \in \mu_{\mathcal{Z}}(U)$, then there is $\langle x, f \rangle$ minimal in $\mathcal{X} \lceil U$ (recall from Definition 2.1.1 (p. 31) that $\mathcal{X} \lceil U := \{\langle x, i \rangle \in \mathcal{X} : x \in U\}$), so $x \in U$, and there is no $\langle x', f' \rangle \prec \langle x, f \rangle$, $x' \in U$, so by $\Pi_{x'} \neq \emptyset$ there is no $x' \in ran(f)$, $x' \in U$, but then $ran(f) \cap U = \emptyset$. "$\Leftarrow$": If $x \in U$, and there is $f \in \Pi_x$, $ran(f) \cap U = \emptyset$, then $\langle x, f \rangle$ is minimal in $\mathcal{X} \lceil U$. (Claim 4.2.2 (p. 95) and Proposition 4.2.1 (p. 94)) □

4.2.2.2 Transitive Preferential Structures

The material in this section is taken from [Sch04], Sect. 3.2.2 there, the result was already shown in [Sch92] with different methods.

We show here:

Proposition 4.2.2 *Let $\mu : \mathcal{Y} \to \mathcal{P}(U)$ satisfy $(\mu \subseteq)$ and (μPR) . Then there is a transitive preferential structure $\mathcal{X}$ s.t. $\mu = \mu_{\mathcal{X}}$. See e.g.*, [Sch04].

Proof

Discussion

The Construction 4.2.1 (p. 95) (also used in [Sch92]) cannot be made transitive as it is, and this will be shown below in Example 4.2.3 (p. 97). The second construction in [Sch92] is a special one, which is transitive, but uses heavily lack of smoothness. (For completeness' sake, we give a similar proof in Proposition 4.2.3 (p. 98).) We present here a more flexible and more adequate construction, which avoids a certain excess in the relation $\prec$ of the construction in Proposition 4.2.3 (p. 98): There, too many elements $\langle y, g \rangle$ are smaller than some $\langle x, f \rangle$, as the relation is independent from g. This excess prevents transitivity.

We refine now the construction of the relation to have better control over successors. Recall that a tree of height $\leq \omega$ seems the right way to encode the successors of an element, as far as transitivity is concerned (which speaks only about finite chains). Now, in the basic construction, different copies have different successors, chosen by different functions (elements of the cartesian product). As it suffices to make one copy of the successor smaller than the element to be minimized, we do the following: Let $\langle x, g \rangle$ with $g \in \Pi\{X : x \in X - f(X)\}$ be one of the elements of the standard construction. Let $\langle x', g' \rangle$ be s.t. $x' \in ran(g)$, then we make again copies $\langle x, g, g' \rangle$, etc. for each such x' and g', and make only $\langle x', g' \rangle$, but not some other $\langle x', g'' \rangle$ smaller than $\langle x, g, g' \rangle$, for some other $g'' \in \Pi\{X' : x' \in X' - f(X')\}$. Thus, we have a much more restricted relation and much better control over it. More precisely, we make trees, where we mark all direct and indirect successors, and each time the choice is made by the appropriate choice functions of the cartesian product. An element with its tree is a successor of another element with its tree, iff the former is an initial segment of the latter – see the definition in Construction 4.2.2 (p. 97).

Recall also that transitivity is for free as we can use the element itself to minimize it. This is made precise by the use of the trees tf_x for a given element x and choice function f_x. But they also serve another purpose. The trees tf_x are constructed as follows: The root is x, the first branching is done according to f_x, and then we continue with constant choice. Let, e.g. $x' \in ran(f_x)$, we can now always choose x', as it will be a legal successor of x' itself, being present in all X' s.t. $x' \in X' - f(X')$. So we have a tree which branches once, directly above the root, and is then constant without branching. Obviously, this is essentially equivalent to the old construction in the not necessarily transitive case. This shows two things: First, the construction with trees gives the same μ as the old construction with simple choice functions. Second, even if we consider successors of successors, nothing changes: We are still with the old x'. Consequently, considering the transitive closure will not change matters, an element $\langle x, tf_x \rangle$ will be minimized by its direct successors iff it will be minimized by direct and indirect successors. If you like, the trees tf_x are the mathematical construction expressing the intuition that we know so little about minimization that we have to consider suicide a serious possibility – the intuitive reason why transitivity imposes no new conditions.

To summarize: Trees seem the right way to encode all the information needed for full control over successors for the transitive case. The special trees tf_x show that we have not changed things substantially, i.e. the new μ-functions in the simple case and for the transitive closure stay the same. We hope that this construction will show its usefulness in other contexts. Its naturalness and generality seem to be a good promise.

We give below the example which shows that the old construction is too brutal for transitivity to hold. Recall that transitivity permits substitution in the following sense: If (the two copies of) x is killed by y_1 and y_2 together, and y_1 is killed by z_1 and z_2 together, then x should be killed by z_1, z_2, and y_2 together. But the old construction substitutes too much: In the old construction, we considered elements $\langle x, f \rangle$, where $f \in \Pi_x$, with $\langle y, g \rangle \prec \langle x, f \rangle$ iff $y \in ran(f)$, independent of g. This construction can, in general, not be made transitive, as Example 4.2.3 (p. 97) below shows.

The new construction avoids this, as it "looks ahead", and not all elements $\langle y_1, t_{y_1} \rangle$ are smaller than $\langle x, t_x \rangle$, where y_1 is a child of x in t_x (or $y_1 \in ran(f)$). The new construction is basically the same as Construction 4.2.1 (p. 95), but avoids to make too many copies smaller than the copy to be killed.

Recall that we need no new properties of μ to achieve transitivity here, as a killed element x might (partially) "commit suicide", i.e. for some i, i' $\langle x, i \rangle \prec \langle x, i' \rangle$, so we cannot substitute x by any set which does not contain x. In this simple situation, if $x \in X - \mu(X)$, we cannot find out whether all copies of x are killed by some $y \neq x$, $y \in X$. We can assume without loss of generality that there is an infinite descending chain of x-copies that are not killed by other elements. Thus, we cannot replace any y_i as above by any set which does not contain y_i, but then substitution becomes trivial, as any set substituting y_i has to contain y_i. Thus, we need no new properties to achieve transitivity.

Example 4.2.3 As we consider only one set in each case, we can index with elements, instead of with functions. So suppose $x, y_1, y_2 \in X$, $y_1, z_1, z_2 \in Y$, $x \notin \mu(X)$, $y_1 \notin \mu(Y)$, and that we need y_1 and y_2 to minimize x, so there are two copies $\langle x, y_1 \rangle$, $\langle x, y_2 \rangle$, likewise we need z_1 and z_2 to minimize y_1. Thus we have $\langle x, y_1 \rangle \succ \langle y_1, z_1 \rangle$, $\langle x, y_1 \rangle \succ \langle y_1, z_2 \rangle$, $\langle x, y_2 \rangle \succ y_2$, $\langle y_1, z_1 \rangle \succ z_1$, $\langle y_1, z_2 \rangle \succ z_2$ (the z_i and y_2 are not killed). If we take the transitive closure, we have $\langle x, y_1 \rangle \succ z_k$ for any i, k, so for any z_k $\{z_k, y_2\}$ will minimize all of x, which is not intended.

The preferential structure is defined in Construction 4.2.2 (p. 97), Claim 4.2.3 (p. 98) shows representation for the simple structure, and Claim 4.2.4 (p. 98) shows representation for the transitive closure of the structure.

The main idea is to use the trees tf_x, whose elements are exactly the elements of the range of the choice function f. This makes Construction 4.2.1 (p. 95) and Construction 4.2.2 (p. 97) basically equivalent, and shows that the transitive case is characterized by the same conditions as the general case. These trees are defined below in Fact 4.2.8 (p. 98), (3), and used in the proofs of Claim 4.2.3 (p. 98) and Claim 4.2.4 (p. 98).

Again, Construction 4.2.2 (p. 97) contains the basic idea for the treatment of the transitive case. It can certainly be reused in other contexts.

Construction 4.2.2

(1) For $x \in Z$, let T_x be the set of trees t_x s.t.

  (a) all nodes are elements of Z,
  (b) the root of t_x is x,
  (c) $height(t_x) \leq \omega$,
  (d) if y is an element in t_x, then there is $f \in \Pi_y := \Pi\{Y \in \mathcal{Y}: y \in Y - \mu(Y)\}$ s.t. the set of children of y is $ran(f)$.

(2) For $x, y \in Z$, $t_x \in T_x$, $t_y \in T_y$, set $t_x \rhd t_y$ iff y is a (direct) child of the root x in t_x and t_y is the subtree of t_x beginning at y.

(3) Let $\mathcal{Z} := \langle \{\langle x, t_x \rangle : x \in Z, t_x \in T_x\}, \langle x, t_x \rangle \succ \langle y, t_y \rangle$ iff $t_x \rhd t_y \rangle$.

Fact 4.2.8

(1) The construction ends at some y iff $\mathcal{Y}_y = \emptyset$, consequently $T_x = \{x\}$ iff $\mathcal{Y}_x = \emptyset$. (We identify the tree of height 1 with its root.)
(2) If $\mathcal{Y}_x \neq \emptyset$, tc_x, the totally ordered tree of height ω, branching with $card = 1$, and with all elements equal to x is an element of T_x. Thus, with (1), $T_x \neq \emptyset$ for any x.
(3) If $f \in \Pi_x$, $f \neq \emptyset$, then the tree tf_x with root x and otherwise composed of the subtrees t_y for $y \in ran(f)$, where $t_y := y$ iff $\mathcal{Y}_y = \emptyset$, and $t_y := tc_y$ iff $\mathcal{Y}_y \neq \emptyset$, is an element of T_x. (Level 0 of tf_x has x as element, the t_y's begin at level 1.)
(4) If y is an element in t_x and t_y the subtree of t_x starting at y, then $t_y \in T_y$.
(5) $\langle x, t_x\rangle \succ \langle y, t_y\rangle$ implies $y \in ran(f)$ for some $f \in \Pi_x$.

Claim 4.2.3 (p. 98) shows basic representation.

Claim 4.2.3 $\forall U \in \mathcal{Y}.\mu(U) = \mu_{\mathcal{Z}}(U)$.

Proof By Claim 4.2.1 (p. 95), it suffices to show that for all $U \in \mathcal{Y}$ $x \in \mu_{\mathcal{Z}}(U) \Leftrightarrow x \in U \wedge \exists f \in \Pi_x.ran(f) \cap U = \emptyset$. Fix $U \in \mathcal{Y}$. "$\Rightarrow$": $x \in \mu_{\mathcal{Z}}(U) \Rightarrow$ ex. $\langle x, t_x\rangle$ minimal in $\mathcal{Z}\lceil U$, thus $x \in U$ and there is no $\langle y, t_y\rangle \in \mathcal{Z}$, $\langle y, t_y\rangle \prec \langle x, t_x\rangle$, $y \in U$. Let f define the set of children of the root x in t_x. If $ran(f) \cap U \neq \emptyset$, if $y \in U$ is a child of x in t_x, and if t_y is the subtree of t_x starting at y, then $t_y \in T_y$ and $\langle y, t_y\rangle \prec \langle x, t_x\rangle$, contradicting minimality of $\langle x, t_x\rangle$ in $\mathcal{Z}\lceil U$. So $ran(f) \cap U = \emptyset$. "$\Leftarrow$": Let $x \in U$. If $\mathcal{Y}_x = \emptyset$, then the tree x has no $\rhd$-successors, and $\langle x, x\rangle$ is $\succ$-minimal in $\mathcal{Z}$. If $\mathcal{Y}_x \neq \emptyset$ and $f \in \Pi_x$ s.t. $ran(f) \cap U = \emptyset$, then $\langle x, tf_x\rangle$ is $\succ$-minimal in $\mathcal{Z}\lceil U$. □

We now consider the transitive closure of $\mathcal{Z}$. (Recall that $\prec^*$ denotes the transitive closure of $\prec$.) Claim 4.2.4 (p. 98) shows that transitivity does not destroy what we have achieved. The trees tf_x will play a crucial role in the demonstration.

Claim 4.2.4 Let $\mathcal{Z}' := \langle\ \{\langle x, t_x\rangle : x \in Z, t_x \in T_x\}, \langle x, t_x\rangle \succ \langle y, t_y\rangle$ iff $t_x \rhd^* t_y\ \rangle$. Then $\mu_{\mathcal{Z}} = \mu_{\mathcal{Z}'}$.

Proof Suppose there is $U \in \mathcal{Y}$, $x \in U$, $x \in \mu_{\mathcal{Z}}(U)$, $x \notin \mu_{\mathcal{Z}'}(U)$. Then there must be an element $\langle x, t_x\rangle \in \mathcal{Z}$ with no $\langle x, t_x\rangle \succ \langle y, t_y\rangle$ for any $y \in U$. Let $f \in \Pi_x$ determine the set of children of x in t_x, then $ran(f) \cap U = \emptyset$, consider tf_x. As all elements $\neq x$ of tf_x are already in $ran(f)$, no element of tf_x is in U. Thus there is no $\langle z, t_z\rangle \prec^* \langle x, tf_x\rangle$ in $\mathcal{Z}$ with $z \in U$, so $\langle x, tf_x\rangle$ is minimal in $\mathcal{Z}'\lceil U$, contradiction. (Claim 4.2.4 (p. 98) and Proposition 4.2.2 (p. 95)) □

We now give the direct proof, which we cannot adapt to the smooth case. Such easy results must be part of the folklore, but we give them for completeness' sake.

Proposition 4.2.3 *In the general case, every preferential structure is equivalent to a transitive one – i.e. they have the same μ-functions.*

Proof If $\langle a, i\rangle \succ \langle b, j\rangle$, we create an infinite descending chain of new copies $\langle b, \langle j, a, i, n\rangle\rangle$, $n \in \omega$, where $\langle b, \langle j, a, i, n\rangle\rangle \succ \langle b, \langle j, a, i, n'\rangle\rangle$ if $n' > n$, and make

$\langle a, i\rangle \succ \langle b, \langle j, a, i, n\rangle\rangle$ for all $n \in \omega$, but cancel the pair $\langle a, i\rangle \succ \langle b, j\rangle$ from the relation (otherwise, we would not have achieved anything), but $\langle b, j\rangle$ stays as element in the set. Now, the relation is trivially transitive, and all these $\langle b, \langle j, a, i, n\rangle\rangle$ just kill themselves, there is no need to minimize them by anything else. We just continued $\langle a, i\rangle \succ \langle b, j\rangle$ in a way it cannot bother us. For the $\langle b, j\rangle$, we do of course the same thing again. So, we have full equivalence, i.e. the μ-functions of both structures are identical (this is trivial to see). □

4.2.2.3 Smooth Structures

Introduction

Cumulativity Without $(\cup)$

We show here that, without sufficient closure properties, there is an infinity of versions of cumulativity, which collapse to usual cumulativity when the domain is closed under finite unions. Closure properties thus reveal themselves as a powerful tool to show independence of properties.

We then show positive results for the smooth and transitive smooth case.

We work in some fixed arbitrary set Z, and all sets considered will be subsets of Z.

Unless said otherwise, we use without further mentioning (μPR) and $(\mu \subseteq)$.

Note that (μPR) and $(\mu \subseteq)$ entail $\mu(A \cup B) \subseteq \mu(A) \cup \mu(B)$ whenever μ is defined for A, B, $A \cup B$. ($\mu(A \cup B) \cap A \subseteq \mu(A)$, $\mu(A \cup B) \cap B \subseteq \mu(B)$ by (μPR), but $\mu(A \cup B) \subseteq A \cup B$ by $(\mu \subseteq)$.)

Definition 4.2.3 For any ordinal α, we define

$(\mu Cum\alpha)$: If for all $\beta \leq \alpha$ $\mu(X_\beta) \subseteq U \cup \bigcup\{X_\gamma : \gamma < \beta\}$ hold, then so does $\bigcap\{X_\gamma : \gamma \leq \alpha\} \cap \mu(U) \subseteq \mu(X_\alpha)$.

$(\mu Cumt\alpha)$: If for all $\beta \leq \alpha$ $\mu(X_\beta) \subseteq U \cup \bigcup\{X_\gamma : \gamma < \beta\}$ hold, then so does $X_\alpha \cap \mu(U) \subseteq \mu(X_\alpha)$.

("t" stands for transitive, see Fact 4.2.9 (p. 103), (2.2) below.)

$(\mu Cum\infty)$ and $(\mu Cumt\infty)$ will be the class of all $(\mu Cum\alpha)$ or $(\mu Cumt\alpha)$ – read their "conjunction", i.e. if we say that $(\mu Cum\infty)$ holds, we mean that all $(\mu Cum\alpha)$ hold.

Note: the first condition thus has the form:

$(\mu Cum0)$ $\mu(X_0) \subseteq U \Rightarrow X_0 \cap \mu(U) \subseteq \mu(X_0)$,

$(\mu Cum1)$ $\mu(X_0) \subseteq U$, $\mu(X_1) \subseteq U \cup X_0 \Rightarrow X_0 \cap X_1 \cap \mu(U) \subseteq \mu(X_1)$,

$(\mu Cum2)$ $\mu(X_0) \subseteq U$, $\mu(X_1) \subseteq U \cup X_0$, $\mu(X_2) \subseteq U \cup X_0 \cup X_1 \Rightarrow$ $X_0 \cap X_1 \cap X_2 \cap \mu(U) \subseteq \mu(X_2)$.

$(\mu Cumt\alpha)$ differs from $(\mu Cum\alpha)$ only in the consequence, the intersection contains only the last X_α – in particular, $(\mu Cum0)$ and $(\mu Cumt0)$ coincide.

Recall that condition $(\mu Cum1)$ is the crucial condition in [Leh92a], which failed, despite (μCUM), but which has to hold in all smooth models. This condition $(\mu Cum1)$ was the starting point of the investigation.

We briefly mention some major results on above conditions, taken from Fact 4.2.9 (p. 103) and shown there – we use the same numbering:

(1.1) $(\mu Cum\alpha) \Rightarrow (\mu Cum\beta)$ for all $\beta \leq \alpha$;
(1.2) $(\mu Cumt\alpha) \Rightarrow (\mu Cumt\beta)$ for all $\beta \leq \alpha$;
(2.1) all $(\mu Cum\alpha)$ hold in smooth preferential structures;
(2.2) all $(\mu Cumt\alpha)$ hold in transitive smooth preferential structures;
(3.1) $(\mu Cum\beta) + (\cup) \Rightarrow (\mu Cum\alpha)$ for all $\beta \leq \alpha$;
(3.2) $(\mu Cumt\beta) + (\cup) \Rightarrow (\mu Cumt\alpha)$ for all $\beta \leq \alpha$;
(5.2) $(\mu Cum\alpha) \Rightarrow (\mu CUM)$ for all α;
(5.3) $(\mu CUM) + (\cup) \Rightarrow (\mu Cum\alpha)$ for all α.

The following inductive definition of $H(U, u)$ and of the property (HU, u) concerns closure under $(\mu Cum\infty)$. Its main property is formulated in Fact 4.2.11 (p. 106), and its main interest is its use in the proof of Proposition 4.2.4 (p. 110).

Definition 4.2.4

$(H(U, u)_\alpha$, $H(U)_\alpha$, (HU, u) , (HU) .)
$H(U, u)_0 := U,$
$H(U, u)_{\alpha+1} := H(U, u)_\alpha \cup \bigcup\{X : u \in X \wedge \mu(X) \subseteq H(U, u)_\alpha\},$
$H(U, u)_\lambda := \bigcup\{H(U, u)_\alpha : \alpha < \lambda\}$ for $limit(\lambda)$,
$H(U, u) := \bigcup\{H(U, u)_\alpha : \alpha < \kappa\}$ for κ sufficiently big ($card(Z)$ suffices, as the procedure trivializes, when we cannot add any new elements).
(HU, u) is the property:
$u \in \mu(U)$, $u \in Y - \mu(Y) \Rightarrow \mu(Y) \not\subseteq H(U, u)$ – of course for all u and U. $(U, Y \in \mathcal{Y})$.
Thus, (HU, u) entails $\mu(U) \subseteq H(U, u)$, $u \in \mu(U) \cap Y \Rightarrow u \in \mu(Y)$.
For the case with $(\cup)$, we further define, independent of u,
$H(U)_0 := U,$
$H(U)_{\alpha+1} := H(U)_\alpha \cup \bigcup\{X : \mu(X) \subseteq H(U)_\alpha\},$
$H(U)_\lambda := \bigcup\{H(U)_\alpha : \alpha < \lambda\}$ for $limit(\lambda)$,
$H(U) := \bigcup\{H(U)_\alpha : \alpha < \kappa\}$ again for κ sufficiently big
(HU) is the property:
$u \in \mu(U)$, $u \in Y - \mu(Y) \Rightarrow \mu(Y) \not\subseteq H(U)$ – of course for all U. $(U, Y \in \mathcal{Y})$.
Thus, (HU) entails $\mu(Y) \subseteq H(U) \Rightarrow \mu(U) \cap Y \subseteq \mu(Y)$.

Obviously, $H(U, u) \subseteq H(U)$, so $(HU) \Rightarrow (HU, u)$.

Example 4.2.4 This important example shows that the conditions $(\mu Cum\alpha)$ and $(\mu Cumt\alpha)$ defined in Definition 4.2.3 (p. 99) are all different in the absence of $(\cup)$, while in its presence they all collapse (see Fact 4.2.9 (p. 103) below). More precisely, the following (class of) examples show that the $(\mu Cum\alpha)$ increases in strength. For any finite or infinite ordinal $\kappa > 0$, we construct an example such that

(a) (μPR) and $(\mu \subseteq)$ hold;
(b) (μCUM) holds;
(c) $(\bigcap)$ holds;
(d) $(\mu Cumt\alpha)$ holds for $\alpha < \kappa$;
(e) $(\mu Cum\kappa)$ fails.

Proof We define a suitable base set and a non-transitive binary relation $\prec$ on this set, as well as a suitable set $\mathcal{X}$ of subsets, closed under arbitrary intersections, but not under finite unions, and define μ on these subsets as usual in preferential structures by $\prec$. Thus, (μPR) and $(\mu \subseteq)$ will hold. It will be immediate that $(\mu Cum\kappa)$ fails and we will show that (μCUM) and $(\mu Cumt\alpha)$ for $\alpha < \kappa$ hold by examining the cases.

For simplicity, we first define a set of generators for $\mathcal{X}$ and close under $(\bigcap)$ afterwards. The set U will have a special position, it is the "useful" starting point to construct chains corresponding to above definitions of $(\mu Cum\alpha)$ and $(\mu Cumt\alpha)$.

In the sequel, i, j will be successor ordinals, λ etc. limit ordinals, α, β, κ any ordinals, thus e.g. $\lambda \leq \kappa$ will imply that λ is a limit ordinal $\leq \kappa$.

The base set and the relation $\prec$:
$\kappa > 0$ is fixed, but arbitrary. We go up to $\kappa > 0$.

The base set is $\{a, b, c\} \cup \{d_\lambda : \lambda \leq \kappa\} \cup \{x_\alpha : \alpha \leq \kappa + 1\} \cup \{x'_\alpha : \alpha \leq \kappa\}$.
$a \prec b \prec c$, $x_\alpha \prec x_{\alpha+1}$, $x_\alpha \prec x'_\alpha$, $x'_0 \prec x_\lambda$ (for any λ) – $\prec$ is NOT transitive.

The generators:

$$U := \{a, c, x_0\} \cup \{d_\lambda : \lambda \leq \kappa\} \text{ – i.e. } \ldots \{d_\lambda : lim(\lambda) \wedge \lambda \leq \kappa\},$$
$$X_i := \{c, x_i, x'_i, x_{i+1}\} \ (i < \kappa),$$
$$X_\lambda := \{c, d_\lambda, x_\lambda, x'_\lambda, x_{\lambda+1}\} \cup \{x'_\alpha : \alpha < \lambda\} \ (\lambda < \kappa),$$
$$X'_\kappa := \{a, b, c, x_\kappa, x'_\kappa, x_{\kappa+1}\} \text{ if } \kappa \text{ is a successor,}$$
$$X'_\kappa := \{a, b, c, d_\kappa, x_\kappa, x'_\kappa, x_{\kappa+1}\} \cup \{x'_\alpha : \alpha < \kappa\} \text{ if } \kappa \text{ is a limit.}$$

Thus, $X'_\kappa = X_\kappa \cup \{a, b\}$ if X_κ were defined.

Note that there is only one X'_κ, and X_α is defined only for $\alpha < \kappa$, so we will not have X_α and X'_α at the same time.

Thus, the values of the generators under μ are the following:

$$\mu(U) = U,$$
$$\mu(X_i) = \{c, x_i\},$$
$$\mu(X_\lambda) = \{c, d_\lambda\} \cup \{x'_\alpha : \alpha < \lambda\},$$
$$\mu(X'_i) = \{a, x_i\} \ (i > 0, \ i \text{ has to be a successor}),$$
$$\mu(X'_\lambda) = \{a, d_\lambda\} \cup \{x'_\alpha : \alpha < \lambda\}.$$

(We do not assume that the domain is closed under μ.)

Intersections: We first consider pairwise intersections:

(1) $U \cap X_0 = \{c, x_0\}$,
(2) $U \cap X_i = \{c\}, i > 0$,
(3) $U \cap X_\lambda = \{c, d_\lambda\}$,
(4) $U \cap X'_i = \{a, c\}\ (i > 0)$,
(5) $U \cap X'_\lambda = \{a, c, d_\lambda\}$,
(6) $X_i \cap X_j$:

 (6.1) $j = i + 1\ \{c, x_{i+1}\}$,
 (6.2) else $\{c\}$,

(7) $X_i \cap X_\lambda$:

 (7.1) $i < \lambda\ \{c, x'_i\}$,
 (7.2) $i = \lambda + 1\ \{c, x_{\lambda+1}\}$,
 (7.3) $i > \lambda + 1\ \{c\}$,

(8) $X_\lambda \cap X_{\lambda'}$: $\{c\} \cup \{x'_\alpha : \alpha \leq min(\lambda, \lambda')\}$.

As X'_κ occurs only once, $X_\alpha \cap X'_\kappa$ etc. give no new results.

Note that μ is constant on all these pairwise intersections.

Iterated intersections: As c is an element of all sets, sets of the type $\{c, z\}$ do not give any new results. The possible subsets of $\{a, c, d_\lambda\}$: $\{c\}$, $\{a, c\}$, $\{c, d_\lambda\}$ exist already. Thus, the only source of new sets via iterated intersections is $X_\lambda \cap X_{\lambda'} = \{c\} \cup \{x'_\alpha : \alpha \leq min(\lambda, \lambda')\}$. But, to intersect them, or with some old sets, will not generate any new sets either. Consequently, the example satisfies $(\bigcap)$ for $\mathcal{X}$ defined by U, X_i $(i < \kappa)$, X_λ $(\lambda < \kappa)$, X'_κ, and above pairwise intersections.

We will now verify the positive properties. This is tedious, but straightforward, and we have to check the different cases.

Validity of (μCUM):

Consider the prerequisite $\mu(X) \subseteq Y \subseteq X$. If $\mu(X) = X$ or if $X - \mu(X)$ is a singleton, X cannot give a violation of (μCUM). So we are left with the following candidates for X:

(1) $X_i := \{c, x_i, x'_i, x_{i+1}\}$, $\mu(X_i) = \{c, x_i\}$;
 Interesting candidates for Y will have three elements, but they will all contain a. (If $\kappa < \omega : U = \{a, c, x_0\}$.)
(2) $X_\lambda := \{c, d_\lambda, x_\lambda, x'_\lambda, x_{\lambda+1}\} \cup \{x'_\alpha : \alpha < \lambda\}$, $\mu(X_\lambda) = \{c, d_\lambda\} \cup \{x'_\alpha : \alpha < \lambda\}$;
 The only sets to contain d_λ are X_λ, U, $U \cap X_\lambda$. But $a \in U$, and $U \cap X_\lambda$ is finite. (X_λ and X'_λ cannot be present at the same time.)
(3) $X'_i := \{a, b, c, x_i, x'_i, x_{i+1}\}$, $\mu(X'_i) = \{a, x_i\}$;
 a is only in U, X'_i, $U \cap X'_i = \{a, c\}$, but $x_i \notin U$, as $i > 0$.
(4) $X'_\lambda := \{a, b, c, d_\lambda, x_\lambda, x'_\lambda, x_{\lambda+1}\} \cup \{x'_\alpha : \alpha < \lambda\}$, $\mu(X'_\lambda) = \{a, d_\lambda\} \cup \{x'_\alpha : \alpha < \lambda\}$
 d_λ is only in X'_λ and U, but U contains no x'_α.

Thus, (μCUM) holds trivially.

$(\mu Cumt\alpha)$ holds for $\alpha < \kappa$:

To simplify language, we say that we reach Y from X iff $X \neq Y$ and there is a sequence X_β, $\beta \leq \alpha$ and $\mu(X_\beta) \subseteq X \cup \bigcup\{X_\gamma : \gamma < \beta\}$, and $X_\alpha = Y$, $X_0 = X$.

Failure of $(\mu Cumt\alpha)$ would then mean that there are X and Y, we can reach Y from X, and $x \in (\mu(X) \cap Y) - \mu(Y)$. Thus, in a counterexample, $Y = \mu(Y)$ is impossible, so none of the intersections can be such Y.

To reach Y from X, we have to get started from X, i.e. there must be Z such that $\mu(Z) \subseteq X$, $Z \not\subseteq X$ (so $\mu(Z) \neq Z$). Inspection of different cases shows that we cannot reach any set Y from any case of the intersections, except from (1), (6.1), and (7.2).

If Y contains a globally minimal element (i.e. there is no smaller element in any set), it can only be reached from any X which already contains this element. The globally minimal elements are a, x_0, and the d_λ, $\lambda \leq \kappa$.

By these observations, we see that X_λ and X'_κ can only be reached from U. From no X_α U can be reached, as the globally minimal a is missing. But U cannot be reached from X'_κ either, as the globally minimal x_0 is missing.

When we look at the relation $\prec$ defining μ, we see that we can reach Y from X only by going upwards, adding bigger elements. Thus, from X_α, we cannot reach any X_β, $\beta < \alpha$, the same holds for X'_κ and X_β, $\beta < \kappa$. Thus, from X'_κ, we cannot go anywhere interesting (recall that the intersections are not candidates for a Y giving a contradiction).

Consider now X_α. We can go up to any $X_{\alpha+n}$, but not to any X_λ, $\alpha < \lambda$, as d_λ is missing, neither to X'_κ, as a is missing, and we will be stopped by the first $\lambda > \alpha$, as x_λ will be missing to go beyond X_λ. Analogous observations hold for the remaining intersections (1), (6.1), and (7.2). In all these sets we can reach, but we will not destroy minimality of any element of X_α (or of the intersections).

Consequently, the only candidates for failure will all start with U. As the only element of U not globally minimal is c, such failure has to have $c \in Y - \mu(Y)$, so Y has to be X'_κ. Suppose we omit one of the X_α in the sequence going up to X'_κ. If $\kappa \geq \lambda > \alpha$, we cannot reach X_λ and beyond, as x'_α will be missing. But we cannot go to $X_{\alpha+n}$ either, as $x_{\alpha+1}$ is missing. So we will be stopped at X_α. Thus, to see failure, we need the full sequence $U = X_0$, $X'_\kappa = Y_\kappa$, $Y_\alpha = X_\alpha$ for $0 < \alpha < \kappa$.

$(\mu Cum\kappa)$ fails: The full sequence $U = X_0$, $X'_\kappa = Y_\kappa$, $Y_\alpha = X_\alpha$ for $0 < \alpha < \kappa$ shows this, as $c \in \mu(U) \cap X'_\kappa$, but $c \notin \mu\left(X'_\kappa\right)$.

Consequently, the example satisfies $(\bigcap)$, (μCUM), $(\mu Cumt\alpha)$ for $\alpha < \kappa$, and $(\mu Cum\kappa)$ fails. □

Fact 4.2.9 We summarize some properties of $(\mu Cum\alpha)$ and $(\mu Cumt\alpha)$ – sometimes with some redundancy. Unless said otherwise, α, β, etc. will be arbitrary ordinals.

For (1)–(6) (μPR) and $(\mu \subseteq)$ are assumed to hold, for (7) only $(\mu \subseteq)$.

(1) Downward:

(1.1) $(\mu Cum\alpha) \Rightarrow (\mu Cum\beta)$ for all $\beta \leq \alpha$;
(1.2) $(\mu Cumt\alpha) \Rightarrow (\mu Cumt\beta)$ for all $\beta \leq \alpha$;

(2) Validity of $(\mu Cum\alpha)$ and $(\mu Cumt\alpha)$:

(2.1) all $(\mu Cum\alpha)$ hold in smooth preferential structures;

(2.2) all $(\mu Cumt\alpha)$ hold in transitive smooth preferential structures;
(2.3) $(\mu Cumt\alpha)$ for $0 < \alpha$ do not necessarily hold in smooth structures without transitivity, even in the presence of $(\bigcap)$.

(3) Upward:

(3.1) $(\mu Cum\beta) + (\cup) \Rightarrow (\mu Cum\alpha)$ for all $\beta \leq \alpha$;
(3.2) $(\mu Cumt\beta) + (\cup) \Rightarrow (\mu Cumt\alpha)$ for all $\beta \leq \alpha$;
(3.3) $\{(\mu Cumt\beta) : \beta < \alpha\} + (\mu CUM) + (\bigcap) \not\Rightarrow (\mu Cum\alpha)$ for $\alpha > 0$.

(4) Connection $(\mu Cum\alpha)/(\mu Cumt\alpha)$:

(4.1) $(\mu Cumt\alpha) \Rightarrow (\mu Cum\alpha)$;
(4.2) $(\mu Cum\alpha) + (\bigcap) \not\Rightarrow (\mu Cumt\alpha)$;
(4.3) $(\mu Cum\alpha) + (\cup) \Rightarrow (\mu Cumt\alpha)$;

(5) (μCUM) and $(\mu Cumi)$:

(5.1) $(\mu CUM) + (\cup)$ entail:

(5.1.1) $\mu(A) \subseteq B \Rightarrow \mu(A \cup B) = \mu(B)$;
(5.1.2) $\mu(X) \subseteq U,\ U \subseteq Y \Rightarrow \mu(Y \cup X) = \mu(Y)$;
(5.1.3) $\mu(X) \subseteq U,\ U \subseteq Y \Rightarrow \mu(Y) \cap X \subseteq \mu(U)$;

(5.2) $(\mu Cum\alpha) \Rightarrow (\mu CUM)$ for all α;
(5.3) $(\mu CUM) + (\cup) \Rightarrow (\mu Cum\alpha)$ for all α;
(5.4) $(\mu CUM) + (\cap) \Rightarrow (\mu Cum0)$;

(6) (μCUM) and $(\mu Cumt\alpha)$:

(6.1) $(\mu Cumt\alpha) \Rightarrow (\mu CUM)$ for all α;
(6.2) $(\mu CUM) + (\cup) \Rightarrow (\mu Cumt\alpha)$ for all α;
(6.3) $(\mu CUM) \not\Rightarrow (\mu Cumt\alpha)$ for all $\alpha > 0$;

(7) $(\mu Cum0) \Rightarrow (\mu PR)$.

Proof We prove these facts in a different order: (1), (2), (5.1), (5.2), (4.1), (6.1), (6.2), (5.3), (3.1), (3.2), (4.2), (4.3), (5.4), (3.3), (6.3), (7).

(1.1): For $\beta < \gamma \leq \alpha$ set $X_\gamma := X_\beta$. Let the prerequisites of $(\mu Cum\beta)$ hold. Then for γ with $\beta < \gamma \leq \alpha$, $\mu(X_\gamma) \subseteq X_\beta$ by $(\mu \subseteq)$, so the prerequisites of $(\mu Cum\alpha)$ hold too, so by $(\mu Cum\alpha)$ $\bigcap\{X_\delta : \delta \leq \beta\} \cap \mu(U) = \bigcap\{X_\delta : \delta \leq \alpha\} \cap \mu(U) \subseteq \mu(X_\alpha) = \mu(X_\beta)$.

(1.2): Analogous.

(2.1): Proof by induction.
$(\mu Cum0)$: Let $\mu(X_0) \subseteq U$, suppose there is $x \in \mu(U) \cap (X_0 - \mu(X_0))$. By smoothness, there is $y \prec x$, $y \in \mu(X_0) \subseteq U$, *contradiction*. (The same arguments work for copies: all copies of x must be minimized by some $y \in \mu(X_0)$, but at least one copy of x has to be minimal in U.)

Suppose $(\mu Cum\beta)$ holds for all $\beta < \alpha$. We show $(\mu Cum\alpha)$. Let the prerequisites of $(\mu Cum\alpha)$ hold, then those for $(\mu Cum\beta)$, $\beta < \alpha$ hold, too. Suppose there is $x \in \mu(U) \cap \bigcap\{X_\gamma : \gamma \leq \alpha\} - \mu(X_\alpha)$. So by $(\mu Cum\beta)$ for $\beta < \alpha$, $x \in \mu(X_\beta)$ for all $\beta < \alpha$; moreover $x \in \mu(U)$. By smoothness, there is $y \in \mu(X_\alpha) \subseteq U \cup \bigcup\{X_{\beta'} : \beta' < \alpha\}$, $y \prec x$, but this is a contradiction. The same argument works again for copies.

(2.2): We use the following fact: Let, in a smooth transitive structure, $\mu(X_\beta) \subseteq U \cup \bigcup\{X_\gamma : \gamma < \beta\}$ for all $\beta \leq \alpha$, and let $x \in \mu(U)$. Then there is no $y \prec x$, $y \in U \cup \bigcup\{X_\gamma : \gamma \leq \alpha\}$.
Proof of the fact by induction: Suppose such $y \in U \cup X_0$ exists. $y \in U$ is impossible. Let $y \in X_0$, by $\mu(X_0) \subseteq U$, $y \in X_0 - \mu(X_0)$, so there is $z \in \mu(X_0)$, $z \prec y$, so $z \prec x$ by transitivity, but $\mu(X_0) \subseteq U$. Let the result holds for all $\beta < \alpha$, but fails for α, so $\neg\exists y \prec x.y \in U \cup \bigcup\{X_\gamma : \gamma < \alpha\}$, but $\exists y \prec x.y \in U \cup \bigcup\{X_\gamma : \gamma \leq \alpha\}$, so $y \in X_\alpha$. If $y \in \mu(X_\alpha)$, then $y \in U \cup \bigcup\{X_\gamma : \gamma < \alpha\}$, but this is impossible, so $y \in X_\alpha - \mu(X_\alpha)$. Let by smoothness $z \prec y$, $z \in \mu(X_\alpha)$, so by transitivity $z \prec x$, *contradiction*. The result is easily modified for the case with copies.
Let the prerequisites of $(\mu Cumt\alpha)$ hold, then those of the fact will hold too. Let now $x \in \mu(U) \cap (X_\alpha - \mu(X_\alpha))$, by smoothness, there must be $y \prec x$, $y \in \mu(X_\alpha) \subseteq U \cup \bigcup\{X_\gamma : \gamma < \alpha\}$, contradicting the fact.

(2.3): Let $\alpha > 0$, and consider the following structure over $\{a, b, c\} : U := \{a, c\}$, $X_0 := \{b, c\}$, $X_\alpha := \ldots := X_1 := \{a, b\}$, and their intersections, $\{a\}$, $\{b\}$, $\{c\}$, $\emptyset$ with the order $c \prec b \prec a$ (without transitivity). This is preferential, so (μPR) and $(\mu \subseteq)$ hold. The structure is smooth for U, all X_β, and their intersections. We have $\mu(X_0) \subseteq U$, $\mu(X_\beta) \subseteq U \cup X_0$ for all $\beta \leq \alpha$, so $\mu(X_\beta) \subseteq U \cup \bigcup\{X_\gamma : \gamma < \beta\}$ for all $\beta \leq \alpha$, but $X_\alpha \cap \mu(U) = \{a\} \not\subseteq \{b\} = \mu(X_\alpha)$ for $\alpha > 0$.

(5.1):

(5.1.1): $\mu(A) \subseteq B \Rightarrow \mu(A \cup B) \subseteq \mu(A) \cup \mu(B) \subseteq B \Rightarrow_{(\mu CUM)} \mu(B) = \mu(A \cup B)$.

(5.1.2): $\mu(X) \subseteq U \subseteq Y \Rightarrow$ (by (5.1.1)) $\mu(Y \cup X) = \mu(Y)$.

(5.1.3): $\mu(Y) \cap X =$ (by (5.1.2)) $\mu(Y \cup X) \cap X \subseteq \mu(Y \cup X) \cap (X \cup U) \subseteq$ (by (μPR)) $\mu(X \cup U) =$ (by (5.1.1)) $\mu(U)$.

(5.2): Using (1.1), it suffices to show $(\mu Cum0) \Rightarrow (\mu CUM)$. Let $\mu(X) \subseteq U \subseteq X$. By $(\mu Cum0)$ $X \cap \mu(U) \subseteq \mu(X)$, so by $\mu(U) \subseteq U \subseteq X \Rightarrow \mu(U) \subseteq \mu(X)$. $U \subseteq X \Rightarrow \mu(X) \cap U \subseteq \mu(U)$ by (μPR), but also $\mu(X) \subseteq U$, so $\mu(X) \subseteq \mu(U)$.

(4.1): Trivial.

(6.1): Follows from (4.1) and (5.2).

(6.2): Let the prerequisites of $(\mu Cumt\alpha)$ hold.
We first show by induction $\mu(X_\alpha \cup U) \subseteq \mu(U)$.
Proof $\alpha = 0$: $\mu(X_0) \subseteq U \Rightarrow \mu(X_0 \cup U) = \mu(U)$ by (5.1.1). Let for all $\beta < \alpha$ $\mu(X_\beta \cup U) \subseteq \mu(U) \subseteq U$. By prerequisite, $\mu(X_\alpha) \subseteq U \cup \bigcup\{X_\beta : \beta < \alpha\}$, thus $\mu(X_\alpha \cup U) \subseteq \mu(X_\alpha) \cup \mu(U) \subseteq \bigcup\{U \cup X_\beta : \beta < \alpha\}$. Moreover for all $\beta < \alpha$, $\mu(X_\beta \cup U) \subseteq U \subseteq X_\alpha \cup U$, so $\mu(X_\alpha \cup U) \cap (U \cup X_\beta) \subseteq \mu(U)$ by (5.1.3), thus $\mu(X_\alpha \cup U) \subseteq \mu(U)$.

Consequently, under the above prerequisites, we have $\mu(X_\alpha \cup U) \subseteq \mu(U) \subseteq U \subseteq U \cup X_\alpha$, so by (μCUM) $\mu(U) = \mu(X_\alpha \cup U)$, and, finally, $\mu(U) \cap X_\alpha = \mu(X_\alpha \cup U) \cap X_\alpha \subseteq \mu(X_\alpha)$ by (μPR).
Note that finite unions take us over the limit step, essentially, as all steps collapse, and $\mu(X_\alpha \cup U)$ will always be $\mu(U)$, so there are no real changes.

(5.3): Follows from (6.2) and (4.1).
(3.1): Follows from (5.2) and (5.3).
(3.2): Follows from (6.1) and (6.2).
(4.2): Follows from (2.3) and (2.1).
(4.3): Follows from (5.2) and (6.2).
(5.4): $\mu(X) \subseteq U \Rightarrow \mu(X) \subseteq U \cap X \subseteq X \Rightarrow \mu(X \cap U) = \mu(X) \Rightarrow X \cap \mu(U) = (X \cap U) \cap \mu(U) \subseteq \mu(X \cap U) = \mu(X)$.
(3.3): See Example 4.2.4 (p. 100).
(6.3): See Example 4.2.4 (p. 100).
(7): Trivial. Let $X \subseteq Y$, so by $(\mu \subseteq)$ $\mu(X) \subseteq X \subseteq Y$, so by $(\mu Cum0)$ $X \cap \mu(Y) \subseteq \mu(X)$. □

Fact 4.2.10 Assume $(\mu \subseteq)$.
We have for $(\mu Cum\infty)$ and (HU, u):

(1) $x \in \mu(Y)$, $\mu(Y) \subseteq H(U, x) \Rightarrow Y \subseteq H(U, x)$;
(2) $(\mu Cum\infty) \Rightarrow (HU, u)$;
(3) $(HU, u) \Rightarrow (\mu Cum\infty)$.

Proof

(1): Trivial by definition of $H(U, x)$.
(2): Let $x \in \mu(U)$, $x \in Y$, $\mu(Y) \subseteq H(U, x)$ (and thus $Y \subseteq H(U, x)$ by definition). Thus, we have a sequence $X_0 := U$, $\mu(X_\beta) \subseteq U \cup \bigcup\{X_\gamma : \gamma < \beta\}$, $x \in X_\beta$, and $Y = X_\alpha$ for some α (after X_0, enumerate arbitrarily $H(U, x)_1$, then $H(U, x)_2$, etc., do nothing at limits). So $x \in \bigcap\{X_\gamma : \gamma \leq \alpha\} \cap \mu(U) \subseteq \mu(X_\alpha) = \mu(Y)$ by $(\mu Cum\infty)$.
Remark: The same argument shows that we can replace "$x \in X$" equivalently by "$x \in \mu(X)$" in the definition of $H(U, x)_{\alpha+1}$, as was done in Definition 3.7.5 in [Sch04].
(3): Suppose $(\mu Cum\alpha)$ fails, we show that then so does (HU, u) for $u = x$. As $(\mu Cum\alpha)$ fails, for all $\beta \leq \alpha$, $\mu(X_\beta) \subseteq U \cup \bigcup\{X_\gamma : \gamma < \beta\}$, but there is $x \in \bigcap\{X_\gamma : \gamma \leq \alpha\} \cap \mu(U)$, $x \notin \mu(X_\alpha)$. Thus for all $\beta \leq \alpha$, $\mu(X_\beta) \subseteq X_\beta \subseteq H(U, x)$. Moreover $x \in \mu(U)$, $x \in X_\alpha - \mu(X_\alpha)$, but $\mu(X_\alpha) \subseteq H(U, x)$, so (HU, u) fails for $u = x$. □

Fact 4.2.11 We continue to show results for $H(U)$ and $H(U, u)$.
Let A, X, U, U', Y, and all A_i be in $\mathcal{Y}$.

(1) $H(U)$ and $H(U, u)$,
(1.1) $H(U, u) \subseteq H(U)$,
(1.2) $(HU) \Rightarrow (HU, u)$,

(1.3) $(\cup)+(\mu PR)$ entail $H(U) \subseteq H(U,u)$ for $u \in \mu(U)$,
(1.4) $(\cup)+(\mu PR)$ entail $(HU,u) \Rightarrow (HU)$,
(2) $(\mu\subseteq)$ and (HU) entail:
(2.1) (μPR),
(2.2) (μCUM),
(3) $(\mu\subseteq)$ and (μPR) entail:
(3.1) $A = \bigcup\{A_i : i \in I\} \Rightarrow \mu(A) \subseteq \bigcup\{\mu(A_i) : i \in I\}$,
(3.2) $U \subseteq H(U)$, and $U \subseteq U' \Rightarrow H(U) \subseteq H(U')$,
(3.3) $\mu(U \cup Y) - H(U) \subseteq \mu(Y)$ – if $\mu(U \cup Y)$ is defined, in particular, if $(\cup)$ holds.
(4) $(\cup)$, $(\mu\subseteq)$, (μPR), and (μCUM) entail:
(4.1) $H(U) = H_1(U)$,
(4.2) $U \subseteq A$, $\mu(A) \subseteq H(U) \Rightarrow \mu(A) \subseteq U$,
(4.3) $\mu(Y) \subseteq H(U) \Rightarrow Y \subseteq H(U)$ and $\mu(U \cup Y) = \mu(U)$,
(4.4) $x \in \mu(U)$, $x \in Y - \mu(Y) \Rightarrow Y \not\subseteq H(U)$ (and thus (HU)),
(4.5) $Y \not\subseteq H(U) \Rightarrow \mu(U \cup Y) \not\subseteq H(U)$.
(5) $(\cup)$, $(\mu\subseteq)$, (HU) entail:
(5.1) $H(U) = H_1(U)$,
(5.2) $U \subseteq A$, $\mu(A) \subseteq H(U) \Rightarrow \mu(A) \subseteq U$,
(5.3) $\mu(Y) \subseteq H(U) \Rightarrow Y \subseteq H(U)$ and $\mu(U \cup Y) = \mu(U)$,
(5.4) $x \in \mu(U)$, $x \in Y - \mu(Y) \Rightarrow Y \not\subseteq H(U)$,
(5.5) $Y \not\subseteq H(U) \Rightarrow \mu(U \cup Y) \not\subseteq H(U)$.

Proof
(1.1) and (1.2) trivial by definition.

(1.3) Proof by induction. $H(U)_0 = H(U,u)_0$ is trivial. Suppose $H(U)_\beta = H(U,u)_\beta$ has been shown for $\beta < \alpha$. The limit step is trivial, so suppose $\alpha = \beta + 1$. Let X be such that $\mu(X) \subseteq H(U)_\beta = H(U,u)_\beta$, so $X \subseteq H(U)_\alpha$. Consider $X \cup U$, so $u \in X \cup U$, $\mu(X \cup U)$ is defined and by (μPR) and $(\mu\subseteq)$, $\mu(X \cup U) \subseteq \mu(X) \cup \mu(U) \subseteq H(U)_\beta = H(U,u)_\beta$, so $X \cup U \subseteq H(U,u)_\alpha$.
(1.4) Immediate by (1.3).
(2.1) By (HU), if $\mu(Y) \subseteq H(U)$, then $\mu(U) \cap Y \subseteq \mu(Y)$. But, if $Y \subseteq U$, then $\mu(Y) \subseteq H(U)$ by $(\mu\subseteq)$.
(2.2) Let $\mu(U) \subseteq X \subseteq U$. Then by (2.1) $\mu(U) = \mu(U) \cap X \subseteq \mu(X)$. By $\mu(U) \subseteq X$ and $(\mu\subseteq)$, $\mu(U) \subseteq U \subseteq H(X)$, so by (HU) and $X \subseteq U$ and $(\mu\subseteq)$, $\mu(X) = \mu(X) \cap U \subseteq \mu(U)$ by $(\mu\subseteq)$.
(3.1) $\mu(A) \cap A_j \subseteq \mu(A_j) \subseteq \bigcup \mu(A_i)$, so by $\mu(A) \subseteq A = \bigcup A_i$ $\mu(A) \subseteq \bigcup \mu(A_i)$.
(3.2) Trivial.
(3.3) $\mu(U \cup Y) - H(U) \subseteq_{(3.2)} \mu(U \cup Y) - U \subseteq$ (by $(\mu\subseteq)$) $\mu(U \cup Y) \cap Y \subseteq_{(\mu PR)} \mu(Y)$.
(4.1) We show that, if $X \subseteq H_2(U)$, then $X \subseteq H_1(U)$; more precisely, if $\mu(X) \subseteq H_1(U)$, then already $X \subseteq H_1(U)$, so the construction stops already at $H_1(U)$. Suppose then $\mu(X) \subseteq \bigcup\{Y : \mu(Y) \subseteq U\}$, and let $A := X \cup U$. We show that $\mu(A) \subseteq U$, so $X \subseteq A \subseteq H_1(U)$. Let $a \in \mu(A)$. By (μPR), $(\mu\subseteq)$, $\mu(A) \subseteq \mu(X) \cup \mu(U)$. If $a \in \mu(U) \subseteq U$, we are done. If $a \in \mu(X)$, there is Y such that $\mu(Y) \subseteq U$ and $a \in Y$, so $a \in \mu(A) \cap Y$. By Fact 4.2.9 (p. 103),

(5.1.3), we have for Y such that $\mu(Y) \subseteq U$ and $U \subseteq A$ $\mu(A) \cap Y \subseteq \mu(U)$. Thus $a \in \mu(U)$, and we are done again.

(4.2) Let $U \subseteq A$, $\mu(A) \subseteq H(U) = H_1(U)$ by (4.1). So $\mu(A) = \bigcup\{\mu(A) \cap Y : \mu(Y) \subseteq U\} \subseteq \mu(U) \subseteq U$, again by Fact 4.2.9 (p. 103), (5.1.3).

(4.3) Let $\mu(Y) \subseteq H(U)$, then by $\mu(U) \subseteq H(U)$ and (μPR), $(\mu \subseteq)$, $\mu(U \cup Y) \subseteq \mu(U) \cup \mu(Y) \subseteq H(U)$, so by (4.2) $\mu(U \cup Y) \subseteq U$ and $U \cup Y \subseteq H(U)$. Moreover, $\mu(U \cup Y) \subseteq U \subseteq U \cup Y \Rightarrow_{(\mu CUM)} \mu(U \cup Y) = \mu(U)$.

(4.4) If not, $Y \subseteq H(U)$, so $\mu(Y) \subseteq H(U)$, so $\mu(U \cup Y) = \mu(U)$ by (4.3), but $x \in Y - \mu(Y) \Rightarrow_{(\mu PR)} x \notin \mu(U \cup Y) = \mu(U)$, *contradiction*.

(4.5) $\mu(U \cup Y) \subseteq H(U) \Rightarrow_{(4.3)} U \cup Y \subseteq H(U)$.

(5) Trivial by (1) and (4). □

The Representation Result

We now turn to the representation result and its proof.

We adapt Proposition 3.7.15 in [Sch04] and its proof. All we need is (HU, u) and $(\mu \subseteq)$. We modify the proof of Remark 3.7.13 (1) in [Sch04] (now Remark 4.2.1 (p. 109)) so we will not need $(\cap)$ anymore. We will give the full proof, although its essential elements have already been published, for three reasons: First, the new version will need less prerequisites than the old proof does (closure under finite intersections is not needed anymore, and replaced by (HU, u)). Second, we will more clearly separate the requirements to do the construction from the construction itself, thus splitting the proof neatly into two parts.

We show how to work with $(\mu \subseteq)$ and (HU, u) only. Thus, once we have shown $(\mu \subseteq)$ and (HU, u), we have finished the substantial side, and enter the administrative part, which will not use any prerequisites about domain closure anymore. At the same time, this gives a uniform proof of the difficult part for the case with and without $(\cup)$, in the former case we can even work with the stronger $H(U)$. The easy direction of the former parts needs a proof of the stronger $H(U)$, but this is easy.

Note that, in the presence of $(\mu \subseteq)$, $(HU, u) \Rightarrow (\mu Cum\infty)$ and $(\mu Cum0) \Rightarrow (\mu PR)$, by Fact 4.2.10 (p. 106), (3) and Fact 4.2.9 (p. 103), (7), so (HU, u) entails (μPR), so we can use it in our context, where (HU, u) will be the central property.

Fact 4.2.12 (HU, u) holds in all smooth models.

Proof Suppose not. So let $x \in \mu(U)$, $x \in Y - \mu(Y)$, $\mu(Y) \subseteq H(U, x)$. By smoothness, there is $x_1 \in \mu(Y)$, $x \succ x_1$, and let κ_1 be the least κ such that $x_1 \in H(U, x)_{\kappa_1}$. κ_1 is not a limit, and $x_1 \in U'_{x_1} - \mu\left(U'_{x_1}\right)$ with $x \in U'_{x_1}$ by definition of $H(U, x)$ for some U'_{x_1}, so as $x_1 \notin \mu\left(U'_{x_1}\right)$, there must be (by smoothness) some other $x_2 \in \mu\left(U'_{x_1}\right) \subseteq H(U, x)_{\kappa_1 - 1}$ with $x \succ x_2$. Continue with x_2, we thus construct a descending chain of ordinals, which cannot be infinite, so there must be $x_n \in \mu\left(U'_{x_n}\right) \subseteq U$, $x \succ x_n$, contradicting minimality of x in U. (More precisely, this works for all copies of x.) □

We first show two basic facts and then turn to the main result, Proposition 4.2.4 (p. 110).

Definition 4.2.5 For $x \in Z$, let
$\mathcal{W}_x := \{\mu(Y): Y \in \mathcal{Y} \wedge x \in Y - \mu(Y)\}$,
$\Gamma_x := \Pi \mathcal{W}_x$, and
$K := \{x \in Z: \exists X \in \mathcal{Y}.x \in \mu(X)\}$.

Remark 4.2.1

(1) $x \in K \Rightarrow \Gamma_x \neq \emptyset$,
(2) $g \in \Gamma_x \Rightarrow ran(g) \subseteq K$.

Proof

(1) We give two proofs, the first uses $(\mu Cum0)$, the second the stronger (by Fact 4.2.10 (p. 106) (3)) (HU, u).
   (a) We have to show that $Y \in \mathcal{Y}$, $x \in Y - \mu(Y) \Rightarrow \mu(Y) \neq \emptyset$. Suppose then $x \in \mu(X)$, this exists, as $x \in K$, and $\mu(Y) = \emptyset$, so $\mu(Y) \subseteq X$, $x \in Y$, so by $(\mu Cum0)$ $x \in \mu(Y)$.
   (b) Consider $H(X, x)$, suppose $\mu(Y) = \emptyset$, $x \in Y$, so $Y \subseteq H(X, x)$, so $x \in \mu(Y)$ by (HU, u).

(2) By definition, $\mu(Y) \subseteq K$ for all $Y \in \mathcal{Y}$. □

Claim 4.2.5 Let $U \in \mathcal{Y}$, $x \in K$. Then

(1) $x \in \mu(U) \Leftrightarrow x \in U \wedge \exists f \in \Gamma_x.ran(f) \cap U = \emptyset$,
(2) $x \in \mu(U) \Leftrightarrow x \in U \wedge \exists f \in \Gamma_x.ran(f) \cap H(U, x) = \emptyset$.

Proof

(1): Case 1: $\mathcal{W}_x = \emptyset$, thus $\Gamma_x = \{\emptyset\}$.
"$\Rightarrow$": Take $f := \emptyset$.
"$\Leftarrow$": $x \in U \in \mathcal{Y}$, $\mathcal{W}_x = \emptyset \Rightarrow x \in \mu(U)$ by definition of $\mathcal{W}_x$.
Case 2: $\mathcal{W}_x \neq \emptyset$.
"$\Rightarrow$": Let $x \in \mu(U) \subseteq U$. Consider $H(U, x)$. If $\mu(Y) \in \mathcal{W}_x$, then $x \in Y - \mu(Y)$, so by (HU, u) for $H(U, x)$ $\mu(Y) - H(U, x) \neq \emptyset$, but $\mu(U) \subseteq U \subseteq H(U, x)$.
"$\Leftarrow$": If $x \in U - \mu(U)$, so $\mu(U) \in \mathcal{W}_x$, moreover $\Gamma_x \neq \emptyset$ by Remark 4.2.1 (p. 109), (1) and thus $\mu(U) \neq \emptyset$, so $\forall f \in \Gamma_x.ran(f) \cap U \neq \emptyset$.
(2): The Case 1 is as for (1).
Case 2: "$\Rightarrow$" was shown already in Case 1.
"$\Leftarrow$": Let $x \in U - \mu(U)$, then by $x \in K$ $\mu(U) \neq \emptyset$ (see proof of Remark 4.2.1 (p. 109)); moreover $\mu(U) \subseteq U \subseteq H(U, x)$, so $\forall f \in \Gamma_x.ran(f) \cap H(U, x) \neq \emptyset$.
(Claim 4.2.5 (p. 109)) □

The following Proposition 4.2.4 (p. 110) is the main positive result of Sect. 4.2.2.3 (p. 99) and shows how to characterize smooth structures in the absence of closure under finite unions. The strategy of the proof follows closely the proof of Proposition 3.3.4 in [Sch04].

Proposition 4.2.4 *Let $\mu : \mathcal{Y} \to \mathcal{P}(Z)$. Then there is a $\mathcal{Y}$-smooth preferential structure $\mathcal{Z}$, such that for all $X \in \mathcal{Y}$ $\mu(X) = \mu_{\mathcal{Z}}(X)$ iff μ satisfies $(\mu \subseteq)$ and (HU, u) above.*

In particular, we need no prerequisites about domain closure.

Proof "⇒" (HU, u) was shown in Fact 4.2.12 (p. 108).

Outline of "⇐": We first define a structure $\mathcal{Z}$ which represents μ, but is not necessarily $\mathcal{Y}$-smooth, refine it to $\mathcal{Z}'$ and show that $\mathcal{Z}'$ represents μ too, and that $\mathcal{Z}'$ is $\mathcal{Y}$-smooth.

In the structure $\mathcal{Z}'$, all pairs destroying smoothness in $\mathcal{Z}$ are successively repaired, by adding minimal elements: If $\langle y, j\rangle$ is not minimal, and has no minimal $\langle x, i\rangle$ below it, we just add one such $\langle x, i\rangle$. As the repair process might itself generate such "bad" pairs, the process may have to be repeated infinitely often. Of course, one has to take care that the representation property is preserved. □

Construction 4.2.3 (Construction of $\mathcal{Z}$)

Let $\mathcal{X} := \{\langle x, g\rangle : x \in K,\ g \in \Gamma_x\}$, $\langle x', g'\rangle \prec \langle x, g\rangle :\Leftrightarrow x' \in ran(g)$, $\mathcal{Z} := \langle \mathcal{X}, \prec\rangle$.

Claim 4.2.6 $\forall U \in \mathcal{Y}.\mu(U) = \mu_{\mathcal{Z}}(U)$

Proof

Case 1: $x \notin K$. Then $x \notin \mu(U)$ and $x \notin \mu_{\mathcal{Z}}(U)$.

Case 2: $x \in K$.

By Claim 4.2.5 (p. 109), (1) it suffices to show that for all $U \in \mathcal{Y}$ $x \in \mu_{\mathcal{Z}}(U) \Leftrightarrow x \in U \wedge \exists f \in \Gamma_x.ran(f) \cap U = \emptyset$. Fix $U \in \mathcal{Y}$.

"⇒": $x \in \mu_{\mathcal{Z}}(U) \Rightarrow$ ex. $\langle x, f\rangle$ minimal in $\mathcal{X}\lceil U$, thus $x \in U$ and there is no $\langle x', f'\rangle \prec \langle x, f\rangle$, $x' \in U$, $x' \in K$. But if $x' \in K$, then by Remark 4.2.1 (p. 109), (1), $\Gamma_{x'} \neq \emptyset$, so we find suitable f'. Thus, $\forall x' \in ran(f).x' \notin U$ or $x' \notin K$. But $ran(f) \subseteq K$, so $ran(f) \cap U = \emptyset$.

"⇐": If $x \in U$, $f \in \Gamma_x$ such that $ran(f) \cap U = \emptyset$, then $\langle x, f\rangle$ is minimal in $\mathcal{X}\lceil U$. (Claim 4.2.6 (p. 110)) □

We will use in the construction of the refined structure $\mathcal{Z}'$ the following definition:

Definition 4.2.6 σ is called x-admissible sequence iff

1. σ is a sequence of length $\leq \omega$, $\sigma = \{\sigma_i : i \in \omega\}$,
2. $\sigma_o \in \Pi\{\mu(Y) : Y \in \mathcal{Y} \wedge x \in Y - \mu(Y)\}$,
3. $\sigma_{i+1} \in \Pi\{\mu(X) : X \in \mathcal{Y} \wedge x \in \mu(X) \wedge ran(\sigma_i) \cap X \neq \emptyset\}$.

By 2, σ_0 minimizes x, and by 3, if $x \in \mu(X)$ and $ran(\sigma_i) \cap X \neq \emptyset$, i.e. we have destroyed minimality of x in X, x will be above some y minimal in X to preserve smoothness.

Let Σ_x be the set of x-admissible sequences, for $\sigma \in \Sigma_x$ let $\widehat{\sigma} := \bigcup\{ran(\sigma_i) : i \in \omega\}$.

Construction 4.2.4 (Construction of $\mathcal{Z}'$)

Note that by Remark 4.2.1 (p. 109), (1), $\Sigma_x \neq \emptyset$, if $x \in K$ (this does σ_0, σ_{i+1} is trivial as by prerequisite $\mu(X) \neq \emptyset$).

Let $\mathcal{X}' := \{\langle x, \sigma \rangle : x \in K \wedge \sigma \in \Sigma_x\}$ and $\langle x', \sigma' \rangle \prec' \langle x, \sigma \rangle :\Leftrightarrow x' \in \widehat{\sigma}$. Finally, let $\mathcal{Z}' := \langle \mathcal{X}', \prec' \rangle$, and $\mu' := \mu_{\mathcal{Z}'}$.

It is now easy to show that $\mathcal{Z}'$ represents μ, and that $\mathcal{Z}'$ is smooth. For $x \in \mu(U)$, we construct a special x-admissible sequence $\sigma^{x,U}$ using the properties of $H(U, x)$ as described at the beginning of this section.

Claim 4.2.7 For all $U \in \mathcal{Y}$, $\mu(U) = \mu_{\mathcal{Z}}(U) = \mu'(U)$.

Proof If $x \notin K$, then $x \notin \mu_{\mathcal{Z}}(U)$ and $x \notin \mu'(U)$ for any U. So assume $x \in K$. If $x \in U$ and $x \notin \mu_{\mathcal{Z}}(U)$, then for all $\langle x, f \rangle \in \mathcal{X}$, there is $\langle x', f' \rangle \in \mathcal{X}$ with $\langle x', f' \rangle \prec \langle x, f \rangle$ and $x' \in U$. Let now $\langle x, \sigma \rangle \in \mathcal{X}'$, then $\langle x, \sigma_0 \rangle \in \mathcal{X}$, and let $\langle x', f' \rangle \prec \langle x, \sigma_0 \rangle$ in $\mathcal{Z}$ with $x' \in U$. As $x' \in K$, $\Sigma_{x'} \neq \emptyset$, let $\sigma' \in \Sigma_{x'}$. Then $\langle x', \sigma' \rangle \prec' \langle x, \sigma \rangle$ in $\mathcal{Z}'$. Thus $x \notin \mu'(U)$. Thus, for all $U \in \mathcal{Y}$, $\mu'(U) \subseteq \mu_{\mathcal{Z}}(U) = \mu(U)$.

It remains to show $x \in \mu(U) \Rightarrow x \in \mu'(U)$.

Assume $x \in \mu(U)$ (so $x \in K$), $U \in \mathcal{Y}$, we will construct minimal σ, i.e. show that there is $\sigma^{x,U} \in \Sigma_x$ such that $\widehat{\sigma^{x,U}} \cap U = \emptyset$. *We construct this* $\sigma^{x,U}$ *inductively, with the* stronger property that $ran\left(\sigma_i^{x,U}\right) \cap H(U, x) = \emptyset$ for all $i \in \omega$.

$\underline{\sigma_0^{x,U}:}$

$x \in \mu(U)$, $x \in Y - \mu(Y) \Rightarrow \mu(Y) - H(U, x) \neq \emptyset$ by (HU, u) for $H(U, x)$. Let $\sigma_0^{x,U} \in \Pi\{\mu(Y) - H(U, x) : Y \in \mathcal{Y}, x \in Y - \mu(Y)\}$, so $ran\left(\sigma_0^{x,U}\right) \cap H(U, x) = \emptyset$.

$\underline{\sigma_i^{x,U} \Rightarrow \sigma_{i+1}^{x,U}:}$

By the induction hypothesis, $ran\left(\sigma_i^{x,U}\right) \cap H(U, x) = \emptyset$. Let $X \in \mathcal{Y}$ be such that $x \in \mu(X)$, $ran\left(\sigma_i^{x,U}\right) \cap X \neq \emptyset$. Thus $X \not\subseteq H(U, x)$, so $\mu(X) - H(U, x) \neq \emptyset$ by Fact 4.2.10 (p. 106), (1). Let $\sigma_{i+1}^{x,U} \in \Pi\left\{\mu(X) - H(U, x) : X \in \mathcal{Y}, x \in \mu(X), ran\left(\sigma_i^{x,U}\right) \cap X \neq \emptyset\right\}$, so $ran\left(\sigma_{i+1}^{x,U}\right) \cap H(U, x) = \emptyset$. The construction satisfies the x-admissibility condition. □

It remains to show:

Claim 4.2.8 $\mathcal{Z}'$ is $\mathcal{Y}$-smooth.

Proof Let $X \in \mathcal{Y}$, $\langle x, \sigma \rangle \in \mathcal{X}' \lceil X$.

Case 1, $x \in X - \mu(X)$: Then $ran(\sigma_0) \cap \mu(X) \neq \emptyset$, let $x' \in ran(\sigma_0) \cap \mu(X)$. Moreover, $\mu(X) \subseteq K$. Then for all $\langle x', \sigma' \rangle \in \mathcal{X}'$ $\langle x', \sigma' \rangle \prec \langle x, \sigma \rangle$. But $\langle x', \sigma^{x',X} \rangle$ as constructed in the proof of Claim 4.2.7 (p. 111) is minimal in $\mathcal{X}' \lceil X$.

Case 2, $x \in \mu(X) = \mu_{\mathcal{Z}}(X) = \mu'(X)$: If $\langle x, \sigma \rangle$ is minimal in $\mathcal{X}' \lceil X$, we are done. So suppose there is $\langle x', \sigma' \rangle \prec \langle x, \sigma \rangle$, $x' \in X$. Thus $x' \in \widehat{\sigma}$. Let $x' \in ran(\sigma_i)$. So $x \in \mu(X)$ and $ran(\sigma_i) \cap X \neq \emptyset$. But $\sigma_{i+1} \in \Pi\{\mu(X') : X' \in \mathcal{Y} \wedge x \in \mu(X') \wedge ran(\sigma_i) \cap X' \neq \emptyset\}$, so X is one of the X', moreover $\mu(X) \subseteq K$, so there is $x'' \in \mu(X) \cap ran(\sigma_{i+1}) \cap K$, so for all $\langle x'', \sigma'' \rangle \in \mathcal{X}'$ $\langle x'', \sigma'' \rangle \prec \langle x, \sigma \rangle$.

But again $\langle x'', \sigma^{x'',X}\rangle$ as constructed in the proof of Claim 4.2.7 (p. 111) is minimal in $\mathcal{X}'\lceil X$. (Claim 4.2.8 (p. 111) and Proposition 4.2.4 (p. 110)) □

We conclude this section by showing that we cannot improve substantially.

Proposition 4.2.5 *There is no fixed size characterization of μ-functions which are representable by smooth structures, if the domain is not closed under finite unions.*

Proof Suppose we have a fixed size characterization, which allows to distinguish μ-functions on domains which are not necessarily closed under finite unions, and which can be represented by smooth structures, from those which cannot be represented in this way. Let the characterization have α parameters for sets, and consider Example 4.2.4 (p. 100) with $\kappa = \beta + 1$, $\beta > \alpha$ (as a cardinal). This structure cannot be represented, as $(\mu Cum\kappa)$ fails – see Fact 4.2.9 (p. 103), (2.1). As we have only α parameters, at least one of the X_γ is not mentioned, say X_δ. Without loss of generality, we may assume that $\delta = \delta' + 1$. We change now the structure, and erase one pair of the relation, $x_\delta \prec x_{\delta+1}$. Thus, $\mu(X_\delta) = \{c, x_\delta, x_{\delta+1}\}$. But now we cannot go anymore from $X_{\delta'}$ to $X_{\delta'+1} = X_\delta$, as $\mu(X_\delta) \not\subseteq X_{\delta'}$. Consequently, the only chain showing that $(\mu Cum\infty)$ fails is interrupted – and we have added no new possibilities, as inspection of cases shows. ($x_{\delta+1}$ is now globally minimal, and increasing $\mu(X)$ cannot introduce new chains, only interrupt chains.) Thus, $(\mu Cum\infty)$ holds in the modified example, and it is thus representable by a smooth structure, as above proposition shows. As we did not touch any of the parameters, the truth value of the characterization is unchanged, which was negative. So the "characterization" cannot be correct. □

4.2.2.4 The Transitive Smooth Case

Unfortunately, $(\mu Cumt\infty)$ is a necessary but not sufficient condition for smooth transitive structures, as can be seen in the following example.

Example 4.2.5 We assume no closure whatever.

$U := \{u_1, u_2, u_3, u_4\}$, $\mu(U) := \{u_3, u_4\}$,
$Y_1 := \{u_4, v_1, v_2, v_3, v_4\}$, $\mu(Y_1) := \{v_3, v_4\}$,
$Y_{2,1} := \{u_2, v_2, v_4\}$, $\mu(Y_{2,1}) := \{u_2, v_2\}$,
$Y_{2,2} := \{u_1, v_1, v_3\}$, $\mu(Y_{2,2}) := \{u_1, v_1\}$.

For no A, B $\mu(A) \subseteq B$ $(A \neq B)$, so the prerequisite of $(\mu Cumt\alpha)$ is false, and $(\mu Cumt\alpha)$ holds, but there is no smooth transitive representation possible: Consider Y_1. If $u_4 \succ v_3$, then $Y_{2,2}$ makes this impossible, if $u_4 \succ v_4$, then $Y_{2,1}$ makes this impossible.

Remark 4.2.2

(1) The situation does not change when we have copies, the same argument will still work: There is a U-minimal copy $\langle u_4, i\rangle$. By smoothness and Y_1, there must be a Y_1-minimal copy, e.g. $\langle v_3, j\rangle \prec \langle u_4, i\rangle$. By smoothness and $Y_{2,2}$, there must be a $Y_{2,2}$-minimal $\langle u_1, k\rangle$ or $\langle v_1, l\rangle$ below $\langle v_3, j\rangle$. But v_1 is in Y_1, contradicting minimality of $\langle v_3, j\rangle$, u_1 is in U, contradicting minimality of $\langle u_4, i\rangle$ by

transitivity. If we choose $\langle v_4, j\rangle$ minimal below $\langle u_4, i\rangle$, we will work with $Y_{2,1}$ instead of $Y_{2,2}$.

(2) We can also close under arbitrary intersections, and the example will still work: We have to consider $U \cap Y_1$, $U \cap Y_{2,1}$, $U \cap Y_{2,2}$, $Y_{2,1} \cap Y_{2,2}$, $Y_1 \cap Y_{2,1}$, $Y_1 \cap Y_{2,2}$, there are no further intersections to consider. We may assume $\mu(A) = A$ for all these intersections (working with copies). But then $\mu(A) \subseteq B$ implies $\mu(A) = A$ for all sets, and all $(\mu Cumt\alpha)$ hold again trivially.

(3) If we had finite unions, we could form $A := U \cup Y_1 \cup Y_{2,1} \cup Y_{2,2}$, then $\mu(A)$ would have to be a subset of $\{u_3\}$ by (μPR), so by (μCUM) $u_4 \not\in \mu(U)$, a contradiction. Finite unions allow us to "look ahead", without $(\cup)$, we see disaster only at the end – and have to backtrack, i.e. try in our example $Y_{2,1}$, once we have seen impossibility via $Y_{2,2}$, and discover impossibility again at the end.

4.2.2.5 General Ranked Structures

Fix $f : \mathcal{Y} \rightarrow \mathcal{P}(Z)$.

The General Case

We summarize in the following Lemma 4.2.2 (p. 113) our results for the general ranked case, many of them trivial.

Lemma 4.2.2 *We assume here for simplicity that all elements occur in the structure.*

(1) *If $\mu(X) = \emptyset$, then each element $x \in X$ either has infinitely many copies, or below each copy of each x, there is an infinite descending chain of other elements.*

(2) *If there is no X such that $x \in \mu(X)$, then we can make infinitely many copies of x.*

(3) *There is no simple way to detect whether there is for all x some X such that $x \in \mu(X)$. More precisely, there is no normal finite characterization of ranked structures, in which each x in the domain occurs in at least one $\mu(X)$.*
Suppose in the sequel that for each x there is some X such that $x \in \mu(X)$. (This is the hard case.)

(4) *If the language is finite, then $X \neq \emptyset$ implies $\mu(X) \neq \emptyset$.*
Suppose now the language to be infinite.

(5) *If we admit all theories, then $\mu(M(T)) = M(T)$ for all complete theories.*

(6) *It is possible to have $\mu(M(\phi)) = \emptyset$ for all formulas ϕ, even though all models occur in exactly one copy.*

(7) *If the domain is sufficiently rich, then we cannot have $\mu(X) = \emptyset$ for "many" X.*

(8) *We see that a small domain (see Case (6)) can have many X with $\mu(X) = \emptyset$, but if the domain is too dense (see Case (7)), then we cannot have many $\mu(X) = \emptyset$. (We do not know any criterion to distinguish poor from rich domains.)*

(9) *If we have all pairs in the domain, we can easily construct the ranking.*

Proof

(1), (2), (4), (5), and (9) are trivial, there is nothing to show for (8).

(3) Suppose there is a normal characterization Φ of such structures, where each element x occurs at least once in a set X such that $x \in \mu(X)$. Such a characterization will be a finite Boolean combination of set expressions Φ, universally quantified, in the spirit of (AND), (RM), etc.

We consider a realistic counterexample – an infinite propositional language and the sets definable by formulas. We do not necessarily assume definability preservation, and work with full equality of results.

Take an infinite propositional language $p_i : i < \omega$. Choose an arbitrary model m, say $m \models p_i : i < \omega$.

Now, determine the height of any model m' as follows: $ht(m') := the$ first p_i such that $m(p_i) \neq m'(p_i)$; in our example then the first p_i is such that $m' \models \neg p_i$. Thus, only m has infinite height. Essentially, the more different m' is from m (in an alphabetical order), the lower it is.

Make now ω many copies of m, in infinite descending order, which you put on top of the rest.

Φ has to fail for some instantiation, as $\mathcal{X}$ does not have the desired property. Write this instantiation of Φ without loss of generality as a disjunction of conjunctions: $\bigvee(\bigwedge \phi_{i,j})$.

Each (consistent or non-empty) component $\phi_{i,j}$ has finite height. More precisely, the minimum of all heights of its models (which is a finite height). Thus, $\mid\!\sim (\phi_{i,j})$ will be just the minimally high models of $\phi_{i,j}$ in this order.

Modify now $\mathcal{X}$ such that m has only one copy, and is just (+1 suffices) above the minimum of all the finitely many $\phi_{i,j}$. Then none of the $\mid\!\sim (\phi_{i,j})$ is affected, and m has now finite height, say h, and is a minimum in any $M(\phi')$, where $\phi' =$ the conjunction of the first h values of m.

(Remark: Obviously, there are two easy generalizations for this ranking: First, we can go beyond ω (but also stay below ω). Second, instead of taking just one m as a scale, and which has maximal height, we can take a set M of models: $ht(m')$ is then the first p_i where $m'(p_i)$ is different from *all* $m \in M$. Note that in this case, in general, not all levels need to be filled. If, e.g. m_0, $m_1 \in M$, and $m_0(p_0) = false$, $m_1(p_0) = true$, then level 0 will be empty.)

(6) Let the p_i again define an infinite language. Denote by p_i^+ the set of all $+p_j$, where $j > i$. Let T be the usual tree of models (each model is a branch) for the p_i, with an artificial root $*$. Let the first model (= branch) be $*^+$, i.e. the leftmost branch in the obvious way of drawing it. Next, we choose $\neg p_0^+$, i.e. we go right, and then all the way left. Next, we consider the four sequences of $+/-p_0$, $+/-p_1$, two of them were done already, both ending in p_1^+, and choose the remaining two, both ending in $\neg p_1^+$, i.e. the positive prolongations of $p_0, \neg p_1$ and $\neg p_0, \neg p_1$. Thus, at each level, we take all possible prolongations. The positive ones were done already, and we count those which begin negatively, and then continue positively. Each formula has in this counting arbitrarily big models.

This is not yet a full enumeration of all models, e.g. the branch with all models negative will never be enumerated. But it suffices for our purposes.

Reverse the order so far constructed, and put the models not enumerated on top. Then all models are considered, and each formula has arbitrarily small models, thus $\mu(\phi) = \emptyset$ for all ϕ.
(7) Let the domain contain all singletons, and let the structure be without copies. The latter can be seen by considering singletons. Suppose now there is a set X in the domain such that $\mu(X) = \emptyset$. Thus, each $x \in X$ must have infinitely many $x' \in X$ $x' \prec x$. Suppose $\mathcal{P}(X)$ is a subset of the domain. Then there must be infinite $Y \in \mathcal{P}(X)$ such that $\mu(Y) \neq \emptyset$: Suppose not. Let $\prec$ be the ranking order. Choose arbitrary $x \in X$. Consider $X' := \{x' \in X : x \prec x'\}$, then $x \in \mu(X')$, and not all such X' can be finite – assuming X is big enough, e.g. uncountable. □

We conclude by giving an example of a definability preserving noncompact preferential logic – in answer to a question by D. Makinson (personal communication):

Example 4.2.6 Take an infinite language, $p_i, i < \omega$. Fix one model, m, which makes p_0 true (and, say, for definiteness, all the others true too), and m' which is just like m, but it makes p_0 false. Well-order all the other p_0-models and all the other $\neg p_0$-models separately.

Construct now the following ranked structure:

On top, put m, directly below it m'. Further down put the block of the other $\neg p_0$-models, and at the bottom the block of the other p_0-models.

As the structure is well-ordered, it is definability preserving (singletons are definable).

Let T be the theory defined by m, m', then $T \mid\sim \neg p_0$.

Let ϕ be such that $M(T) \subseteq M(\phi)$, then $M(\phi)$ contains a p_0-model other than m, so $\phi \mid\sim p_0$.

4.2.2.6 Smooth-Ranked Structures

We assume that all elements occur in the structure, so smoothness and $\mu(X) \neq \emptyset$ for $X \neq \emptyset$ coincide.
The following abstract definition is motivated by

$\approx (u)$, the set of $u' \in W$ which have same rank as u,
$\prec (u)$, the set of $u' \in W$ which have lower rank than u,
$\succ (u)$, the set of $u' \in W$ which have higher rank than u.

All other $u' \in W$ will by default have unknown rank in comparison.
We can diagnose, e.g. $u' \in\approx (u)$ if $u, u' \in \mu(X)$ for some X, and $u' \in\succ (u)$ if $u \in \mu(X)$ and $u' \in X - \mu(X)$ for some X.
If we sometimes do not know more, we will have to consider also $\preceq (u)$ and $\succeq (u)$ – this will be needed in Sect. 8.1.2 (p. 219), where we will have only incomplete information, due to hidden dimensions.
All other $u' \in W$ will by default have unknown rank in comparison.

Definition 4.2.7

(1) Define for each $u \in W$ three subsets of $W \approx (u)$, $\prec (u)$, and $\succ (u)$. Let $\mathcal{O}$ be the set of all these subsets, i.e. $\mathcal{O} := \{\approx (u), \prec (u), \succ (u) : u \in W\}$.

(2) We say that $\mathcal{O}$ is generated by a choice function f iff

(1) $\forall U \in \mathcal{Y} \forall x, x' \in f(U)\, x' \in\approx (x)$,
(2) $\forall U \in \mathcal{Y} \forall x \in f(U) \forall x' \in U - f(U)\, x' \in\succ (x)$.

(3) $\mathcal{O}$ is said to be representable by a ranking iff there is a function $f : W \rightarrow \langle O, \blacktriangleleft \rangle$ into a total order $\langle O, \blacktriangleleft \rangle$ such that

(1) $u' \in\approx (u) \Rightarrow f(u') = f(u)$,
(2) $u' \in\prec (u) \Rightarrow f(u') \blacktriangleleft f(u)$,
(3) $u' \in\succ (u) \Rightarrow f(u') \blacktriangleright f(u)$.

(4) Let $\mathcal{C}(\mathcal{O})$ be the closure of $\mathcal{O}$ under the following operations:

- $u \in\approx (u)$,
- if $u' \in\approx (u)$, then $\approx (u) =\approx (u')$, $\prec (u) =\prec (u')$, $\succ (u) =\succ (u')$,
- $u' \in\prec (u)$ iff $u \in\succ (u')$,
- $u \in\prec (u')$, $u' \in\prec (u'') \Rightarrow u \in\prec (u'')$,
  or, equivalently,
- $u \in\prec (u') \Rightarrow \prec (u') \subseteq\prec (u)$.

Note that we will generally loose much ignorance in applying the next two facts.

Fact 4.2.13 A partial (strict) order on W can be extended to a total (strict) order.

Proof Take an arbitrary enumeration of all pairs a, b of $W : \langle a, b\rangle_i : i \in \kappa$. Suppose all $\langle a, b\rangle_j$ for $j < i$ have been ordered, and we have no information if $a \prec b$ or $a \approx b$ or $a \succ b$. Choose arbitrarily $a \prec b$. A contradiction would be a (finite) cycle involving $\prec$. But then we would have known already that $b \preceq a$. □

We now use a generalized abstract nonsense result, taken from [LMS01], which must also be part of the folklore:

Fact 4.2.14 Given a set X and a binary relation R on X, there exists a total preorder (i.e. a total, reflexive, transitive relation) S on X that extends R such that

$$\forall x, y \in X (xSy, ySx \Rightarrow xR^*y),$$

where R^* is the reflexive and transitive closure of R.

Proof Define $x \equiv y$ iff xR^*y and yR^*x. The relation $\equiv$ is an equivalence relation. Let $[x]$ be the equivalence class of x under $\equiv$. Define $[x] \preceq [y]$ iff xR^*y. The definition of $\preceq$ does not depend on the representatives x and y chosen. The relation $\preceq$ on equivalence classes is a partial order.

Let $\leq$ be any total order on these equivalence classes that extends $\preceq$ by above Fact 4.2.13 (p. 116).
Define xSy iff $[x] \leq [y]$. The relation S is total (since $\leq$ is total) and transitive (since $\leq$ is transitive), and is therefore a total preorder. It extends R by the definition of $\preceq$ and the fact that $\leq$ extends $\preceq$. Suppose now xSy and ySx. We have $[x] \leq [y]$ and $[y] \leq [x]$, and therefore $[x] = [y]$ by antisymmetry. Therefore, $x \equiv y$ and xR^*y. □

Fact 4.2.15 $\mathcal{O}$ can be represented by a ranking iff in $\mathcal{C}(\mathcal{O})$ the sets $\approx (u)$, $\prec (u)$, and $\succ (u)$ are pairwise disjoint.

Proof (Outline) By the construction of $\mathcal{C}(\mathcal{O})$ and disjointness, there are no cycles involving $\prec$. Extend the relation by Fact 4.2.14 (p. 116). Let $\approx (u)$ be the equivalence classes. Define $\approx (u) \blacktriangleleft \approx (u')$ iff $u \in \prec (u')$. □

Proposition 4.2.6 *Let* $f : \mathcal{Y} \rightarrow \mathcal{P}(W)$. f *is representable by a smooth-ranked structure iff in* $\mathcal{C}(\mathcal{O})$ *the sets* $\approx (u)$, $\prec (u)$, $\succ (u)$ *are pairwise disjoint, where* $\mathcal{O}$ *is the system generated by* f, *as in Definition* 4.2.7 (p. 116).

Proof If the sets are not pairwise disjoint, we have a cycle. If not, use Fact 4.2.15 (p. 117). □

Chapter 5
Preferential Structures – Part II

5.1 Simplifications by Domain Conditions, Logical Properties

5.1.1 Introduction

We examine here simplifications made possible by stronger closure conditions of the domain $\mathcal{Y}$, in particular $(\cup)$.

For general preferential structures, there is nothing to show – there were no prerequisites about closure of the domain.

The smooth case is more interesting. The work for the not necessarily transitive case was done already, and as we did not know how to do better, we give now directly the result for the smooth transitive case, using an essential way $(\cup)$.

5.1.2 Smooth Structures

For completeness' sake and for the reader's convenience, we will just repeat here our result from [Sch04], with the slight improvement that we do not need $(\cap)$ any more, and the codomain need not be $\mathcal{Y}$ any more. The central condition is, of course, $(\cup)$, which we use now as we prepare the classical propositional case, where we have $\vee$.

5.1.2.1 Discussion of the Smooth and Transitive Case

In a certain way, it is not surprising that transitivity does not impose stronger conditions in the smooth case either. Smoothness is itself a weak kind of transitivity: If an element is not minimal, then there is a minimal element below it, i.e. $x \succ y$ with y not minimal is possible, there is $z' \prec y$, but then there is z minimal with $x \succ z$. This is "almost" $x \succ z'$, transitivity.

To obtain representation, we will combine here the ideas of the smooth, but not necessarily transitive case with those of the general transitive case – as the reader will have suspected. Thus, we will index again with trees and work with (suitably adapted) admissible sequences for the construction of the trees. In the construction of the admissible sequences, we were careful to repair all damage done in previous

D.M. Gabbay, K. Schlechta, *Logical Tools for Handling Change in Agent-Based Systems*, Cognitive Technologies, DOI 10.1007/978-3-642-04407-6_5,

steps. We have to add now reparation of all damage done by using transitivity, i.e. the transitivity of the relation might destroy minimality, and we have to construct minimal elements below all elements for which we thus destroyed minimality. Both cases are combined by considering immediately all Y such that $x \in Y - H(U)$. Of course, the properties described in Fact 4.2.11 (p. 106) play again a central role.

The (somewhat complicated) construction will be commented on in more detail below.

Note that even beyond Fact 4.2.11 (p. 106), closure of the domain under finite unions is used in the construction of the trees. This – or something like it – is necessary, as we have to respect the hulls of all elements treated so far (the predecessors), and not only of the first element, because of transitivity. For the same reason, we need more bookkeeping, to annotate all the hulls (or the union of the respective U's) of all predecessors to be respected. One can perhaps do with a weaker operation than union – i.e. just look at the hulls of all U's separately, to obtain a transitive construction where unions are lacking, and see the case of plausibility logic below – but we have not investigated this problem.

To summarize, we combine the ideas from the transitive general case and the simple smooth case, using the crucial Fact 4.2.11 (p. 106) to show that the construction goes through. The construction leaves still some freedom, and modifications are possible as indicated below in the course of the proof. The construction is perhaps the most complicated in the entire book, as it combines several ideas, some of which are already somewhat involved. If necessary, the proof can certainly still be elaborated, and its main points (use of a suitable $H(U)$ to avoid U, successive repair of damage done in the construction, trees as indexing) may probably be used in other contexts too.

5.1.2.2 The Construction

Recall that $\mathcal{Y}$ will be closed under finite unions in this section, and let again $\mu : \mathcal{Y} \to \mathcal{P}(Z)$. Proposition 5.1.1 (p. 120) is the representation result for the smooth transitive case.

Proposition 5.1.1 *Let $\mathcal{Y}$ be closed under finite unions, and $\mu : \mathcal{Y} \to \mathcal{P}(Z)$. Then there is a $\mathcal{Y}$-smooth transitive preferential structure $\mathcal{Z}$, such that for all $X \in \mathcal{Y}$ $\mu(X) = \mu_{\mathcal{Z}}(X)$ iff μ satisfies $(\mu \subseteq)$, (μPR), (μCUM).*

Proof

5.1.2.3 The Idea

We have to adapt Construction 4.2.4 (p. 110) (using x-admissible sequences) to the transitive situation and to our construction with trees. If $\langle \emptyset, x\rangle$ is the root, $\sigma_0 \in \Pi\{\mu(Y) : x \in Y - \mu(Y)\}$ determines some children of the root. To preserve smoothness, we have to compensate and add other children by the σ_{i+1} : $\sigma_{i+1} \in \Pi\{\mu(X) : x \in \mu(X), ran(\sigma_i) \cap X \neq \emptyset\}$. On the other hand, we have to pursue the same construction for the children so constructed. Moreover, these indirect children have to be added to those children of the root, which have to be compensated (as the first children are compensated by σ_1) to preserve smoothness. Thus, we build the tree in a simultaneous vertical and horizontal induction.

This construction can be simplified, by considering immediately all $Y \in \mathcal{Y}$ such that $x \in Y \not\subseteq H(U)$ – independent of whether $x \notin \mu(Y)$ (as done in σ_0), or whether $x \in \mu(Y)$, and some child y constructed before is in Y (as done in the σ_{i+1}), or whether $x \in \mu(Y)$, and some indirect child y of x is in Y (to take care of transitivity, as indicated above). We make this simplified construction.

There are two ways to proceed. First, we can take as $\lhd^*$ in the trees the transitive closure of $\lhd$. Second, we can deviate from the idea that children are chosen by selection functions f, and take nonempty subsets of elements instead, making more elements children than in the first case. We take the first alternative, as it is more in the spirit of the construction.

We will suppose for simplicity that $Z = K$ – the general case is easy to obtain by a technique similar to that in Sect. 4.2.2.1 (p. 94), but complicates the picture. For each $x \in Z$, we construct trees t_x, which will be used to index different copies of x, and control the relation $\prec$.

These trees t_x will have the following form:

(a) the root of t is $\langle \emptyset, x\rangle$ or $\langle U, x\rangle$ with $U \in \mathcal{Y}$ and $x \in \mu(U)$,
(b) all other nodes are pairs $\langle Y, y\rangle$, $Y \in \mathcal{Y}$, $y \in \mu(Y)$,
(c) $ht(t) \leq \omega$,
(d) if $\langle Y, y\rangle$ is an element in t_x, then there is some $\mathcal{Y}(y) \subseteq \{W \in \mathcal{Y} : y \in W\}$, and $f \in \Pi\{\mu(W) : W \in \mathcal{Y}(y)\}$ such that the set of children of $\langle Y, y\rangle$ is $\{\langle Y \cup W, f(W)\rangle : W \in \mathcal{Y}(y)\}$.

The first coordinate is used for bookkeeping when constructing children, in particular for condition (d).

The relation $\prec$ will essentially be determined by the subtree relation.

We first construct the trees t_x for those sets U where $x \in \mu(U)$, and then take care of the others. In the construction for the minimal elements, at each level $n > 0$, we may have several ways to choose a selection function f_n, and each such choice leads to the construction of a different tree – we construct all these trees. (We could also construct only one tree, but then the choice would have to be made coherently for different x, U. It is simpler to construct more trees than necessary.)

We control the relation by indexing with trees, just as it was done in the not necessarily smooth case before.

Definition 5.1.1 If t is a tree with root $\langle a, b\rangle$, then t/c will be the same tree, only with the root $\langle c, b\rangle$.

Construction 5.1.1 (A) The set T_x of trees t for fixed x:

(1) Construction of the set $T\mu_x$ of trees for those sets $U \in \mathcal{Y}$, where $x \in \mu(U)$:

Let $U \in \mathcal{Y}$, $x \in \mu(U)$. The trees $t_{U,x} \in T\mu_x$ are constructed inductively, observing simultaneously:
If $\langle U_{n+1}, x_{n+1} \rangle$ is a child of $\langle U_n, x_n \rangle$, then (a) $x_{n+1} \in \mu(U_{n+1}) - H(U_n)$ and (b) $U_n \subseteq U_{n+1}$.
Set $U_0 := U$, $x_0 := x$.

Level 0: $\langle U_0, x_0 \rangle$.

Level $n \to n+1$: Let $\langle U_n, x_n \rangle$ be in level n. Suppose $Y_{n+1} \in \mathcal{Y}$, $x_n \in Y_{n+1}$, and $Y_{n+1} \not\subseteq H(U_n)$. Note that $\mu(U_n \cup Y_{n+1}) - H(U_n) \neq \emptyset$ by Fact 4.2.11 (p. 106), (5.5) and $\mu(U_n \cup Y_{n+1}) - H(U_n) \subseteq \mu(Y_{n+1})$ by Fact 4.2.11 (p. 106), (3.3). Choose $f_{n+1} \in \Pi\{\mu(U_n \cup Y_{n+1}) - H(U_n) : Y_{n+1} \in \mathcal{Y}, x_n \in Y_{n+1} \not\subseteq H(U_n)\}$ (for the construction of this tree, at this element), and let the set of children of $\langle U_n, x_n \rangle$ be $\{\langle U_n \cup Y_{n+1}, f_{n+1}(Y_{n+1}) \rangle : Y_{n+1} \in \mathcal{Y}, x_n \in Y_{n+1} \not\subseteq H(U_n)\}$. (If there is no such Y_{n+1}, $\langle U_n, x_n \rangle$ has no children.) Obviously, (a) and (b) hold.

We call such trees U, x-trees.

(2) Construction of the set T'_x of trees for the nonminimal elements. Let $x \in Z$. Construct the tree t_x as follows (here, one tree per x suffices for all U):

Level 0: $\langle \emptyset, x \rangle$

Level 1: Choose arbitrary $f \in \Pi\{\mu(U) : x \in U \in \mathcal{Y}\}$. Note that $U \neq \emptyset \to \mu(U) \neq \emptyset$ by $Z = K$ (by Remark 4.2.1 (p. 109), (1)). Let $\{\langle U, f(U) \rangle : x \in U \in \mathcal{Y}\}$ be the set of children of $< \emptyset, x >$. This assures that the element will be nonminimal.

Level > 1: Let $\langle U, f(U) \rangle$ be an element of level 1, as $f(U) \in \mu(U)$, there is a $t_{U,f(U)} \in T\mu_{f(U)}$. Graft one of these trees $t_{U,f(U)} \in T\mu_{f(U)}$ at $\langle U, f(U) \rangle$ on the level 1. This assures that a minimal element will be below it to guarantee smoothness.

Finally, let $T_x := T\mu_x \cup T'_x$.

(B) The relation $\lhd$ between trees: For $x, y \in Z$, $t \in T_x$, $t' \in T_y$, set $t \rhd t'$ iff for some Y $\langle Y, y \rangle$ is a child of the root $\langle X, x \rangle$ in t, and t' is the subtree of t beginning at this $\langle Y, y \rangle$.

(C) The structure $\mathcal{Z}$:
Let $\mathcal{Z} := < \{\langle x, t_x \rangle : x \in Z, t_x \in T_x\}, \langle x, t_x \rangle \succ \langle y, t_y \rangle$ iff $t_x \rhd^* t_y >$.

The rest of the proof are simple observations.

Fact 5.1.1

(1) If $t_{U,x}$ is a U, x-tree, $\langle U_n, x_n \rangle$ an element of $t_{U,x}$, $\langle U_m, x_m \rangle$ a direct or indirect child of $\langle U_n, x_n \rangle$, then $x_m \notin H(U_n)$.
(2) Let $\langle Y_n, y_n \rangle$ be an element in $t_{U,x} \in T\mu_x$, t' the subtree starting at $\langle Y_n, y_n \rangle$, then t' is a Y_n, y_n-tree.
(3) $\prec$ is free from cycles.
(4) If $t_{U,x}$ is a U, x-tree, then $\langle x, t_{U,x} \rangle$ is $\prec$ −minimal in $\mathcal{Z} \lceil U$.
(5) No $\langle x, t_x \rangle$, $t_x \in T'_x$ is minimal in any $\mathcal{Z} \lceil U$, $U \in \mathcal{Y}$.
(6) Smoothness is respected for the elements of the form $\langle x, t_{U,x} \rangle$.

(7) Smoothness is respected for the elements of the form $\langle x, t_x \rangle$ with $t_x \in T'_x$.
(8) $\mu = \mu_{\mathcal{Z}}$.

Proof

(1) Trivial by (a) and (b).
(2) Trivial by (a).
(3) Note that no $\langle x, t_x \rangle$ $t_x \in T'_x$ can be smaller than any other element (smaller elements require $U \neq \emptyset$ at the root). So no cycle involves any such $\langle x, t_x \rangle$. Consider now $\langle x, t_{U,x} \rangle$, $t_{U,x} \in T\mu_x$. For any $\langle y, t_{V,y} \rangle \prec \langle x, t_{U,x} \rangle$, $y \notin H(U)$ by (1), but $x \in \mu(U) \subseteq H(U)$, so $x \neq y$.
(4) This is trivial by (1).
(5) Let $x \in U \in \mathcal{Y}$, then f as used in the construction of level 1 of t_x chooses $y \in \mu(U) \neq \emptyset$, and some $\langle y, t_{U,y} \rangle$ is in $\mathcal{Z}\lceil U$ and below $\langle x, t_x \rangle$.
(6) Let $x \in A \in \mathcal{Y}$, we have to show that either $\langle x, t_{U,x} \rangle$ is minimal in $\mathcal{Z}\lceil A$, or that there is $\langle y, t_y \rangle \prec \langle x, t_{U,x} \rangle$ minimal in $\mathcal{Z}\lceil A$. Case 1, $A \subseteq H(U)$: Then $\langle x, t_{U,x} \rangle$ is minimal in $\mathcal{Z}\lceil A$, again by (1). Case 2, $A \not\subseteq H(U)$: Then A is one of the Y_1 considered for level 1. So there is $\langle U \cup A, f_1(A) \rangle$ in level 1 with $f_1(A) \in \mu(A) \subseteq A$ by Fact 4.2.11 (p. 106), (3.3). But note that by (1) all elements below $\langle U \cup A, f_1(A) \rangle$ avoid $H(U \cup A)$. Let t be the subtree of $t_{U,x}$ beginning at $\langle U \cup A, f_1(A) \rangle$, then by (2) t is one of the $U \cup A$, $f_1(A)$-trees, and $\langle f_1(A), t \rangle$ is minimal in $\mathcal{Z}\lceil U \cup A$ by (4), so in $\mathcal{Z}\lceil A$, and $\langle f_1(A), t \rangle \prec \langle x, t_{U,x} \rangle$.
(7) Let $x \in A \in \mathcal{Y}$, $\langle x, t_x \rangle$, $t_x \in T'_x$, and consider the subtree t beginning at $\langle A, f(A) \rangle$, then t is one of the A, $f(A)$-trees, and $\langle f(A), t \rangle$ is minimal in $\mathcal{Z}\lceil A$ by (4).
(8) Let $x \in \mu(U)$. Then any $\langle x, t_{U,x} \rangle$ is $\prec$ $-$minimal in $\mathcal{Z}\lceil U$ by (4), so $x \in \mu_{\mathcal{Z}}(U)$. Conversely, let $x \in U - \mu(U)$. By (5), no $\langle x, t_x \rangle$ is minimal in U. Consider now some $\langle x, t_{V,x} \rangle \in \mathcal{Z}$, so $x \in \mu(V)$. As $x \in U - \mu(U)$, $U \not\subseteq H(V)$ by Fact 4.2.11 (p. 106), (5.4). Thus U was considered in the construction of level 1 of $t_{V,x}$. Let t be the subtree of $t_{V,x}$ beginning at $\langle V \cup U, f_1(U) \rangle$, by $\mu(V \cup U) - H(V) \subseteq \mu$ (U) (Fact 4.2.11 (p. 106), (3.3)) $f_1(U) \in \mu(U) \subseteq U$ and $\langle f_1(U), t \rangle \prec \langle x, t_{V,x} \rangle$. (Fact 5.1.1 (p. 122) and Proposition 5.1.1 (p. 120).) □

5.1.3 Ranked Structures

We summarize for completeness' sake results from [Sch04]:

First two results for the case without copies (Proposition 5.1.2 (p. 123) and Proposition 5.1.3 (p. 124)).

Proposition 5.1.2 *Let $\mathcal{Y} \subseteq \mathcal{P}(U)$ be closed under finite unions. Then $(\mu \subseteq)$, $(\mu\emptyset)$, $(\mu =)$ characterize ranked structures for which for all $X \in \mathcal{Y}$, $X \neq \emptyset \Rightarrow \mu_<(X) \neq \emptyset$ hold, i.e. $(\mu \subseteq)$, $(\mu\emptyset)$, $(\mu =)$ hold in such structures for $\mu_<$, and if they hold for some μ, we can find a ranked relation $<$ on U such that $\mu = \mu_<$. Moreover, the structure can be choosen $\mathcal{Y}$-smooth.*

For the following representation result, we assume only $(\mu\emptyset fin)$, but the domain has to contain singletons.

Proposition 5.1.3 *Let $\mathcal{Y} \subseteq \mathcal{P}(U)$ be closed under finite unions and contain singletons. Then $(\mu \subseteq)$, $(\mu\emptyset fin)$, $(\mu =)$, $(\mu \in)$ characterize ranked structures for which for all finite $X \in \mathcal{Y}$, $X \neq \emptyset \Rightarrow \mu_<(X) \neq \emptyset$ hold, i.e. $(\mu \subseteq)$, $(\mu\emptyset fin)$, $(\mu =)$, $(\mu \in)$ hold in such structures for $\mu_<$, and if they hold for some μ, we can find a ranked relation $<$ on U such that $\mu = \mu_<$.*

Note that the prerequisites of Proposition 5.1.3 (p. 124) hold in particular in the case of ranked structures without copies, where all elements of U are present in the structure – we need infinite descending chains to have $\mu(X) = \emptyset$ for $X \neq \emptyset$.

We turn now to the general case, where every element may occur in several copies.

Fact 5.1.2

(1) $(\mu \subseteq) + (\mu PR) + (\mu =) + (\mu\cup) + (\mu \in)$ do not imply representation by a ranked structure.
(2) The infinitary version of $(\mu \parallel)$
$(\mu \parallel \infty)$ $\mu(\bigcup\{A_i : i \in I\}) = \bigcup\{\mu(A_i) : i \in I'\}$ for some $I' \subseteq I$.
will not always hold in ranked structures.

We assume again the existence of singletons for the following representation result.

Proposition 5.1.4 *Let $\mathcal{Y}$ be closed under finite unions and contain singletons. Then $(\mu \subseteq) + (\mu PR) + (\mu \parallel) + (\mu\cup) + (\mu \in)$ characterize ranked structures.*

5.1.4 The Logical Properties with Definability Preservation

We repeat for completeness' sake:

Proposition 5.1.5 *Let $\mid\sim$ be a logic for $\mathcal{L}$. Set $T^{\mathcal{M}} := Th(\mu_{\mathcal{M}}(M(T)))$, where $\mathcal{M}$ is a preferential structure.*

(1) *Then there is a (transitive) definability preserving classical preferential model $\mathcal{M}$ such that $\overline{\overline{T}} = T^{\mathcal{M}}$ iff*
(LLE) $\overline{T} = \overline{T'} \Rightarrow \overline{\overline{T}} = \overline{\overline{T'}}$,
(CCL) $\overline{\overline{T}}$ *is classically closed,*
(SC) $T \subseteq \overline{\overline{T}}$,
(PR) $\overline{\overline{\overline{T \cup T'}}} \subseteq \overline{\overline{\overline{\overline{T}} \cup T'}}$,
for all $T, T' \subseteq \mathcal{L}$.
(2) *The structure can be chosen smooth, iff, in addition*
(CUM) $T \subseteq \overline{T'} \subseteq \overline{\overline{T}} \Rightarrow \overline{\overline{T}} = \overline{\overline{T'}}$
holds.

The proof is an immediate consequence of Proposition 5.1.6 (p. 125) and Proposition 5.1.1 (p. 120).

Proposition 5.1.6 *Consider for a logic* $\mid\!\sim$ *on* $\mathcal{L}$ *the properties*

(LLE) $\overline{T} = \overline{T'} \Rightarrow \overline{\overline{T}} = \overline{\overline{T'}}$,

(CCL) $\overline{\overline{T}}$ is classically closed,

(SC) $T \subseteq \overline{\overline{T}}$,

(PR) $\overline{\overline{T \cup T'}} \subseteq \overline{\overline{\overline{T}} \cup T'}$,

(CUM) $T \subseteq \overline{T'} \subseteq \overline{\overline{T}} \Rightarrow \overline{\overline{T}} = \overline{\overline{T'}}$,

for all $T, T' \subseteq \mathcal{L}$,

and for a function $\mu : \mathbf{D}_{\mathcal{L}} \rightarrow \mathcal{P}(M_{\mathcal{L}})$ *the properties*

(μdp) μ *is definability preserving, i.e.* $\mu(M(T)) = M(T')$ *for some* T'

$(\mu \subseteq)$ $\mu(X) \subseteq X$,

(μPR) $X \subseteq Y \Rightarrow \mu(Y) \cap X \subseteq \mu(X)$,

(μCUM) $\mu(X) \subseteq Y \subseteq X \Rightarrow \mu(X) = \mu(Y)$,

for all $X, Y \in \mathbf{D}_{\mathcal{L}}$.

It then holds:

(a) *If* μ *satisfies* (μdp), $(\mu \subseteq)$, (μPR), *then* $\mid\!\sim$ *defined by* $\overline{\overline{T}} := T^{\mu} := Th(\mu(M(T)))$ *satisfies (LLE), (CCL), (SC), (PR). If* μ *satisfies in addition* (μCUM), *then* (CUM) *will hold too.*

(b) *If* $\mid\!\sim$ *satisfies (LLE), (CCL), (SC), (PR), then there is* $\mu : \mathbf{D}_{\mathcal{L}} \rightarrow \mathcal{P}(M_{\mathcal{L}})$ *such that* $\overline{\overline{T}} = T^{\mu}$ *for all* $T \subseteq \mathcal{L}$ *and* μ *satisfies* (μdp), $(\mu \subseteq)$, (μPR). *If, in addition,* (CUM) *holds, then* (μCUM) *will hold too.*

The proof follows from Proposition 2.3.1 (p. 45).

The logical properties of definability preserving ranked structures are straightforward now, and left to the reader.

5.2 $\mathcal{A}$-Ranked Structures

We do now the completeness proofs for the preferential part of hierarchical conditionals. All motivation, etc. will be found in Sect. 7.3 (p. 205).

First, the basic semantical definition.

Definition 5.2.1 Let $\mathbf{A}$ be a fixed set, and $\mathcal{A}$ a finite, totally ordered (by $<$) disjoint cover by nonempty subsets of $\mathbf{A}$.
For $x \in \mathbf{A}$, let $rg(x)$ the unique $A \in \mathcal{A}$ such that $x \in A$, so $rg(x) < rg(y)$ is defined in the natural way.

A preferential structure $\langle \mathcal{X}, \prec \rangle$ ($\mathcal{X}$ a set of pairs $\langle x, i \rangle$) is called $\mathcal{A}$-ranked iff for all x, x', $rg(x) < rg(x')$ implies $\langle x, i \rangle \prec \langle x', i' \rangle$ for all $\langle x, i \rangle, \langle x', i' \rangle \in \mathcal{X}$.

5.2.1 Representation Results for $\mathcal{A}$-Ranked Structures

5.2.1.1 Discussion

The not necessarily smooth and the smooth case will be treated differently.

Strangely, the smooth case is simpler, as an added new layer in the proof settles it. Yet, this is not surprising when looking closer, as minimal elements never have higher rank, and we know from (μCUM) that minimizing by minimal elements suffices. All we have to add that any element in the minimal layer minimizes any element higher up.

In the simple, not necessarily smooth, case, we have to go deeper into the original proof to obtain the result. The following idea, inspired by the treatment of the smooth case, will not work: Instead of minimizing by arbitrary elements, minimize only by elements of minimal rank, as the following example shows. If it worked, we might add just another layer to the original proof without $(\mu\mathcal{A})$ (see Definition 5.2.2 (p. 126)), as in the smooth case.

Example 5.2.1 Consider the base set $\{a, b, c\}$, $\mu(\{a, b, c\}) = \{b\}$, $\mu(\{a, b\}) = \{a, b\}$, $\mu(\{a, c\}) = \emptyset$, $\mu(\{b, c\}) = \{b\}$, $\mathcal{A}$ defined by $\{a, b\} < \{c\}$.
Obviously, $(\mu\mathcal{A})$ is satisfied. μ can be represented by the (not transitive!) relation $a \prec c \prec a$, $b \prec c$, which is $\mathcal{A}$-ranked. But trying to minimize a in $\{a, b, c\}$ in the minimal layer will lead to $b \prec a$, and thus $a \notin \mu(\{a, b\})$, which is wrong.

The proofs of the general and transitive general case are (minor) adaptations of the proofs in Sect. 4.2.2 (p. 94). For the smooth case, we only have to add a supplementary layer in the end (Fact 5.2.2 (p. 129)), which will make the construction $\mathcal{A}$-ranked.

In the following, we will assume the partition $\mathcal{A}$ to be given. We could also construct it from the properties of μ, but this would need stronger closure properties of the domain. The construction of $\mathcal{A}$ is more difficult than the construction of the ranking in fully ranked structures, as $x \in \mu(X)$, $y \in X - \mu(X)$ will guarantee only $rg(x) \leq rg(y)$, and not $rg(x) < rg(y)$, as is the case in the latter situation. This corresponds to the separate treatment of the α and other formulas in the logical version, discussed in Sect. 5.2.1.4 (p. 130).

5.2.1.2 $\mathcal{A}$-Ranked General and Transitive Structures

Introduction

We will show here the following representation result:

Let $\mathcal{A}$ be given. An operation $\mu : \mathcal{Y} \to \mathcal{P}(Z)$ is representable by an $\mathcal{A}$-ranked preferential structure iff μ satisfies $(\mu \subseteq)$, (μPR), $(\mu\mathcal{A})$ (Proposition 5.2.1 (p. 127)), and, moreover, the structure can be chosen transitive (Proposition 5.2.2 (p. 127)). Note that we carefully avoid any unnecessary assumptions about the domain $\mathcal{Y} \subseteq \mathcal{P}(Z)$ of the function μ.

Definition 5.2.2 We define a new condition:

Let $\mathcal{A}$ be given as defined in Definition 5.2.1 (p. 125).

$(\mu\mathcal{A})$ If $X \in \mathcal{Y}$, $A, A' \in \mathcal{A}$, $A < A'$, $X \cap A \neq \emptyset$, $X \cap A' \neq \emptyset$, then $\mu(X) \cap A' = \emptyset$.

This new condition will be central for the modified representation.

The Basic, Not Necessarily Transitive, Case

Corollary 5.2.1 *Let $\mu : \mathcal{Y} \to \mathcal{P}(Z)$ satisfy $(\mu \subseteq)$, (μPR), $(\mu \mathcal{A})$, and let $U \in \mathcal{Y}$.*
If $x \in U$ and $\exists x' \in U, rg(x') < rg(x)$, then $\forall f \in \Pi_x, ran(f) \cap U \neq \emptyset$.

Proof By $(\mu \mathcal{A})$ $x \notin \mu(U)$, thus by Claim 4.2.1 (p. 95) $\forall f \in \Pi_x, ran(f) \cap U \neq \emptyset$. □

Proposition 5.2.1 *Let $\mathcal{A}$ be given.*
An operation $\mu : \mathcal{Y} \to \mathcal{P}(Z)$ is representable by an $\mathcal{A}$-ranked preferential structure iff μ satisfies $(\mu \subseteq)$, (μPR), $(\mu \mathcal{A})$.

Proof One direction is trivial. The central argument is: If $a \prec b$ in X, and $X \subseteq Y$, then $a \prec b$ in Y too.

We turn to the other direction. The preferential structure is defined in Construction 5.2.1 (p. 127), and Claim 5.2.1 (p. 127) shows representation. □

Construction 5.2.1 Let $\mathcal{X} := \{\langle x, f \rangle : x \in Z \wedge f \in \Pi_x\}$, and $\langle x', f' \rangle \prec \langle x, f \rangle$ $:\leftrightarrow x' \in ran(f)$ or $rg(x') < rg(x)$. Note that, as $\mathcal{A}$ is given, we also know $rg(x)$.

Let $\mathcal{Z} := \langle \mathcal{X}, \prec \rangle$. Obviously, $\mathcal{Z}$ is $\mathcal{A}$-ranked.

Claim 5.2.1 For $U \in \mathcal{Y}$, $\mu(U) = \mu_{\mathcal{Z}}(U)$.

Proof By Claim 4.2.1 (p. 95), it suffices to show that for all $U \in \mathcal{Y}, x \in \mu_{\mathcal{Z}}(U) \leftrightarrow x \in U$ and $\exists f \in \Pi_x, ran(f) \cap U = \emptyset$. So let $U \in \mathcal{Y}$.

"$\Rightarrow$": If $x \in \mu_{\mathcal{Z}}(U)$, then there is $\langle x, f \rangle$ minimal in $\mathcal{X} \lceil U$ – where $\mathcal{X} \lceil U := \{\langle x, i \rangle \in \mathcal{X} : x \in U\}$), so $x \in U$, and there is no $\langle x', f' \rangle \prec \langle x, f \rangle$, $x' \in U$, so by $\Pi_{x'} \neq \emptyset$ there is no $x' \in ran(f)$, $x' \in U$, but then $ran(f) \cap U = \emptyset$.
"$\Leftarrow$": If $x \in U$ and there is $f \in \Pi_x, ran(f) \cap U = \emptyset$, then by Corollary 5.2.1 (p. 127), there is no $x' \in U$, $rg(x') < rg(x)$, so $\langle x, f \rangle$ is minimal in $\mathcal{X} \lceil U$.
(Claim 5.2.1 (p. 127) and Proposition 5.2.1 (p. 127).) □

The Transitive Case

Proposition 5.2.2 *Let $\mathcal{A}$ be given.*
An operation $\mu : \mathcal{Y} \to \mathcal{P}(Z)$ is representable by an $\mathcal{A}$-ranked transitive preferential structure iff μ satisfies $(\mu \subseteq)$, (μPR), $(\mu \mathcal{A})$.

Construction 5.2.2 (1) For $x \in Z$, let T_x be the set of trees t_x s.t.

(a) all nodes are elements of Z,
(b) the root of t_x is x,
(c) $height(t_x) \leq \omega$, and
(d) if y is an element in t_x, then there is $f \in \Pi_y := \Pi\{Y \in \mathcal{Y}: y \in Y - \mu(Y)\}$ s.t. the set of children of y is $ran(f) \cup \{y' \in Z : rg(y') < rg(y)\}$.

(2) For $x, y \in Z$, $t_x \in T_x$, $t_y \in T_y$, set $t_x \rhd t_y$ iff y is a (direct) child of the root x in t_x, and t_y is the subtree of t_x beginning at y.
(3) Let $\mathcal{Z} := \langle \{\langle x, t_x \rangle : x \in Z, t_x \in T_x\}, \langle x, t_x \rangle \succ \langle y, t_y \rangle$ iff $t_x \rhd t_y$ $\rangle$.

Fact 5.2.1

(1) The construction ends at some y iff $\mathcal{Y}_y = \emptyset$ and there is no y' s.t. $rg(y') < rg(y)$, consequently $T_x = \{x\}$ iff $\mathcal{Y}_x = \emptyset$ and there are no x' with lesser rang. (We identify the tree of height 1 with its root.)
(2) We define a special tree tc_x for all x: For all nodes y in tc_x, the successors are as follows:

$$\text{if } \mathcal{Y}_y \neq \emptyset, \text{ then } z \text{ is a successor iff } z = y \text{ or } rg(z) < rg(y),$$
$$\text{if } \mathcal{Y}_y = \emptyset, \text{ then } z \text{ is a successor iff } rg(z) < rg(y).$$

(In the first case, we make $f \in \mathcal{Y}_y$ always choose y itself.) tc_x is an element of T_x. Thus, with (1), $T_x \neq \emptyset$ for any x. Note: $tc_x = x$ iff $\mathcal{Y}_x = \emptyset$ and x has minimal rang.
(3) If $f \in \Pi_x$, then the tree tf_x with root x and otherwise composed of the subtrees tc_y for $y \in ran(f) \cup \{y' : rg(y') < rg(y)\}$ is an element of T_x. (Level 0 of tf_x has x as element, the $t'_y s$ begin at level 1.)
(4) If y is an element in t_x and t_y the subtree of t_x starting at y, then $t_y \in T_y$.
(5) $\langle x, t_x \rangle \succ \langle y, t_y \rangle$ implies $y \in ran(f) \cup \{x' : rg(x') < rg(x)\}$ for some $f \in \Pi_x$.

Claim 5.2.2 (p. 128) shows basic representation.

Claim 5.2.2 $\forall U \in \mathcal{Y}, \mu(U) = \mu_{\mathcal{Z}}(U)$

Proof By Claim 4.2.1 (p. 95), it suffices to show that for all $U \in \mathcal{Y}$, $x \in \mu_{\mathcal{Z}}(U) \leftrightarrow x \in U \wedge \exists f \in \Pi_x . ran(f) \cap U = \emptyset$.
Fix $U \in \mathcal{Y}$.

"$\Rightarrow$": $x \in \mu_{\mathcal{Z}}(U) \rightarrow$ ex. $\langle x, t_x \rangle$ minimal in $\mathcal{Z} \lceil U$, thus $x \in U$ and there is no $\langle y, t_y \rangle \in \mathcal{Z}$, $\langle y, t_y \rangle \prec \langle x, t_x \rangle$, $y \in U$. Let f define the first part of the set of children of the root x in t_x. If $ran(f) \cap U \neq \emptyset$, if $y \in U$ is a child of x in t_x, and if t_y is the subtree of t_x starting at y, then $t_y \in T_y$ and $\langle y, t_y \rangle \prec \langle x, t_x \rangle$, contradicting minimality of $\langle x, t_x \rangle$ in $\mathcal{Z} \lceil U$. So $ran(f) \cap U = \emptyset$.
"$\Leftarrow$": Let $x \in U$ and $\exists f \in \Pi_x$, $ran(f) \cap U = \emptyset$. By Corollary 5.2.1 (p. 127), there is no $x' \in U$, $rg(x') < rg(x)$. If $\mathcal{Y}_x = \emptyset$, then the tree tc_x has no $\rhd$-successors in U, and $\langle x, tc_x \rangle$ is $\succ$ $-$minimal in $\mathcal{Z} \lceil U$. If $\mathcal{Y}_x \neq \emptyset$ and $f \in \Pi_x$ s.t. $ran(f) \cap U = \emptyset$, then $\langle x, tf_x \rangle$ is again $\succ$ $-$minimal in $\mathcal{Z} \lceil U$. □

We consider now the transitive closure of $\mathcal{Z}$. (Recall that $\prec^*$ denotes the transitive closure of $\prec$.) Claim 5.2.3 (p. 128) shows that transitivity does not destroy what we have achieved. The trees tf_x play a crucial role in the demonstration.

Claim 5.2.3 Let $\mathcal{Z}' := \langle \{\langle x, t_x \rangle : x \in Z, t_x \in T_x\}, \langle x, t_x \rangle \succ \langle y, t_y \rangle \text{ iff } t_x \rhd^* t_y \rangle$. Then $\mu_{\mathcal{Z}} = \mu_{\mathcal{Z}'}$.

Proof Suppose there is $U \in \mathcal{Y}$, $x \in U$, $x \in \mu_{\mathcal{Z}}(U)$, $x \notin \mu_{\mathcal{Z}'}(U)$. Then there must be an element $\langle x, t_x \rangle \in \mathcal{Z}$ with no $\langle x, t_x \rangle \succ \langle y, t_y \rangle$ for any $y \in U$. Let $f \in \Pi_x$

determine the first part of the set of children of x in t_x, then $ran(f) \cap U = \emptyset$, consider tf_x. All elements $w \neq x$ of tf_x are already in $ran(f)$, or $rg(w) < rg(x)$ holds. (Note that the elements chosen by rang in tf_x continue by themselves or by another element of even smaller rang, but the rang order is transitive.) But all w s.t. $rg(w) < rg(x)$ were already successors at level 1 of x in tf_x. By Corollary 5.2.1 (p. 127), there is no $w \in U$, $rg(w) < rg(x)$. Thus, no element $\neq x$ of tf_x is in U. Thus there is no $\langle z, t_z \rangle \prec^* \langle x, tf_x \rangle$ in $\mathcal{Z}$ with $z \in U$, so $\langle x, tf_x \rangle$ is minimal in $\mathcal{Z}' \lceil U$, contradiction. (Claim 5.2.3 (p. 128) and Proposition 5.2.2 (p. 127)) □

5.2.1.3 $\mathcal{A}$-Ranked Smooth Structures

All smooth cases have a simple solution. We use one of our existing proofs for the not necessarily $\mathcal{A}$-ranked case, and add one litte result:

Fact 5.2.2 Let $(\mu\mathcal{A})$ hold, and let $\mathcal{Z} = \langle \mathcal{X}, \prec \rangle$ be a smooth preferential structure representing μ, i.e. $\mu = \mu_{\mathcal{Z}}$.
Suppose that
$\langle x, i \rangle \prec \langle y, j \rangle$ implies $rg(x) \leq rg(y)$.
Define $\mathcal{Z}' := \langle \mathcal{X}, \sqsubset \rangle$ where $\langle x, i \rangle \sqsubset \langle y, j \rangle$ iff $\langle x, i \rangle \prec \langle y, j \rangle$ or $rg(x) < rg(y)$.
Then $\mathcal{Z}'$ is $\mathcal{A}$-ranked.
$\mathcal{Z}'$ is smooth too, and $\mu_{\mathcal{Z}} = \mu_{\mathcal{Z}'} =: \mu'$.
In addition, if $\prec$ is free from cycles, so is $\sqsubset$, if $\prec$ is transitive, so is $\sqsubset$.

Proof $\mathcal{A}$-rankedness is trivial.
Suppose $\langle x, i \rangle$ is $\prec$-minimal, but not $\sqsubset$-minimal. Then there must be $\langle y, j \rangle \sqsubset \langle x, i \rangle$, $\langle y, j \rangle \not\prec \langle x, i \rangle$, $y \in X$, so $rg(y) < rg(x)$. By $(\mu\mathcal{A})$, all $x \in \mu(X)$ have minimal $\mathcal{A}$-rang among the elements of X, so this is impossible. Thus, μ-minimal elements stay μ'-minimal, so smoothness will also be preserved – remember that we increased the relation.
By prerequisite, there cannot be any cycle involving only $\prec$, but the rang order is free from cycles too, and $\prec$ respects the rang order, so $\sqsubset$ is free from cycles.
Let $\prec$ be transitive, so is the rang order. But if $\langle x, i \rangle \prec \langle y, j \rangle$ and $rg(y) < rg(z)$ for some $\langle z, k \rangle$, then by prerequisite $rg(x) \leq rg(y)$, so $rg(x) < rg(z)$, so $\langle x, i \rangle \sqsubset \langle z, k \rangle$ by definition. Likewise for $rg(x) < rg(y)$ and $\langle y, j \rangle \prec \langle z, k \rangle$. □

All that remains to show then is that our constructions of smooth and of smooth and transitive structures satisfy the condition
$\langle x, i \rangle \prec \langle y, j \rangle$ implies $rg(x) \leq rg(y)$.

Proposition 5.2.3 *Let $\mathcal{A}$ be given.*
Let – for simplicity – $\mathcal{Y}$ be closed under finite unions and $\mu : \mathcal{Y} \to \mathcal{P}(Z)$. Then there is a $\mathcal{Y}$-smooth $\mathcal{A}$-ranked preferential structure $\mathcal{Z}$, s.t. for all $X \in \mathcal{Y}$, $\mu(X) = \mu_{\mathcal{Z}}(X)$ iff μ satisfies $(\mu\subseteq)$, (μPR), (μCUM), $(\mu\mathcal{A})$.

Proof Consider the construction in the proof of Proposition 4.2.4 (p. 110). We have to show that it respects the rang order with respect to $\mathcal{A}$, i.e. that $\langle x', \sigma' \rangle \prec' \langle x, \sigma \rangle$ implies $rg(x') \leq rg(x)$. This is easy: By definition, $x' \in \bigcup\{ran(\sigma_i) : i \in \omega\}$. If

$x' \in ran(\sigma_0)$, then for some Y $x' \in \mu(Y)$, $x \in Y - \mu(Y)$, so $rg(x') \leq rg(x)$ by $(\mu\mathcal{A})$. If $x' \in ran(\sigma_i)$, $i > 0$, then for some X $x', x \in \mu(X)$, so $rg(x) = rg(x')$ by $(\mu\mathcal{A})$. (Proposition 5.2.3 (p. 129)) □

Proposition 5.2.4 *Let $\mathcal{A}$ be given.*

Let – for simplicity – $\mathcal{Y}$ be closed under finite unions and $\mu : \mathcal{Y} \to \mathcal{P}(Z)$. Then there is a $\mathcal{Y}$-smooth $\mathcal{A}$-ranked transitive preferential structure $\mathcal{Z}$, s.t. for all $X \in \mathcal{Y}$, $\mu(X) = \mu_{\mathcal{Z}}(X)$ iff μ satisfies $(\mu \subseteq)$, (μPR), (μCUM), $(\mu\mathcal{A})$.

Proof Consider the construction in the proof of Proposition 5.1.1 (p. 120).

Thus, we only have to show that in $\mathcal{Z}$ defined by

$\mathcal{Z} := \langle \{\langle x, t_x\rangle : x \in Z, t_x \in T_x\}, \langle x, t_x\rangle \succ \langle y, t_y\rangle$ iff $t_x \rhd^* t_y \rangle$, $t_x \rhd t_y$ implies $rg(y) \leq rg(x)$.

But by construction of the trees, $x_n \in Y_{n+1}$ and $x_{n+1} \in \mu(U_n \cup Y_{n+1})$, so $rg(x_{n+1}) \leq rg(x_n)$. (Proposition 5.2.4 (p. 130)) □

5.2.1.4 The Logical Properties with Definability Preservation

First, a small fact about $\mathcal{A}$.

Fact 5.2.3 Let $\mathcal{A}$ be as above (and thus finite). Then each A_i is equivalent to a formula α_i.

Proof We use the standard topology and its compactness. By definition, each $M(A_i)$ is closed. By finiteness all unions of such $M(A_i)$ are closed too, so $\mathbf{C}(M(A_i))$ is closed. By compactness, each open cover $X_j : j \in J$ of the open $M(A_i)$ contains a finite subcover, so also $\bigcup\{M(A_j) : j \neq i\}$ has a finite open cover. [But the $M(\phi)$, ϕ a formula form a basis of the closed sets, so we are done.] □

Proposition 5.2.5 *Let $\mid\!\sim$ be a logic for $\mathcal{L}$. Set $T^{\mathcal{M}} := Th(\mu_{\mathcal{M}}(M(T)))$, where $\mathcal{M}$ is a preferential structure.*

(1) *Then there is a (transitive) definability preserving classical preferential model $\mathcal{M}$ s.t. $\overline{\overline{T}} = T^{\mathcal{M}}$ iff*
(LLE), (CCL), (SC), (PR) hold for all $T, T' \subseteq \mathcal{L}$.
(2) *The structure can be chosen smooth, iff, in addition (CUM) holds.*
(3) *The structure can be chosen $\mathcal{A}$-ranked, iff, in addition*
$(\mathcal{A}$-min$)$ $T \not\vdash \neg\alpha_i$ and $T \not\vdash \neg\alpha_j$, $i < j$ implies $\overline{\overline{T}} \vdash \neg\alpha_j$ holds.

The proof is an immediate consequence of Proposition 5.2.6 (p. 130) and the respective above results. This proposition (or its analogue) was mostly already shown in [Sch92] and [Sch96-1] and is repeated here for completeness' sake, but with a new and partly stronger proof.

Proposition 5.2.6 *Consider for a logic $\mid\!\sim$ on $\mathcal{L}$ the properties*
(LLE), (CCL), (SC), (PR), (CUM), $(\mathcal{A}$-min$)$ hold for all $T, T' \subseteq \mathcal{L}$,
and for a function $\mu : \mathbf{D}_{\mathcal{L}} \to \mathcal{P}(M_{\mathcal{L}})$ the properties
(μdp) μ is definability preserving, i.e. $\mu(M(T)) = M(T')$ for some T'

$(\mu \subseteq)$, (μPR), (μCUM), $(\mu\mathcal{A})$
for all $X, Y \in \boldsymbol{D}_{\mathcal{L}}$.
It then holds:

(a) *If* μ *satisfies* (μdp), $(\mu \subseteq)$, (μPR), *then* $\mid\!\sim$ *defined by* $\overline{\overline{T}} := T^{\mu} := Th(\mu(M(T)))$ *satisfies* (LLE), (CCL), (SC), (PR). *If* μ *satisfies in addition* (μCUM), *then (CUM) will hold too. If* μ *satisfies in addition* $(\mu\mathcal{A})$, *then* ($\mathcal{A}$*-min*) *will hold too.*

(b) *If* $\mid\!\sim$ *satisfies* (LLE), (CCL), (SC), (PR), *then there is* $\mu : \boldsymbol{D}_{\mathcal{L}} \rightarrow \mathcal{P}(M_{\mathcal{L}})$ *s.t.* $\overline{\overline{T}} = T^{\mu}$ *for all* $T \subseteq \mathcal{L}$ *and* μ *satisfies* (μdp), $(\mu \subseteq)$, (μPR). *If, in addition,* (CUM) *holds, then* (μCUM) *will hold too. If, in addition,* ($\mathcal{A}$*-min*) *holds, then* $(\mu\mathcal{A})$ *will hold too.*

Proof All properties except ($\mathcal{A}$-min) and $(\mu\mathcal{A})$ are shown in Proposition 2.3.1 (p. 45). But the remaining two are trivial. □

5.3 Two-Sequent Calculi

5.3.1 Introduction

This section serves mainly as a posteriori motivation for our examination of weak closure conditions of the domain. The second author realized first when looking at Lehmann's plausibility logic that absence of $(\cup)$ might be a problem for representation – see [Sch96-3] or [Sch04].

Beyond motivation, the reader will see here two "real-life" examples where closure under $(\cup)$ is not given, and thus problems arise. So this is also a warning against a too naive treatment of representation problems, neglecting domain closure issues.

5.3.2 Plausibility Logic

5.3.2.1 Discussion of Plausibility Logic

Plausibility logic was introduced by Lehmann [Leh92a], [Leh92b] as a sequent calculus in a propositional language without connectives. Thus, a plausibility logic language $\mathcal{L}$ is just a set, whose elements correspond to propositional variables, and a sequent has the form $X \mid\!\sim Y$, where X, Y are *finite* subsets of $\mathcal{L}$. Thus, in the intuitive reading, $\bigwedge X \mid\!\sim \bigvee Y$. (We use $\mid\!\sim$ instead of the $\vdash$ used in [Leh92a], [Leh92b] and continue to reserve $\vdash$ for classical logic.)

5.3.2.2 The Details

Notation 5.3.1 We abuse notation, and write $X \mid\!\sim a$ for $X \mid\!\sim \{a\}$, $X, a \mid\!\sim Y$ for $X \cup \{a\} \mid\!\sim Y$, $ab \mid\!\sim Y$ for $\{a, b\} \mid\!\sim Y$, etc. When discussing plausibility logic, X, Y, etc. will denote finite subsets of $\mathcal{L}$, a, b, etc., elements of $\mathcal{L}$.

We first define the logical properties that we will examine.

Definition 5.3.1 X and Y will be finite subsets of $\mathcal{L}$, a, etc., elements of $\mathcal{L}$. The base axiom and rules of plausibility logic are (we use the prefix "*Pl*" to differentiate them from the usual ones) the following:

(PlI) (Inclusion): $X \mid\sim a$ for all $a \in X$,

$(PlRM)$ (Right monotony): $X \mid\sim Y \Rightarrow X \mid\sim a, Y$,

$(PlCLM)$ (Cautious left monotony): $X \mid\sim a$, $X \mid\sim Y \Rightarrow X, a \mid\sim Y$,

$(PlCC)$ (Cautious cut): $X, a_1 \ldots a_n \mid\sim Y$, and for all $1 \leq i \leq n$, $X \mid\sim a_i, Y \Rightarrow X \mid\sim Y$, and as a special case of $(PlCC)$:

$(PlUCC)$ (Unit cautious cut): $X, a \mid\sim Y, X \mid\sim a, Y \Rightarrow X \mid\sim Y$.

We denote by PL, for plausibility logic, the full system, i.e. $(PlI) + (PlRM) + (PlCLM) + (PlCC)$.

We now adapt the definition of a preferential model to plausibility logic. This is the central definition on the semantic side.

Definition 5.3.2 Fix a plausibility logic language $\mathcal{L}$. A model for $\mathcal{L}$ is then just an arbitrary subset of $\mathcal{L}$.
If $\mathcal{M} := \langle M, \prec \rangle$ is a preferential model such that M is a set of (indexed) $\mathcal{L}$-models, then for a finite set $X \subseteq \mathcal{L}$ (to be imagined on the left hand side of $\mid\sim$!), we define

(a) $m \models X$ iff $X \subseteq m$,
(b) $M(X) := \{m\colon \langle m, i \rangle \in M$ for some i and $m \models X\}$,
(c) $\mu(X) := \{m \in M(X)\colon \exists \langle m, i \rangle \in M\ \neg\exists \langle m', i' \rangle \in M\ (m' \in M(X) \wedge \langle m', i' \rangle \prec \langle m, i \rangle)\}$,
(d) $X \models_{\mathcal{M}} Y$ iff $\forall m \in \mu(X), m \cap Y \neq \emptyset$.

(a) reflects the intuitive reading of X as $\bigwedge X$ and (d) that of Y as $\bigvee Y$ in $X \mid\sim Y$. Note that X is a set of "formulas" and $\mu(X) = \mu_{\mathcal{M}}(M(X))$.

We note as trivial consequences of the definition.

Fact 5.3.1

(a) $a \models_{\mathcal{M}} b$ iff for all $m \in \mu(a)$, $b \in m$,
(b) $X \models_{\mathcal{M}} Y$ iff $\mu(X) \subseteq \bigcup\{M(b) : b \in Y\}$,
(c) $m \in \mu(X) \wedge X \subseteq X' \wedge m \in M(X') \Rightarrow m \in \mu(X')$.

We note without proof: $(PlI) + (PlRM) + (PlCC)$ is complete (and sound) for preferential models – see [Sch96-3] or [Sch04] for a proof.

We note the following fact for smooth preferential models:

Fact 5.3.2 Let U, X, Y be any sets, $\mathcal{M}$ be smooth for at least $\{Y, X\}$, and let $\mu(Y) \subseteq U \cup X$, $\mu(X) \subseteq U$, then $X \cap Y \cap \mu(U) \subseteq \mu(Y)$. (This is, of course, a special case of $(\mu Cum1)$.)

Example 5.3.1 Let $\mathcal{L} := \{a, b, c, d, e, f\}$ and $\mathcal{X} := \{a \mid\sim b, b \mid\sim a, a \mid\sim c, a \mid\sim fd, dc \mid\sim ba, dc \mid\sim e, fcba \mid\sim e\}$. We show that $\mathcal{X}$ does not have a smooth representation.

Fact 5.3.3 $\mathcal{X}$ does not entail $a \mid\!\sim e$.

See [Sch96-3] or [Sch04] for a proof.

Suppose now that there is a smooth preferential model $\mathcal{M} = \langle M, \prec \rangle$ for plausibility logic which represents $\mid\!\sim$, i.e. for all X, Y finite subsets of $\mathcal{L}$, $X \mid\!\sim Y$ iff $X \models_{\mathcal{M}} Y$. (See Definition 5.3.2 (p. 132) and Fact 5.3.1 (p. 132).)

$a \mid\!\sim a$, $a \mid\!\sim b$, $a \mid\!\sim c$ implies for $m \in \mu(a)$ $a, b, c \in m$. Moreover, as $a \mid\!\sim df$, then also $d \in m$ or $f \in m$. As $a \not\mid\!\sim e$, there must be $m \in \mu(a)$ such that $e \notin m$. Suppose now $m \in \mu(a)$ with $f \in m$. So $a, b, c, f \in m$, thus by $m \in \mu(a)$ and Fact 5.3.1 (p. 132), $m \in \mu(a, b, c, f)$. But $fcba \mid\!\sim e$, so $e \in m$. We thus have shown that $m \in \mu(a)$ and $f \in m$ implies $e \in m$. Consequently, there must be $m \in \mu(a)$ such that $d \in m$, $e \notin m$. Thus, in particular, as $cd \mid\!\sim e$, there is $m \in \mu(a)$, $a, b, c, d \in m$, $m \notin \mu(cd)$. But by $cd \mid\!\sim ab$, and $b \mid\!\sim a$, $\mu(cd) \subseteq M(a) \cup M(b)$ and $\mu(b) \subseteq M(a)$ by Fact 5.3.1 (p. 132). Let now $T := M(cd)$, $R := M(a)$, $S := M(b)$, and $\mu_{\mathcal{M}}$ be the choice function of the minimal elements in the structure $\mathcal{M}$, we then have by $\mu(S) = \mu_{\mathcal{M}}(M(S))$:

1. $\mu_{\mathcal{M}}(T) \subseteq R \cup S$,
2. $\mu_{\mathcal{M}}(S) \subseteq R$,
3. there is $m \in S \cap T \cap \mu_{\mathcal{M}}(R)$, but $m \notin \mu_{\mathcal{M}}(T)$,

but this contradicts above Fact 5.3.2 (p. 132).

5.3.3 A Comment on the Work by Arieli and Avron

We turn to a similar case, published in [AA00]. Definitions are due to [AA00], and for motivation the reader is referred there.

Definition 5.3.3 (1) A Scott consequence relation, abbreviated scr, is a binary relation $\vdash$ between sets of formulas that satisfies the following conditions:

$(s-R)$ if $\Gamma \cap \Delta \neq \emptyset$, then $\Gamma \vdash \Delta$,

(M) if $\Gamma \vdash \Delta$ and $\Gamma \subseteq \Gamma'$, $\Delta \subseteq \Delta'$, then $\Gamma' \vdash \Delta'$,

(C) if $\Gamma \vdash \psi, \Delta$ and $\Gamma', \psi \vdash \Delta'$, then $\Gamma, \Gamma' \vdash \Delta, \Delta'$.

(2) A Scott cautious consequence relation, abbreviated sccr, is a binary relation $\mid\!\sim$ between nonempty sets of formulas that satisfies the following conditions:

$(s-R)$ if $\Gamma \cap \Delta \neq \emptyset$, then $\Gamma \mid\!\sim \Delta$,

(CM) if $\Gamma \mid\!\sim \Delta$ and $\Gamma \mid\!\sim \psi$, then $\Gamma, \psi \mid\!\sim \Delta$,

(CC) if $\Gamma \mid\!\sim \psi$ and $\Gamma, \psi \mid\!\sim \Delta$, then $\Gamma \mid\!\sim \Delta$.

Example 5.3.2 We have two consequence relations, $\vdash$ and $\mid\!\sim$.

The rules to consider are

$$LCC^n \frac{\Gamma \mid\!\sim \psi_1, \Delta \ldots \Gamma \mid\!\sim \psi_n, \Delta \Gamma, \psi_1, \ldots, \psi_n \mid\!\sim \Delta}{\Gamma \mid\!\sim \Delta},$$

$$RW^n \frac{\Gamma \mid\!\sim \psi_i, \Delta i = 1 \ldots n\Gamma, \psi_1, \ldots, \psi_n \vdash \phi}{\Gamma \mid\!\sim \phi, \Delta},$$
$$Cum\ \Gamma, \Delta \neq \emptyset, \Gamma \vdash \Delta \Rightarrow \Gamma \mid\!\sim \Delta,$$
$$RM\ \Gamma \mid\!\sim \Delta \Rightarrow \Gamma \mid\!\sim \psi, \Delta,$$
$$CM \frac{\Gamma \mid\!\sim \psi \Gamma \mid\!\sim \Delta}{\Gamma, \psi \mid\!\sim \Delta},$$
$$s\text{–}R\ \Gamma \cap \Delta \neq \emptyset \Rightarrow \Gamma \mid\!\sim \Delta,$$
$$M\ \Gamma \vdash \Delta, \Gamma \subseteq \Gamma', \Delta \subseteq \Delta' \Rightarrow \Gamma' \vdash \Delta',$$
$$C \frac{\Gamma_1 \vdash \psi, \Delta_1 \Gamma_2, \psi \vdash \Delta_2}{\Gamma_1, \Gamma_2 \vdash \Delta_1, \Delta_2}.$$

Let $\mathcal{L}$ be any set. Define now $\Gamma \vdash \Delta$ iff $\Gamma \cap \Delta \neq \emptyset$. Then $s-R$ and M for $\vdash$ are trivial. For C: If $\Gamma_1 \cap \Delta_1 \neq \emptyset$ or $\Gamma_1 \cap \Delta_1 \neq \emptyset$, the result is trivial. If not, $\psi \in \Gamma_1$ and $\psi \in \Delta_2$, which implies the result. So $\vdash$ is a scr.

Consider now the rules for a sccr which is $\vdash$-plausible for this $\vdash$. Cum is equivalent to $s-R$, which is essentially (PlI) of plausibility logic. Consider RW^n. If ϕ is one of the ψ_i, then the consequence $\Gamma \mid\!\sim \phi, \Delta$ is a case of one of the other hypotheses. If not, $\phi \in \Gamma$, so $\Gamma \mid\!\sim \phi$ by $s-R$, so $\Gamma \mid\!\sim \phi, \Delta$ by RM (if Δ is finite). So, for this $\vdash$, RW^n is a consequence of $s-R + RM$.

We are left with LCC^n, RM, CM, $s-R$. It was shown in [Sch04] and [Sch96-3] that this does not suffice to guarantee smooth representability, by failure of $(\mu Cum1)$.

5.4 Blurred Observation – Absence of Definability Preservation

5.4.1 Introduction

Lack of definability preservation results in uncertainty. We do not know exactly the result, but only that it is not too far away from what we (seem to) observe.

Thus, we pretend to know more than we know, and, according to our general policy of not neglecting ignorance, we should branch here into a multitude of solutions.

We take here a seemingly different way, but, as a matter of fact, we just describe the boundaries of what is permitted. So, everything which lies in those boundaries is a possible solution, and every such solution should be considered as equal, and, again, we should not pretend to know more than we actually do.

5.4.1.1 General Remarks, Affected Conditions

We assume now – unless explicitly stated otherwise – $\mathcal{Y} \subseteq \mathcal{P}(Z)$ to be closed under arbitrary intersections (this is used for the definition of $\widehat{.}$) and finite unions, and $\emptyset, Z \in \mathcal{Y}$. This holds, of course, for $\mathcal{Y} = \boldsymbol{D}_{\mathcal{L}}$, $\mathcal{L}$ any propositional language.

The aim of Sect. 5.4 (p. 134) is to present the results of [Sch04] connected to problems of definability preservation in a uniform way, stressing the crucial condition $\widehat{X} \cap \widehat{Y} = \widehat{X \cap Y}$. This presentation shall help and guide future research concerning similar problems.

For motivation, we first consider the problem with definability preservation for the rules

$$(PR)\overline{\overline{T \cup T'}} \subseteq \overline{\overline{\overline{\overline{T}} \cup T'}} \text{ and}$$

$$(\mid\!\sim=) T \vdash T', Con(\overline{\overline{T'}}, T) \Rightarrow \overline{\overline{T}} = \overline{\overline{\overline{\overline{T'}} \cup T}} \text{ holds,}$$

which are consequences of

$$(\mu PR) X \subseteq Y \Rightarrow \mu(Y) \cap X \subseteq \mu(X) \text{ or}$$

$$(\mu =) X \subseteq Y, \mu(Y) \cap X \neq \emptyset \Rightarrow \mu(Y) \cap X = \mu(X), \text{ respectively,}$$

and definability preservation.

Example 4.2.1 (p. 87) showed that in the general case without definability preservation, (PR) fails, and the following Example 5.4.1 (p. 135) shows that in the ranked case, $(\mid\!\sim=)$ may fail. So failure is not just a consequence of the very liberal definition of general preferential structures.

Example 5.4.1 Take $\{p_i : i \in \omega\}$ and put $m := m_{\bigwedge p_i}$, the model which makes all p_i true, in the top layer, and all the others in the bottom layer. Let $m' \neq m$, $T' := \emptyset$, $T := Th(m, m')$. Then $\overline{\overline{T'}} = T'$, so $Con(\overline{\overline{T'}}, T)$, $\overline{\overline{T}} = Th(m')$, $\overline{\overline{\overline{\overline{T'}} \cup T}} = T$.

We remind the reader of Definition 2.1.5 (p. 32) and Fact 2.1.1 (p. 32), partly taken from [Sch04].

We turn to the central condition.

5.4.1.2 The Central Condition

We analyse the problem of (PR), seen in Example 5.4.1 (p. 135) (1) above, working in the intended application.

(PR) is equivalent to $M(\overline{\overline{T}} \cup T') \subseteq M(\overline{\overline{\overline{\overline{T}} \cup T'}})$. To show (PR) from (μPR), we argue as follows, the crucial point is marked by "?":

$M(\overline{\overline{\overline{\overline{T}} \cup T'}}) = M(Th(\mu(M_{T \cup T'}))) = \widehat{\mu(M_{T \cup T'})} \supseteq \mu(M_{T \cup T'}) = \mu(M_T \cap M_{T'}) \supseteq$ (by (μPR)) $\mu(M_T) \cap M_{T'} ? \widehat{\mu(M_T)} \cap M_{T'} = M(Th(\mu(M_T))) \cap M_{T'} = M(\overline{\overline{T}}) \cap M_{T'}$ $= M(\overline{\overline{T}} \cup T')$. If μ is definability preserving, then $\mu(M_T) = \widehat{\mu(M_T)}$, so "?" above is equality, and everything is fine. In general, however, we have only $\mu(M_T) \subseteq \widehat{\mu(M_T)}$, and the argument collapses.

But it is not necessary to impose $\mu(M_T) = \widehat{\mu(M_T)}$, as we still have room to move: $\widehat{\mu(M_{T \cup T'})} \supseteq \mu(M_{T \cup T'})$. (We do not consider here $\mu(M_T \cap M_{T'}) \supseteq \mu(M_T) \cap M_{T'}$ as room to move, as we are now interested only in questions related to definability preservation.) If we had $\widehat{\mu(M_T)} \cap M_{T'} \subseteq \widehat{\mu(M_T) \cap M_{T'}}$, we could use $\mu(M_T) \cap M_{T'} \subseteq \mu(M_T \cap M_{T'}) = \mu(M_{T \cup T'})$ and monotony of $\widehat{\cdot}$ to obtain $\widehat{\mu(M_T)} \cap M_{T'} \subseteq$

$\widehat{\mu(M_T) \cap M_{T'}} \subseteq \widehat{\mu(M_T \cap M_{T'})} = \widehat{\mu(M_{T \cup T'})}$. If, for instance, $T' = \{\psi\}$, we have $\widehat{\mu(M_T)} \cap M_{T'} = \widehat{\mu(M_T) \cap M_{T'}}$ by Fact 2.1.1 (p. 32) $(Cl \cap +)$. Thus, definability preservation is not the only solution to the problem.

We have seen in Fact 2.1.1 (p. 32) that $\widehat{X \cup Y} = \widehat{X} \cup \widehat{Y}$; moreover $X - Y = X \cap \mathbf{C}Y$ ($\mathbf{C}Y$ the set complement of Y), so, when considering Boolean expressions of model sets (as we do in usual properties describing logics), the central question is whether

$(\sim \cap)$ $\widehat{X \cap Y} = \widehat{X} \cap \widehat{Y}$ holds.

We take a closer look at this question.

$\widehat{X \cap Y} \subseteq \widehat{X} \cap \widehat{Y}$ holds by Fact 2.1.1 (p. 32) (6). Using $(Cl\cup)$ and monotony of $\widehat{\cdot}$, we have $\widehat{X} \cap \widehat{Y} = \widehat{((X \cap Y) \cup (X - Y))} \cap \widehat{((X \cap Y) \cup (Y - X))} = ((\widehat{X \cap Y}) \cup (\widehat{X - Y})) \cap ((\widehat{X \cap Y}) \cup (\widehat{Y - X})) = \widehat{X \cap Y} \cup (\widehat{X - Y} \cap \widehat{Y - X})$, thus $\widehat{X} \cap \widehat{Y} \subseteq \widehat{X \cap Y}$ iff

$(\sim \cap')$ $\widehat{Y - X} \cap \widehat{X - Y} \subseteq \widehat{X \cap Y}$ holds.

Intuitively speaking, the condition holds iff we cannot approximate any element both from $Y - X$ and $[X - Y]$, which cannot be approximated from $X \cap Y$, too.

Note that in the above Example 5.4.1 (p. 135) (1) $X := \mu(M_T) = M_{\mathcal{L}} - \{n'\}$, $Y := M_{T'} = \{n, n'\}$, $\widehat{X - Y} = M_{\mathcal{L}}$, $\widehat{Y - X} = \{n'\}$, $\widehat{X \cap Y} = \{n\}$, and $\widehat{X} \cap \widehat{Y} = \{n, n'\}$.

We consider now particular cases:

(1) If $X \cap Y = \emptyset$, then by $\emptyset \in \mathcal{Y}$, $(\sim \cap)$ holds iff $\widehat{X} \cap \widehat{Y} = \emptyset$.

(2) If $X \in \mathcal{Y}$ and $Y \in \mathcal{Y}$, then $\widehat{X - Y} \subseteq X$ and $\widehat{Y - X} \subseteq Y$, so $\widehat{X - Y} \cap \widehat{Y - X} \subseteq X \cap Y \subseteq \widehat{X \cap Y}$ and $(\sim \cap)$ trivially holds.

(3) $X \in \mathcal{Y}$ and $\mathbf{C}X \in \mathcal{Y}$ together also suffice – in these cases $\widehat{Y - X} \cap \widehat{X - Y} = \emptyset$: $\widehat{Y - X} = \widehat{Y \cap \mathbf{C}X} \subseteq \mathbf{C}X$, and $\widehat{X - Y} \subseteq X$, so $\widehat{Y - X} \cap \widehat{X - Y} \subseteq X \cap \mathbf{C}X = \emptyset \subseteq \widehat{X \cap Y}$. (The same holds, of course, for Y.) (In the intended application, such X will be $M(\phi)$ for some formula ϕ. But, a warning, $\mu(M(\phi))$ need not again be the $M(\psi)$ for some ψ.)

We turn to the properties of various structures and apply our results.

5.4.1.3 Application to Various Structures

We now take a look at other frequently used logical conditions. First, in the context on nonmonotonic logics, the following rules will always hold in smooth preferential structures, even if we consider full theories, and not necessarily definability preserving structures:

Fact 5.4.1 Also for full theories, and not necessarily, definability preserving structures hold:

(1) (LLE), (RW), (AND), (REF), by definition and $(\mu \subseteq)$,
(2) (OR),
(3) (CM) in smooth structures,
(4) the infinitary version of (CUM) in smooth structures. In definability preserving structures, but also when considering only formulas, the following two conditions hold:
(5) (PR),
(6) $(\mid\!\sim=)$ in ranked structures.

Proof We use the corresponding algebraic properties. The result then follows from Proposition 2.3.1 (p. 45). □

We turn to theory revision. The following definition and example, taken from [Sch04], show that the usual AGM axioms for theory revision fail in distance-based structures in the general case, unless we require definability preservation. See Sect. 8.2 (p. 227) for discussion and motivation.

Definition 5.4.1 We summarize the AGM postulates $(K*7)$ and $(K*8)$ in $(*4)$:
$(*4)$ If $T*T'$ is consistent with T'', then $T*(T' \cup T'') = \overline{(T*T') \cup T''}$.

Example 5.4.2 Consider an infinite propositional language $\mathcal{L}$.

Let X be an infinite set of models, and m, m_1, m_2 be models for $\mathcal{L}$. Arrange the models of $\mathcal{L}$ in the real plane such that all $x \in X$ have the same distance < 2 (in the real plane) from m, m_2 has distance 2 from m, and m_1 has distance 3 from m.

Let T, T_1, T_2 be complete (consistent) theories, T' a theory with infinitely many models, $M(T) = \{m\}$, $M(T_1) = \{m_1\}$, $M(T_2) = \{m_2\}$. $M(T') = X \cup \{m_1, m_2\}$, $M(T'') = \{m_1, m_2\}$.

Assume $Th(X) = T'$, so X will not be definable by a theory. Then $M(T) \mid M(T') = X$, but $T*T' = Th(X) = T'$. So $T*T'$ is consistent with T'', and $\overline{(T*T') \cup T''} = T''$. But $T' \cup T'' = T''$, and $T*(T' \cup T'') = T_2 \neq T''$, contradicting $(*4)$.

We now show that the version with formulas only holds here, too, just as does above (PR), when we consider formulas only – this is needed below for T'' only. This was already shown in [Sch04]; we now give a proof based on our new principles.

Fact 5.4.2 $(*4)$ holds when considering only formulas.

Proof When we fix the left hand side, the structure is ranked, so $Con(T*T', T'')$ implies $(M_T \mid M_{T'}) \cap M_{T''} \neq \emptyset$ by $T'' = \{\psi\}$ and thus $M_T \mid \overbrace{M_{T' \cup T''}} = \underbrace{M_T \mid (M_{T'} \cap M_{T''}) = (M_T \mid M_{T'}) \cap M_{T''}}$. So $M(T*(T' \cup T'')) = \overline{M_T \mid M_{T' \cup T''}} = \overbrace{(M_T \mid M_{T'}) \cap M_{T''}} =$ (by $T'' = \{\psi\}$, see above) $\overline{\underbrace{(M_T \mid M_{T'})} \cap \underbrace{M_{T''}}} = \overline{(M_T \mid M_{T'})} \cap M_{T''} = M((T*T') \cup T'')$, and $T*(T' \cup T'') = \overline{(T*T') \cup T''}$. □

5.4.2 General and Smooth Structures Without Definability Preservation

5.4.2.1 Introduction

Note that in Sects. 3.2 and 3.3 of [Sch04], as well as in Proposition 4.2.2 of [Sch04] we have characterized $\mu : \mathcal{Y} \to \mathcal{Y}$ or $\mid : \mathcal{Y} \times \mathcal{Y} \to \mathcal{Y}$, but a closer inspection of the proofs shows that the destination can as well be assumed $\mathcal{P}(Z)$, consequently we can simply re use above algebraic representation results also for the not definability preserving case. (Note that the easy direction of all these results work for destination $\mathcal{P}(Z)$, too.) In particular, also the proof for the not definability preserving case of revision in [Sch04] can be simplified – but we will not go into details here.

($\cup$) and ($\bigcap$) are again assumed to hold now – we need ($\bigcap$) for $\overbrace{\cdot}$.

The central functions and conditions to consider are summarized in the following definition.

Definition 5.4.2 Let $\mu : \mathcal{Y} \to \mathcal{Y}$; we define $\mu_i : \mathcal{Y} \to \mathcal{P}(Z)$:

$\mu_0(U) := \{x \in U : \neg\exists Y \in \mathcal{Y}(Y \subseteq U \text{ and } x \in Y - \mu(Y))\}$,

$\mu_1(U) := \{x \in U : \neg\exists Y \in \mathcal{Y}(\mu(Y) \subseteq U \text{ and } x \in Y - \mu(Y))\}$,

$\mu_2(U) := \{x \in U : \neg\exists Y \in \mathcal{Y}(\mu(U \cup Y) \subseteq U \text{ and } x \in Y - \mu(Y))\}$,

(note that we use ($\cup$) here),

$\mu_3(U) := \{x \in U : \forall y \in U, x \in \mu(\{x, y\})\}$,

(we use here ($\cup$) and that singletons are in $\mathcal{Y}$).

"Small" is now in the sense of Definition 2.1.5 (p. 32).

$(\mu PR0)$ $\mu(U) - \mu_0(U)$ is small,

$(\mu PR1)$ $\mu(U) - \mu_1(U)$ is small,

$(\mu PR2)$ $\mu(U) - \mu_2(U)$ is small,

$(\mu PR3)$ $\mu(U) - \mu_3(U)$ is small.

$(\mu PR0)$ with its function will be the one to consider for general preferential structures, and $(\mu PR2)$ the one for smooth structures.

A Nontrivial Problem

Unfortunately, we cannot use $(\mu PR0)$ in the smooth case, too, as Example 5.4.4 (p. 141) below will show. This sheds some doubt on the possibility to find an easy common approach to all cases of not definability preserving preferential, and perhaps other, structures. The next best guess, $(\mu PR1)$ will not work either, as the same example shows – or by Fact 5.4.3 (p. 140) (10), if μ satisfies (μCum), then $\mu_0(U) = \mu_1(U)$. $(\mu PR3)$ and μ_3 are used for ranked structures.

We will now see that this first impression of a difficult situation is indeed well founded. First, note that in our context, μ will not necessarily respect (μPR). Thus, if, e.g. $x \in Y - \mu(Y)$, and $\mu(Y) \subseteq U$, we cannot necessarily conclude that $x \notin \mu(U \cup Y)$ – the fact that x is minimized in $U \cup Y$ might be hidden by the bigger $\mu(U \cup Y)$.

Consequently, we may have to work with small sets (Y in the case of μ_2 above) to see the problematic elements – recall that the smaller the set $\mu(X)$ is, the less it can "hide" missing elements – but will need bigger sets ($U \cup Y$ in above example) to recognize the contradiction.

Second, "problematic" elements are those involved in a contradiction, i.e. contradicting the representation conditions. Now, a negation of a conjunction is a disjunction of negations. So, generally, we will have to look at various possibilities of violated conditions. But the general situation is much worse, still.

Example 5.4.3 Look at the ranked case and assume no closure properties of the domain. Recall that we might be unable to see $\mu(X)$, but see only $\widehat{\mu(X)}$. Suppose we have $\widehat{\mu(X_1)} \cap (X_2 - \widehat{\mu(X_2)}) \neq \emptyset$, $\widehat{\mu(X_2)} \cap (X_3 - \widehat{\mu(X_3)}) \neq \emptyset$, $\widehat{\mu(X_{n-1})} \cap (X_n - \widehat{\mu(X_n)}) \neq \emptyset$, $\widehat{\mu(X_n)} \cap (X_1 - \widehat{\mu(X_1)}) \neq \emptyset$, which seems to be a contradiction. (It is only a real contradiction if it still holds without the closures.) But, we do not know where the contradiction is situated. It might well be that for all but one i really $\mu(X_i) \cap (X_{i+1} - \mu(X_{i+1})) \neq \emptyset$, and not only that for the closure $\widehat{\mu(X_i)}$ of $\mu(X_i)$ $\widehat{\mu(X_i)} \cap (X_{i+1} - \widehat{\mu(X_{i+1})}) \neq \emptyset$, but we might be unable to find this out. So we have to branch into all possibilities, i.e. for one, or several i, $\widehat{\mu(X_i)} \cap (X_{i+1} - \widehat{\mu(X_{i+1})}) \neq \emptyset$, but $\mu(X_i) \cap (X_{i+1} - \mu(X_{i+1})) = \emptyset$.

The situation might even be worse, when those $\widehat{\mu(X_i)} \cap (X_{i+1} - \widehat{\mu(X_{i+1})}) \neq \emptyset$ are involved in several cycles, etc. Consequently, it seems very difficult to describe all possible violations in one concise condition, and thus we will examine here only some specific cases, and do not pretend that they are the only ones, that other cases are similar, or that our solutions (which depend on closure conditions) are the best ones.

Outline of Our Solutions in Some Particular Cases

The strategy of representation without definability preservation will in all cases be very simple: Under sufficient conditions, among them smallness (μPRi) as described above, the corresponding function μ_i has all the properties to guarantee representation by a corresponding structures, and we can just take our representation theorems for the dp case to show this. Using smallness again, we can show that we have obtained a sufficient approximation – see Proposition 5.4.1 (p. 142), Proposition 5.4.2 (p. 143), and Proposition 5.4.3 (p. 145).

We first show some properties for the μ_i, $i = 0, 1, 2$. A corresponding result for μ_3 is given in Fact 5.4.5 (p. 144) below. (The conditions and results are sufficiently different for μ_3 to make a separation more natural.)

Property (9) of the following Fact 5.4.3 (p. 140) fails for μ_0 and μ_1, as Example 5.4.4 (p. 141) below will show. We will therefore work in the smooth case with μ_2.

5.4.2.2 Results

Fact 5.4.3 (This is partly Fact 5.2.6 in [Sch04].)

Recall that $\mathcal{Y}$ is closed under $(\cup)$, and $\mu : \mathcal{Y} \to \mathcal{Y}$. Let A, B, U, U', X, Y be elements of $\mathcal{Y}$ and the μ_i be defined from μ as in Definition 5.4.2 (p. 138). i will here be 0, 1, or 2, but not 3.

(1) Let μ satisfy $(\mu \subseteq)$, then $\mu_1(X) \subseteq \mu_0(X)$ and $\mu_2(X) \subseteq \mu_0(X)$,

(2) Let μ satisfy $(\mu \subseteq)$ and (μCum), then $\mu(U \cup U') \subseteq U \Leftrightarrow \mu(U \cup U') = \mu(U)$,

(3) Let μ satisfy $(\mu \subseteq)$, then $\mu_i(U) \subseteq \mu(U)$ and $\mu_i(U) \subseteq U$,

(4) Let μ satisfy $(\mu \subseteq)$ and one of the (μPRi), then $\mu(A \cup B) \subseteq \mu(A) \cup \mu(B)$,

(5) Let μ satisfy $(\mu \subseteq)$ and one of the (μPRi), then $\mu_2(X) \subseteq \mu_1(X)$,

(6) Let μ satisfy $(\mu \subseteq)$, (μPRi), then $\mu_i(U) \subseteq U' \Leftrightarrow \mu(U) \subseteq U'$,

(7) Let μ satisfy $(\mu \subseteq)$ and one of the (μPRi), then $X \subseteq Y$, $\mu(X \cup U) \subseteq X \Rightarrow \mu(Y \cup U) \subseteq Y$,

(8) Let μ satisfy $(\mu \subseteq)$ and one of the (μPRi), then $X \subseteq Y \Rightarrow X \cap \mu_i(Y) \subseteq \mu_i(X)$ – so (μPR) holds for μ_i, (more precisely, only for μ_2 we need the prerequisites, in the other cases the definition suffices),

(9) Let μ satisfy $(\mu \subseteq)$, $(\mu PR2)$, (μCum), then $\mu_2(X) \subseteq Y \subseteq X \Rightarrow \mu_2(X) = \mu_2(Y)$ – so (μCum) holds for μ_2.

(10) $(\mu \subseteq)$ and (μCum) for μ entail $\mu_0(U) = \mu_1(U)$.

Proof

(1) $\mu_1(X) \subseteq \mu_0(X)$ follows from $(\mu \subseteq)$ for μ. For μ_2: By $Y \subseteq U$, $U \cup Y = U$, so $\mu(U) \subseteq U$ by $(\mu \subseteq)$.

(2) $\mu(U \cup U') \subseteq U \subseteq U \cup U' \Rightarrow_{(\mu CUM)} \mu(U \cup U') = \mu(U)$.

(3) $\mu_i(U) \subseteq U$ by definition. To show $\mu_i(U) \subseteq \mu(U)$, take in all three cases $Y := U$, and use for $i = 1, 2$ $(\mu \subseteq)$.

(4) By definition of μ_0, we have $\mu_0(A \cup B) \subseteq A \cup B$, $\mu_0(A \cup B) \cap (A - \mu(A)) = \emptyset$, $\mu_0(A \cup B) \cap (B - \mu(B)) = \emptyset$, so $\mu_0(A \cup B) \cap A \subseteq \mu(A)$, $\mu_0(A \cup B) \cap B \subseteq \mu(B)$, and $\mu_0(A \cup B) \subseteq \mu(A) \cup \mu(B)$. By $\mu : \mathcal{Y} \to \mathcal{Y}$ and $(\cup)$, $\mu(A) \cup \mu(B) \in \mathcal{Y}$. Moreover, by (3) $\mu_0(A \cup B) \subseteq \mu(A \cup B)$, so $\mu_0(A \cup B) \subseteq (\mu(A) \cup \mu(B)) \cap \mu(A \cup B)$, so by (1) $\mu_i(A \cup B) \subseteq (\mu(A) \cup \mu(B)) \cap \mu(A \cup B)$ for $i = 0, \ldots, 2$. If $\mu(A \cup B) \not\subseteq \mu(A) \cup \mu(B)$, then $(\mu(A) \cup \mu(B)) \cap \mu(A \cup B) \subset \mu(A \cup B)$, contradicting (μPRi).

(5) Let $Y \in \mathcal{Y}$, $\mu(Y) \subseteq U$, $x \in Y - \mu(Y)$, then (by (4)) $\mu(U \cup Y) \subseteq \mu(U) \cup \mu(Y) \subseteq U$.

(6) "$\Leftarrow$" by (3). "$\Rightarrow$": By (μPRi), $\mu(U) - \mu_i(U)$ is small, so there is no $X \in \mathcal{Y}$ such that $\mu_i(U) \subseteq X \subset \mu(U)$. If there were $U' \in \mathcal{Y}$ such that $\mu_i(U) \subseteq U'$, but $\mu(U) \not\subseteq U'$, then for $X := U' \cap \mu(U) \in \mathcal{Y}$, $\mu_i(U) \subseteq X \subset \mu(U)$, *contradiction.*

(7) $\mu(Y \cup U) = \mu(Y \cup X \cup U) \subseteq_{(4)} \mu(Y) \cup \mu(X \cup U) \subseteq Y \cup X = Y$.

(8) For $i = 0, 1$: Let $x \in X - \mu_0(X)$, then there is A such that $A \subseteq X$, $x \in A - \mu(A)$, so $A \subseteq Y$. The case $i = 1$ is similar. We need here only the definitions. For $i = 2$: Let $x \in X - \mu_2(X)$, A such that $x \in A - \mu(A)$, $\mu(X \cup A) \subseteq X$, then by (7) $\mu(Y \cup A) \subseteq Y$.

(9) "$\subseteq$": Let $x \in \mu_2(X)$, so $x \in Y$, and $x \in \mu_2(Y)$ by (8). "$\supseteq$": Let $x \in \mu_2(Y)$, so $x \in X$. Suppose $x \notin \mu_2(X)$, so there is $U \in \mathcal{Y}$ such that $x \in U - \mu(U)$ and $\mu(X \cup U) \subseteq X$. Note that by $\mu(X \cup U) \subseteq X$ and (2), $\mu(X \cup U) = \mu(X)$. Now, $\mu_2(X) \subseteq Y$, so by (6) $\mu(X) \subseteq Y$, thus $\mu(X \cup U) = \mu(X) \subseteq Y \subseteq Y \cup U \subseteq X \cup U$, so $\mu(Y \cup U) = \mu(X \cup U) = \mu(X) \subseteq Y$ by (μCum), so $x \notin \mu_2(Y)$, *contradiction*.

(10) $\mu_1(U) \subseteq \mu_0(U)$ by (1). Let Y such that $\mu(Y) \subseteq U$, $x \in Y - \mu(Y)$, $x \in U$. Consider $Y \cap U$, $x \in Y \cap U$, $\mu(Y) \subseteq Y \cap U \subseteq Y$, so $\mu(Y) = \mu(Y \cap U)$ by (μCum), and $x \notin \mu(Y \cap U)$. Thus, $\mu_0(U) \subseteq \mu_1(U)$. □

Fact 5.4.4 In the presence of $(\mu \subseteq)$, (μCum) for μ, we have: $(\mu PR0) \Leftrightarrow (\mu PR1)$ and $(\mu PR2) \Rightarrow (\mu PR1)$.

If (μPR) also holds for μ, then so will $(\mu PR1) \Rightarrow (\mu PR2)$.

(Recall that $(\cup)$ and $(\cap)$ are assumed to hold.)

Proof

$(\mu PR0) \Leftrightarrow (\mu PR1)$: By Fact 5.4.3 (p. 140), (10), $\mu_0(U) = \mu_1(U)$ if (μCum) holds for μ.

$(\mu PR2) \Rightarrow (\mu PR1)$: Suppose $(\mu PR2)$ holds. By $(\mu PR2)$ and (5), $\mu_2(U) \subseteq \mu_1(U)$, so $\mu(U) - \mu_1(U) \subseteq \mu(U) - \mu_2(U)$. By $(\mu PR2)$, $\mu(U) - \mu_2(U)$ is small, then so is $\mu(U) - \mu_1(U)$, so $(\mu PR1)$ holds.

$(\mu PR1) \Rightarrow (\mu PR2)$: Suppose $(\mu PR1)$ holds and $(\mu PR2)$ fails. By failure of $(\mu PR2)$, there is $X \in \mathcal{Y}$ such that $\mu_2(U) \subseteq X \subset \mu(U)$. Let $x \in \mu(U) - X$, as $x \notin \mu_2(U)$, there is Y such that $\mu(U \cup Y) \subseteq U$, $x \in Y - \mu(Y)$. Let $Z := U \cup Y \cup X$. By (μPR), $x \notin \mu(U \cup Y)$ and $x \notin \mu(U \cup Y \cup X)$. Moreover, $\mu(U \cup X \cup Y) \subseteq \mu(U \cup Y) \cup \mu(X)$ by Fact 5.4.3 (p. 140) (4), $\mu(U \cup Y) \subseteq U$, $\mu(X) \subseteq X \subseteq \mu(U) \subseteq U$ by prerequisite, so $\mu(U \cup X \cup Y) \subseteq U \subseteq U \cup Y \subseteq U \cup X \cup Y$, so $\mu(U \cup X \cup Y) = \mu(U \cup Y) \subseteq U$. Thus, $x \notin \mu_1(U)$, and $\mu_1(U) \subseteq X$, too, a contradiction. □

Here is an example which shows that Fact 5.4.3 (p. 140), (9) may fail for μ_0 and μ_1.

Example 5.4.4 Consider $\mathcal{L}$ with $v(\mathcal{L}) := \{p_i : i \in \omega\}$. Let $m \not\models p_0$; let $m' \in M(p_0)$ be arbitrary. Make for each $n \in M(p_0) - \{m'\}$ one copy of m, likewise of m', set $\langle m, n\rangle \prec \langle m', n\rangle$ for all n, and $n \prec \langle m, n\rangle$, $n \prec \langle m', n\rangle$ for all n. The resulting structure $\mathcal{Z}$ is smooth and transitive.

Let $\mathcal{Y} := \boldsymbol{D}_{\mathcal{L}}$, define $\mu(X) := \widehat{\mu_{\mathcal{Z}}(X)}$ for $X \in \mathcal{Y}$.

Let $m' \in X - \mu_{\mathcal{Z}}(X)$. Then $m \in X$ or $M(p_0) \subseteq X$. In the latter case, as all m'' such that $m'' \neq m'$, $m'' \models p_0$ are minimal, $M(p_0) - \{m'\} \subseteq \mu_{\mathcal{Z}}(X)$, so $m' \in \widehat{\mu_{\mathcal{Z}}(X)} = \mu(X)$. Thus, as $\mu_{\mathcal{Z}}(X) \subseteq \mu(X)$, if $m' \in X - \mu(X)$, then $m \in X$.

Define now $X := M(p_0) \cup \{m\}$, $Y := M(p_0)$.

We first show that μ_0 does not satisfy (μCum). $\mu_0(X) := \{x \in X : \neg\exists A \in \mathcal{Y}(A \subseteq X : x \in A - \mu(A))\}$. $m \notin \mu_0(X)$, as $m \notin \mu(X) = \widehat{\mu_{\mathcal{Z}}(X)}$. Moreover, $m' \notin \mu_0(X)$, as $\{m, m'\} \in \mathcal{Y}$, $\{m, m'\} \subseteq X$, and $\mu(\{m, m'\}) = \mu_{\mathcal{Z}}(\{m, m'\}) = \{m\}$. So $\mu_0(X) \subseteq Y \subseteq X$. Consider now $\mu_0(Y)$. As $m \notin Y$, for any $A \in \mathcal{Y}$, $A \subseteq Y$, if

$m' \in A$, then $m' \in \mu(A)$, too, by above argument, so $m' \in \mu_0(Y)$, and μ_0 does not satisfy (μCum).

We turn to μ_1. By Fact 5.4.3 (p. 140) (1), $\mu_1(X) \subseteq \mu_0(X)$, so $m, m' \notin \mu_1(X)$, and again $\mu_1(X) \subseteq Y \subseteq X$. Consider again $\mu_1(Y)$. As $m \notin Y$, for any $A \in \mathcal{Y}$, $\mu(A) \subseteq Y$, if $m' \in A$, then $m' \in \mu(A)$, too: If $M(p_0) - \{m'\} \subseteq A$, then $m' \in \overbrace{\mu_{\mathcal{Z}}(A)}$; if $M(p_0) - \{m'\} \not\subseteq A$, but $m' \in A$, then either $m' \in \mu_{\mathcal{Z}}(A)$ or $m \in \mu_{\mathcal{Z}}(A) \subseteq \mu(A)$, but $m \notin Y$. Thus, (μCum) fails for μ_1, too.

It remains to show that μ satisfies $(\mu \subseteq)$, (μCum), $(\mu PR0)$, $(\mu PR1)$. Note that by Fact 4.2.10 (p. 106) (3) and Proposition 4.2.4 (p. 110) $\mu_{\mathcal{Z}}$ satisfies (μCum), as $\mathcal{Z}$ is smooth. $(\mu \subseteq)$ is trivial. We show (μPRi) for $i = 0, 1$. As $\mu_{\mathcal{Z}}(A) \subseteq \mu(A)$, by (μPR) and (μCum) for $\mu_{\mathcal{Z}}$, $\mu_{\mathcal{Z}}(X) \subseteq \mu_0(X)$ and $\mu_{\mathcal{Z}}(X) \subseteq \mu_1(X)$: To see this, we note $\mu_{\mathcal{Z}}(X) \subseteq \mu_0(X)$. Let $x \in X - \mu_0(X)$, then there is Y such that $x \in Y - \mu(Y)$. $Y \subseteq X$, but $\mu_{\mathcal{Z}}(Y) \subseteq \mu(Y)$, so by $Y \subseteq X$ and (μPR) for $\mu_{\mathcal{Z}}$ $x \notin \mu_{\mathcal{Z}}(X)$. $\mu_{\mathcal{Z}}(X) \subseteq \mu_1(X)$: Let $x \in X - \mu_1(X)$, then there is Y such that $x \in Y - \mu(Y)$, $\mu(Y) \subseteq X$, so $x \in Y - \mu_{\mathcal{Z}}(Y)$ and $\mu_{\mathcal{Z}}(Y) \subseteq X$. $\mu_{\mathcal{Z}}(X \cup Y) \subseteq \mu_{\mathcal{Z}}(X) \cup \mu_{\mathcal{Z}}(Y) \subseteq X \subseteq X \cup Y$, so $\mu_{\mathcal{Z}}(X \cup Y) = \mu_{\mathcal{Z}}(X)$ by (μCum) for $\mu_{\mathcal{Z}}$. $x \in Y - \mu_{\mathcal{Z}}(Y) \Rightarrow x \notin \mu_{\mathcal{Z}}(X \cup Y)$ by (μPR) for $\mu_{\mathcal{Z}}$, so $x \notin \mu_{\mathcal{Z}}(X)$.

But by Fact 5.4.3 (p. 140), (3) $\mu_i(X) \subseteq \mu(X)$. As by definition, $\mu(X) - \mu_{\mathcal{Z}}(X)$ is small, (μPRi) hold for $i = 0, 1$. It remains to show (μCum) for μ. Let $\mu(X) \subseteq Y \subseteq X$, then $\mu_{\mathcal{Z}}(X) \subseteq \mu(X) \subseteq Y \subseteq X$, so by (μCum) for $\mu_{\mathcal{Z}}$ $\mu_{\mathcal{Z}}(X) = \mu_{\mathcal{Z}}(Y)$, so by definition of μ, $\mu(X) = \mu(Y)$.

(Note that by Fact 5.4.3 (page 140) (10), $\mu_0 = \mu_1$ follows from (μCum) for μ, so we could have demonstrated part of the properties also differently.)

By Fact 5.4.3 (p. 140), (3) and (8) and Proposition 4.2.2 (p. 95), μ_0 has a representation by a (transitive) preferential structure, if $\mu : \mathcal{Y} \to \mathcal{Y}$ satisfies $(\mu \subseteq)$ and $(\mu PR0)$, and μ_0 is defined as in Definition 5.4.2 (p. 138).

We thus have (taken from [Sch04], Proposition 5.2.5 there):

Proposition 5.4.1 *Let Z be an arbitrary set, $\mathcal{Y} \subseteq \mathcal{P}(Z)$, $\mu : \mathcal{Y} \to \mathcal{Y}$, $\mathcal{Y}$ closed under arbitrary intersections and finite unions, and $\emptyset, Z \in \mathcal{Y}$, and let $\overbrace{.}$ be defined with respect to $\mathcal{Y}$.*

(a) *If μ satisfies $(\mu \subseteq)$, $(\mu PR0)$, then there is a transitive preferential structure $\mathcal{Z}$ over Z such that for all $U \in \mathcal{Y}$, $\mu(U) = \overbrace{\mu_{\mathcal{Z}}(U)}$.*

(b) *If $\mathcal{Z}$ is a preferential structure over Z and $\mu : \mathcal{Y} \to \mathcal{Y}$ such that for all $U \in \mathcal{Y}$, $\mu(U) = \overbrace{\mu_{\mathcal{Z}}(U)}$, then μ satisfies $(\mu \subseteq)$, $(\mu PR0)$.*

Proof (a) Let μ satisfy $(\mu \subseteq)$, $(\mu PR0)$. μ_0 as defined in Definition 5.4.2 (p. 138) satisfies properties $(\mu \subseteq)$, (μPR) by Fact 5.4.3 (p. 140), (3) and (8). Thus, by Proposition 4.2.2 (p. 95), there is a transitive structure $\mathcal{Z}$ over Z such that $\mu_0 = \mu_{\mathcal{Z}}$, but by $(\mu PR0)$ $\mu(U) = \overbrace{\mu_0(U)} = \overbrace{\mu_{\mathcal{Z}}(U)}$ for $U \in \mathcal{Y}$.

(b) $(\mu \subseteq) : \mu_{\mathcal{Z}}(U) \subseteq U$, so by $U \in \mathcal{Y}$ $\mu(U) = \overbrace{\mu_{\mathcal{Z}}(U)} \subseteq U$.

$(\mu PR0)$: If $(\mu PR0)$ is false, there is $U \in \mathcal{Y}$ such that for $U' := \bigcup\{Y' - \mu(Y') : Y' \in \mathcal{Y}, Y' \subseteq U\}$ $\overbrace{\mu(U) - U'} \subset \mu(U)$. By $\mu_{\mathcal{Z}}(Y') \subseteq \mu(Y')$, $Y' - \mu(Y') \subseteq Y' -$

$\mu_{\mathcal{Z}}(Y')$. No copy of any $x \in Y' - \mu_{\mathcal{Z}}(Y')$ with $Y' \subseteq U$, $Y' \in \mathcal{Y}$ can be minimal in $\mathcal{Z}\lceil U$. Thus, by $\mu_{\mathcal{Z}}(U) \subseteq \mu(U)$, $\mu_{\mathcal{Z}}(U) \subseteq \mu(U) - U'$, so $\overbrace{\mu_{\mathcal{Z}}(U)} \subseteq \overbrace{\mu(U) - U'} \subset \mu(U)$, *contradiction*. □

We turn to the smooth case.

If $\mu : \mathcal{Y} \to \mathcal{Y}$ satisfies $(\mu \subseteq)$, $(\mu PR2)$, (μCUM) and μ_2 is defined from μ as in Definition 5.4.2 (p. 138), then μ_2 satisfies $(\mu \subseteq)$, (μPR), (μCum) by Fact 5.4.3 (p. 140) (3), (8), and (9), and can thus be represented by a (transitive) smooth structure, by Proposition 5.1.1 (p. 120), and we finally have (taken from [Sch04], Proposition 5.2.9 there):

Proposition 5.4.2 *Let Z be an arbitrary set, $\mathcal{Y} \subseteq \mathcal{P}(Z)$, $\mu : \mathcal{Y} \to \mathcal{Y}$, $\mathcal{Y}$ closed under arbitrary intersections and finite unions, and $\emptyset, Z \in \mathcal{Y}$, and let $\overbrace{\ .\ }$ be defined with respect to $\mathcal{Y}$.*

(a) *If μ satisfies $(\mu \subseteq)$, $(\mu PR2)$, (μCUM), then there is a transitive smooth preferential structure $\mathcal{Z}$ over Z such that for all $U \in \mathcal{Y}$, $\mu(U) = \overbrace{\mu_{\mathcal{Z}}(U)}$.*

(b) *If $\mathcal{Z}$ is a smooth preferential structure over Z and $\mu : \mathcal{Y} \to \mathcal{Y}$ such that for all $U \in \mathcal{Y}$, $\mu(U) = \overbrace{\mu_{\mathcal{Z}}(U)}$, then μ satisfies $(\mu \subseteq)$, $(\mu PR2)$, (μCUM).*

Proof (a) If μ satisfies $(\mu \subseteq)$, $(\mu PR2)$, (μCUM), then μ_2 defined from μ as in Definition 5.4.2 (p. 138) satisfies $(\mu \subseteq)$, (μPR), (μCUM) by Fact 5.4.3 (p. 140) (3), (8), and (9). Thus, by Proposition 5.1.1 (p. 120), there is a smooth transitive preferential structure $\mathcal{Z}$ over Z such that $\mu_2 = \mu_{\mathcal{Z}}$, but by $(\mu PR2)$ $\mu(U) = \overbrace{\mu_2(U)} = \overbrace{\mu_{\mathcal{Z}}(U)}$.

(b) $(\mu \subseteq)$: $\mu_{\mathcal{Z}}(U) \subseteq U \Rightarrow \mu(U) = \overbrace{\mu_{\mathcal{Z}}(U)} \subseteq U$ by $U \in \mathcal{Y}$.

$(\mu PR2)$: If $(\mu PR2)$ fails, then there is $U \in \mathcal{Y}$ such that for $U' := \bigcup\{Y' - \mu(Y') : Y' \in \mathcal{Y}, \mu(U \cup Y') \subseteq U\}$ $\overbrace{\mu(U) - U'} \subset \mu(U)$.

By $\mu_{\mathcal{Z}}(Y') \subseteq \mu(Y')$, $Y' - \mu(Y') \subseteq Y' - \mu_{\mathcal{Z}}(Y')$. But no copy of any $x \in Y' - \mu_{\mathcal{Z}}(Y')$ with $\mu_{\mathcal{Z}}(U \cup Y') \subseteq \mu(U \cup Y') \subseteq U$ can be minimal in $\mathcal{Z}\lceil U$: As $x \in Y' - \mu_{\mathcal{Z}}(Y')$, if $\langle x, i\rangle$ is any copy of x, then there is $\langle y, j\rangle \prec \langle x, i\rangle$, $y \in Y'$. Consider now $U \cup Y'$. As $\langle x, i\rangle$ is not minimal in $\mathcal{Z}\lceil U \cup Y'$, by smoothness of $\mathcal{Z}$ there must be $\langle z, k\rangle \prec \langle x, i\rangle$, $\langle z, k\rangle$ minimal in $\mathcal{Z}\lceil U \cup Y'$. But all minimal elements of $\mathcal{Z}\lceil U \cup Y'$ must be in $\mathcal{Z}\lceil U$, so there must be $\langle z, k\rangle \prec \langle x, i\rangle$, $z \in U$, thus $\langle x, i\rangle$ is not minimal in $\mathcal{Z}\lceil U$. Thus by $\mu_{\mathcal{Z}}(U) \subseteq \mu(U)$, $\mu_{\mathcal{Z}}(U) \subseteq \mu(U) - U'$, so $\overbrace{\mu_{\mathcal{Z}}(U)} \subseteq \overbrace{\mu(U) - U'} \subset \mu(U)$, *contradiction*. (μCUM): Let $\mu(X) \subseteq Y \subseteq X$. Now $\mu_{\mathcal{Z}}(X) \subseteq \overbrace{\mu_{\mathcal{Z}}(X)} = \mu(X)$, so by smoothness of $\mathcal{Z}$ $\mu_{\mathcal{Z}}(Y) = \mu_{\mathcal{Z}}(X)$, thus $\mu(X) = \overbrace{\mu_{\mathcal{Z}}(X)} = \overbrace{\mu_{\mathcal{Z}}(Y)} = \mu(Y)$. □

5.4.3 Ranked Structures

We recall from [Sch04] and Sect. 5.1.3 (p. 123) above the basic properties of ranked structures. We give now an easy version of representation results for ranked structures without definability preservation.

Notation 5.4.1 We abbreviate $\mu(\{x, y\})$ by $\mu(x, y)$ etc.

Fact 5.4.5 Let the domain contain singletons and be closed under $(\cup)$.
Let for $\mu : \mathcal{Y} \to \mathcal{Y}$ the following properties hold:
$(\mu =)$ for finite sets, $(\mu \in)$, $(\mu PR3)$, $(\mu\emptyset fin)$.
Then the following properties hold for μ_3 as defined in Definition 5.4.2 (p. 138):

(1) $\mu_3(X) \subseteq \mu(X)$,
(2) for finite X, $\mu(X) = \mu_3(X)$,
(3) $(\mu \subseteq)$,
(4) (μPR),
(5) $(\mu\emptyset fin)$,
(6) $(\mu =)$,
(7) $(\mu \in)$,
(8) $\mu(X) = \widehat{\mu_3(X)}$.

Proof (1) Suppose not, so $x \in \mu_3(X)$, $x \in X - \mu(X)$, so by $(\mu \in)$ for μ, there is $y \in X$, $x \notin \mu(x, y)$, $contradiction$.
(2) By $(\mu PR3)$ for μ and (1), for finite U $\mu(U) = \mu_3(U)$.
(3) $(\mu \subseteq)$ is trivial for μ_3.
(4) Let $X \subseteq Y$, $x \in \mu_3(Y) \cap X$, suppose $x \in X - \mu_3(X)$, so there is $y \in X \subseteq Y$, $x \notin \mu(x, y)$, so $x \notin \mu_3(Y)$.
(5) $(\mu\emptyset fin)$ for μ_3 follows from $(\mu\emptyset fin)$ for μ and (2).
(6) Let $X \subseteq Y$, $y \in \mu_3(Y) \cap X$, $x \in \mu_3(X)$, we have to show $x \in \mu_3(Y)$. By (4), $y \in \mu_3(X)$. Suppose $x \notin \mu_3(Y)$. So there is $z \in Y$, $x \notin \mu(x, z)$. As $y \in \mu_3(Y)$, $y \in \mu(y, z)$. As $x \in \mu_3(X)$, $x \in \mu(x, y)$, as $y \in \mu_3(X)$, $y \in \mu(x, y)$. Consider $\{x, y, z\}$. Suppose $y \notin \mu(x, y, z)$, then by $(\mu \in)$ for μ, $y \notin \mu(x, y)$ or $y \notin \mu(y, z)$, $contradiction$. Thus $y \in \mu(x, y, z) \cap \mu(x, y)$. As $x \in \mu(x, y)$, and $(\mu =)$ for μ and finite sets, $x \in \mu(x, y, z)$. Recall that $x \notin \mu(x, z)$. But for finite sets $\mu = \mu_3$, and by (4) (μPR) holds for μ_3, so it holds for μ and finite sets, $contradiction$.
(7) Let $x \in X - \mu_3(X)$, so there is $y \in X$ s.t. $x \notin \mu(x, y) = \mu_3(x, y)$.
(8) As $\mu(X) \in \mathcal{Y}$, and $\mu_3(X) \subseteq \mu(X)$, $\widehat{\mu_3(X)} \subseteq \mu(X)$, so by $(\mu PR3)$ $\widehat{\mu_3(X)} = \mu(X)$. □

Fact 5.4.6 If $\mathcal{Z}$ is ranked, and we define $\mu(X) := \widehat{\mu_{\mathcal{Z}}(X)}$, and $\mathcal{Z}$ has no copies, then the following hold:

(1) $\mu_{\mathcal{Z}}(X) = \{x \in X : \forall y \in X, x \in \mu(x, y)\}$, so $\mu_{\mathcal{Z}}(X) = \mu_3(X)$ for $X \in \mathcal{Y}$,
(2) $\mu(X) = \mu_{\mathcal{Z}}(X)$ for finite X,
(3) $(\mu =)$ for finite sets for μ,
(4) $(\mu \in)$ for μ,
(5) $(\mu\emptyset fin)$ for μ,
(6) $(\mu PR3)$ for μ.

Proof (1) holds for ranked structures.
(2) and (6) are trivial. (3) and (5) hold for $\mu_{\mathcal{Z}}$, so by (2) for μ.

(4) If $x \notin \mu(X)$, then $x \notin \mu_{\mathcal{Z}}(X)$, $(\mu \in)$ holds for $\mu_{\mathcal{Z}}$, so there is $y \in X$ such that $x \notin \mu_{\mathcal{Z}}(x, y) = \mu(x, y)$ by (2). □

We summarize:

Proposition 5.4.3 *Let Z be an arbitrary set,* $\mathcal{Y} \subseteq \mathcal{P}(Z)$, $\mu : \mathcal{Y} \to \mathcal{Y}$, $\mathcal{Y}$ *closed under arbitrary intersections and finite unions, contain singletons, and* $\emptyset, Z \in \mathcal{Y}$, *and let* $\overbrace{.}$ *be defined with respect to* $\mathcal{Y}$.

(a) *If* μ *satisfies* $(\mu =)$ *for finite sets,* $(\mu \in)$, $(\mu PR3)$, $(\mu\emptyset fin)$, *then there is a ranked preferential structure* $\mathcal{Z}$ *without copies over Z such that for all* $U \in \mathcal{Y}$ $\mu(U) = \overbrace{\mu_{\mathcal{Z}}(U)}$.

(b) *If* $\mathcal{Z}$ *is a ranked preferential structure over Z without copies and* $\mu : \mathcal{Y} \to \mathcal{Y}$ *such that for all* $U \in \mathcal{Y}$ $\mu(U) = \overbrace{\mu_{\mathcal{Z}}(U)}$, *then* μ *satisfies* $(\mu =)$ *for finite sets,* $(\mu \in)$, $(\mu PR3)$, $(\mu\emptyset fin)$.

Proof (a) Let μ satisfy $(\mu =)$ for finite sets, $(\mu \in)$, $(\mu PR3)$, $(\mu\emptyset fin)$, then μ_3 as defined in Definition 5.4.2 (p. 138) satisfies properties $(\mu \subseteq)$, $(\mu\emptyset fin)$, $(\mu =)$, $(\mu \in)$ by Fact 5.4.5 (p. 144). Thus, by Proposition 5.1.3 (p. 124), there is a transitive structure $\mathcal{Z}$ over Z such that $\mu_3 = \mu_{\mathcal{Z}}$, but by Fact 5.4.5 (p. 144) (8) $\mu(U) = \overbrace{\mu_3(U)} = \overbrace{\mu_{\mathcal{Z}}(U)}$ for $U \in \mathcal{Y}$.

(b) This was shown in Fact 5.4.6 (p. 144). □

5.5 The Limit Variant

5.5.1 Introduction

Distance-based semantics give perhaps the clearest motivation for the limit variant. For instance, the Stalnaker/Lewis semantics for counterfactual conditionals define $\phi > \psi$ to hold in a (classical) model m iff in those models of ϕ, which are closest to m, ψ holds. For this to make sense, we need, of course, a distance d on the model set. We call this approach the minimal variant. Usually, one makes a limit assumption: The set of ϕ-models closest to m is not empty if ϕ is consistent – i.e. the ϕ-models are not arranged around m in a way that they come closer and closer, without a minimal distance. This is, of course, a very strong assumption, and which is probably difficult to justify philosophically. It seems to have its only justification in the fact that it avoids degenerate cases, where, in above example, for consistent ϕ $m \models \phi > FALSE$ holds. As such, this assumption is unsatisfactory.

Our aim here is to analyse the limit version more closely, in particular, to see criteria whether the much more complex limit version can be reduced to the simpler minimal variant. In the limit version, roughly, ψ is a consequence of ϕ, if ψ holds "in the limit" in all ϕ-models. That is, iff, "going sufficiently far down", ψ will become and stay true.

The problem is not simple, as there are two sides which come into play, and sometimes we need both to cooperate to achieve a satisfactory translation. The first component is what we call the "algebraic limit", i.e. we stipulate that the limit version should have properties which correspond to the algebraic properties of the minimal variant. An exact correspondence cannot always be achieved, and we give a translation which seems reasonable. But once the translation is done, even if it is exact, there might still be problems linked to translation to logic.

(1) The structural limit: It is a natural and much more convincing solution to the problem described above to modify the basic definition, and work without the rather artificial assumption that the closest world exists. We adopt what we call a "limit approach", and define $m \models \phi > \psi$ iff there is a distance d' such that for all $m' \models \phi$ and $d(m, m') \leq d'$ $m' \models \psi$. Thus, from a certain point onward, ψ becomes and stays true. We will call this definition the structural limit, as it is based directly on the structure (the distance on the model set).
(2) The algebraic limit: The model sets to consider are spheres around m, $S := \{m' \in M(\phi) : d(m, m') \leq d'\}$ for some d', such that $S \neq \emptyset$. The system of such S is nested, i.e. totally ordered by inclusion; and if $m \models \phi$, it has a smallest element $\{m\}$, etc. When we forget the underlying structure, and consider just the properties of these systems of spheres around different m, and for different ϕ, we obtain what we call the algebraic limit.
(3) The logical limit: The logical limit speaks about the logical properties which hold "in the limit", i.e. finally in all such sphere systems.

The interest to investigate this algebraic limit is twofold: First, we shall see (for other kinds of structures) that there are reasonable and not so reasonable algebraic limits. Second, this distinction permits us to separate algebraic from logical problems, which have to do with definability of model sets, in short definability problems. We will see that we find common definability problems and also common solutions in the usual minimal and limit variant.

In particular, the decomposition into three layers on both sides (minimal and limit version) can reveal that a (seemingly) natural notion of structural limit results in algebraic properties which have not much to do anymore with the minimal variant. So, to speak about a limit variant, we will demand that this variant is not only a natural structural limit, but results in a natural abstract limit, too. Conversely, if the algebraic limit preserves the properties of the minimal variant, there is hope that it preserves the logical properties, too – not more than hope, however, due to definability problems.

We now give the basic definitions for preferential and ranked preferential structures.

Definition 5.5.1

(1) General preferential structures:

(1.1) The version without copies:

Let $\mathcal{M} := \langle U, \prec \rangle$. Define

$Y \subseteq X \subseteq U$ Y is a minimizing initial segment, or MISE, of X iff

(a) $\forall x \in X, \exists x \in Y$ s.t. $y \preceq x$ – where $y \preceq x$ stands for $x \prec y$ or $x = y$ (i.e. Y is minimizing) and

(b) $\forall y \in Y, \forall x \in X\ (x \prec y \Rightarrow x \in Y)$ (i.e. Y is downward closed or an initial part).

(1.2) The version with copies:

Let $\mathcal{M} := \langle \mathcal{U}, \prec \rangle$ be as above. Define for $Y \subseteq X \subseteq \mathcal{U}$, Y is a minimizing initial segment, or MISE, of X iff

(a) $\forall \langle x, i \rangle \in X\ \exists \langle y, j \rangle \in Y$ s.t. $\langle y, j \rangle \preceq \langle x, i \rangle$

and

(b) $\forall \langle y, j \rangle \in Y, \forall \langle x, i \rangle \in X\ (\langle x, i \rangle \prec \langle y, j \rangle \Rightarrow \langle x, i \rangle \in Y)$.

(1.3) For $X \subseteq \mathcal{U}$, let $\Lambda(X)$ be the set of MISE of X.

(1.4) We say that a set $\mathcal{X}$ of MISE is cofinal in another set of MISE $\mathcal{X}'$ (for the same base set X) iff for all $Y' \in \mathcal{X}'$, there is $Y \in \mathcal{X}$, $Y \subseteq Y'$.

(1.5) A MISE X is called definable iff $\{x : \exists i, \langle x, i \rangle \in X\} \in \boldsymbol{D}_{\mathcal{L}}$.

(1.6) $T \models_{\mathcal{M}} \phi$ iff there is $Y \in \Lambda(\mathcal{U} \lceil M(T))$ such that $Y \models \phi$.

($\mathcal{U} \lceil M(T) := \{\langle x, i \rangle \in \mathcal{U} : x \in M(T)\}$ – if there are no copies, we simplify in the obvious way.)

(2) Ranked preferential structures:

In the case of ranked structures, we may assume without loss of generality that the MISE sets have a particularly simple form:

For $X \subseteq U$, $A \subseteq X$ is MISE iff $X \neq \emptyset$ and $\forall a \in A\ \forall x \in X\ (x \prec a \vee x \perp a \Rightarrow x \in A)$. ($A$ is downward and horizontally closed.)

(3) Theory revision:

Recall that we have a distance d on the model set, and are interested in $y \in Y$ which are close to X.

Thus, given X, Y, we define analogously:

$B \subseteq Y$ is MISE iff

(1) $B \neq \emptyset$

(2) there is d' such that $B := \{y \in Y : \exists x \in X, d(x, y) \leq d'\}$ (we could also have chosen $d(x, y) < d'$, this is not important).

And we define $\phi \in T * T'$ iff there is $B \in \Lambda(M(T), M(T'))$ $B \models \phi$.

5.5.2 The Algebraic Limit

There are basic problems with the algebraic limit in general preferential structures.

Example 5.5.1 Let $a \prec b$, $a \prec c$, $b \prec d$, $c \prec d$ (but $\prec$ not transitive!), then $\{a, b\}$ and $\{a, c\}$ are such S and S', but there is no $S'' \subseteq S \cap S'$ which is an initial segment. If, for instance, in a and b ψ holds, in a and c ψ', then "in the limit" ψ and ψ' will hold, but not $\psi \wedge \psi'$. This does not seem right. We should not be obliged to give up ψ to obtain ψ'.

When we look at the system of such S generated by a preferential structure and its algebraic properties, we will therefore require it to be closed under finite intersections, or at least, that if S, S' are such segments, then there must be $S'' \subseteq S \cap S'$ which is also such a segment.

We make this official. Let $\Lambda(X)$ be the set of initial segments of X, then we require:

$(\Lambda\cap)$ If $A, B \in \Lambda(X)$ then there is $C \subseteq A \cap B$, $C \in \Lambda(X)$. More precisely, a limit should be a structural limit in a reasonable sense – whatever the underlying structure is – and the resulting algebraic limit should respect $(\Lambda\cap)$.

We should not demand too much, either. It would be wrong to demand closure under arbitrary intersections, as this would mean that there is an initial segment which makes all consequences true – trivializing the very idea of a limit. But we can make our requirements more precise, and bind the limit variant closely to the minimal variant, by looking at the algebraic version of both.

Before we look at deeper problems, we show some basic facts about the algebraic limit.

Fact 5.5.1 (Taken from [Sch04], Fact 3.4.3, Proposition 3.10.16 there.)

Let the relation $\prec$ be transitive. The following hold in the limit variant of general preferential structures:

(1) If $A \in \Lambda(Y)$, and $A \subseteq X \subseteq Y$, then $A \in \Lambda(X)$.

(2) If $A \in \Lambda(Y)$, and $A \subseteq X \subseteq Y$, and $B \in \Lambda(X)$, then $A \cap B \in \Lambda(Y)$.

(3) If $A \in \Lambda(Y)$, $B \in \Lambda(X)$, then there is $Z \subseteq A \cup B$, $Z \in \Lambda(Y \cup X)$.

The following hold in the limit variant of ranked structures without copies, where the domain is closed under finite unions and contains all finite sets.

(4) $A, B \in \Lambda(X) \Rightarrow A \subseteq B$ or $B \subseteq A$,

(5) $A \in \Lambda(X)$, $Y \subseteq X$, $Y \cap A \neq \emptyset \Rightarrow Y \cap A \in \Lambda(Y)$,

(6) $\Lambda' \subseteq \Lambda(X)$, $\bigcap \Lambda' \neq \emptyset \Rightarrow \bigcap \Lambda' \in \Lambda(X)$,

(7) $X \subseteq Y$, $A \in \Lambda(X) \Rightarrow \exists B \in \Lambda(Y)$, $B \cap X = A$.

Proof (1) Trivial.

(2.1) $A \cap B$ is closed in Y: Let $\langle x, i\rangle \in A \cap B$, $\langle y, j\rangle \prec \langle x, i\rangle$, then $\langle y, j\rangle \in A$. If $\langle y, j\rangle \notin X$, then $\langle y, j\rangle \notin A$, *contradiction*. So $\langle y, j\rangle \in X$, but then $\langle y, j\rangle \in B$.

(2.2) $A \cap B$ minimizes Y: Let $\langle a, i\rangle \in Y$.

(a) If $\langle a, i\rangle \in A - B \subseteq X$, then there is $\langle y, j\rangle \prec \langle a, i\rangle$, $\langle y, j\rangle \in B$. By closure of A, $\langle y, j\rangle \in A$.

(b) If $\langle a, i\rangle \notin A$, then there is $\langle a', i'\rangle \in A \subseteq X$, $\langle a', i'\rangle \prec \langle a, i\rangle$, continue by (a).

(3) Let $Z := \{\langle x, i\rangle \in A: \neg\exists\langle b, j\rangle \preceq \langle x, i\rangle.\langle b, j\rangle \in X - B\} \cup \{\langle y, j\rangle \in B: \neg\exists\langle a, i\rangle \preceq \langle y, j\rangle.\langle a, i\rangle \in Y - A\}$, where $\preceq$ stands for $\prec$ or $=$.

(3.1) Z minimizes $Y \cup X$: We consider Y, X is symmetrical.

(a) We first show: If $\langle a, k\rangle \in A - Z$, then there is $\langle y, i\rangle \in Z$, $\langle a, k\rangle \succ \langle y, i\rangle$. Proof: If $\langle a, k\rangle \in A - Z$, then there is $\langle b, j\rangle \preceq \langle a, k\rangle$, $\langle b, j\rangle \in X - B$. Then there is $\langle y, i\rangle \prec \langle b, j\rangle$, $\langle y, i\rangle \in B$. But $\langle y, i\rangle \in Z$ too: If not, there would be $\langle a', k'\rangle \preceq \langle y, i\rangle$, $\langle a', k'\rangle \in Y - A$, but $\langle a', k'\rangle \prec \langle a, k\rangle$, contradicting closure of A.

(b) If $\langle a'', k''\rangle \in Y - A$, there is $\langle a, k\rangle \in A$, $\langle a, k\rangle \prec \langle a'', k''\rangle$. If $\langle a, k\rangle \notin Z$, continue with (a).

(3.2) Z is closed in $Y \cup X$: Let then $\langle z, i\rangle \in Z$, $\langle u, k\rangle \prec \langle z, i\rangle$, $\langle u, k\rangle \in Y \cup X$. Suppose $\langle z, i\rangle \in A$ – the case $\langle z, i\rangle \in B$ is symmetrical.

(a) $\langle u, k\rangle \in Y - A$ cannot be, by closure of A.

(b) $\langle u, k\rangle \in X - B$ cannot be, as $\langle z, i\rangle \in Z$, and by definition of Z.

(c) If $\langle u,k\rangle \in A - Z$, then there is $\langle v,l\rangle \preceq \langle u,k\rangle$, $\langle v,l\rangle \in X - B$, so $\langle v,l\rangle \prec \langle z,i\rangle$, contradicting (b).

(d) If $\langle u,k\rangle \in B - Z$, then there is $\langle v,l\rangle \preceq \langle u,k\rangle$, $\langle v,l\rangle \in Y - A$, contradicting (a).

(4) Suppose not, so there are $a \in A - B$, $b \in B - A$. But if $a \perp b$, $a \in B$ and $b \in A$, similarly if $a \prec b$ or $b \prec a$.

(5) As $A \in \Lambda(X)$ and $Y \subseteq X$, $Y \cap A$ is downward and horizontally closed. As $Y \cap A \neq \emptyset$, $Y \cap A$ minimizes Y.

(6) $\bigcap \Lambda'$ is downward and horizontally closed, as all $A \in \Lambda'$ are. As $\bigcap \Lambda' \neq \emptyset$, $\bigcap \Lambda'$ minimizes X.

(7) Set $B := \{b \in Y : \exists a \in A . a \perp b \text{ or } b \leq a\}$ □

We have as immediate logical consequence the following fact:

Fact 5.5.2 (Fact 3.4.4 of [Sch04].)

If $\prec$ is transitive, then in the limit variant the following properties hold:

(1) (AND),

(2) (OR).

Proof Let $\mathcal{Z}$ be the structure.

(1) Immediate by Fact 5.5.1 (p. 148), (2) – set $A = B$.

(2) Immediate by Fact 5.5.1 (p. 148). □

5.5.3 The Logical Limit

5.5.3.1 Translation Between the Minimal and the Limit Variant

A good example for problems linked to the translation from the algebraic limit to the logical limit is the property $(\mu =)$ of ranked structures:

$$(\mu =) X \subseteq Y, \mu(Y) \cap X \neq \emptyset \Rightarrow \mu(Y) \cap X = \mu(X)$$

or its logical form

$$(\mid\sim =) T \vdash T', Con(\overline{\overline{T'}}, T) \Rightarrow \overline{\overline{T}} = \overline{\overline{\overline{T'}} \cup T}.$$

$\mu(Y)$ or its analogue $\overline{\overline{T'}}$ (set $X := M(T)$, $Y := M(T')$) speak about the limit, the "ideal", and this, of course, is not what we have in the limit version. This limit version was introduced precisely to avoid speaking about the ideal.

So, first, we have to translate $\mu(Y) \cap X \neq \emptyset$ to something else, and the natural candidate seems to be

$$\forall B \in \Lambda(Y). B \cap X \neq \emptyset.$$

In logical terms, we have replaced the set of consequences of Y by some $Th(B)$ where $T' \subseteq Th(B) \subseteq \overline{\overline{T'}}$. The conclusion can now be translated in a similar way to

$\forall B \in \Lambda(Y)\, \exists A \in \Lambda(X)\; A \subseteq B \cap X$ and $\forall A \in \Lambda(X)\, \exists B \in \Lambda(Y)$ s.t. $B \cap X \subseteq A$.
The total translation reads now:

($\Lambda =$) Let $X \subseteq Y$. Then

$$\Big(\forall B \in \Lambda(Y), B \cap X \neq \emptyset\Big) \Rightarrow \Big(\forall B \in \Lambda(Y)\, \exists A \in \Lambda(X)$$

$$A \subseteq B \cap X \text{ and } \forall A \in \Lambda(X)\, \exists B \in \Lambda(Y) \text{ s.t. } B \cap X \subseteq A\Big).$$

By Fact 5.5.1 (p. 148), (5) and (7), we see that this holds in ranked structures. Thus, the limit reading seems to provide a correct algebraic limit.

Yet, Example 5.5.2 (p. 150) below shows the following:

Let $m' \neq m$ be arbitrary. For $T' := Th(\{m, m'\})$, $T := \emptyset$, we have $T' \vdash T$, $\overline{\overline{\overline{T'}}} = Th(\{m'\})$, $\overline{\overline{\overline{T}}} = Th(\{m\})$, $Con(\overline{\overline{T}}, T')$, but $Th(\{m\}) = \overline{\overline{\overline{T \cup T'}}} \neq \overline{\overline{\overline{T'}}}$.

Thus:

(1) The prerequisite holds, though usually for $A \in \Lambda(T)$, $A \cap M(T') = \emptyset$.
(2) (PR) fails, which is independent of the prerequisite $Con(\overline{\overline{T}}, T')$, so the problem is not just due to the prerequisite.
(3) Both inclusions of ($\mid\!\sim=$) fail.

We will see below in Corollary 5.5.1 (p. 154) a sufficient condition to make ($\mid\!\sim=$) hold in ranked structures. It has to do with definability or formulas, more precisely, the crucial property is to have sufficiently often $\widehat{A} \cap \widehat{M(T')} = \widehat{A \cap M(T')}$ for $A \in \Lambda(T)$ – see Sect. 5.4.1 (p. 134) for reference.

Example 5.5.2 (Taken from [Sch04], Example 3.10.1 (1) there.)

Take an infinite propositional language $p_i : i \in \omega$. We have ω_1 models (assume for simplicity CH).

Take the model m which makes all p_i true, and put it on top. Next, going down, take all models which make p_0 false, and then all models which make p_0 true, but p_1 false, etc. in a ranked construction. So, successively more p_i will become (and stay) true. Consequently, $\emptyset \models_\Lambda p_i$ for all i. But the structure has no minimum, and the "logical" limit m is not in the setwise limit. Let $T := \emptyset$ and $m' \neq m$, $T' := Th(\{m, m'\})$, then $\overline{\overline{\overline{T}}} = Th(\{m\})$, $\overline{\overline{\overline{T'}}} = Th(\{m'\})$, and $\overline{\overline{\overline{T' \cup T}}} = \overline{\overline{\overline{T'}}} = Th(\{m'\})$ and $\overline{\overline{\overline{T \cup T'}}} = \overline{\overline{\overline{T}}} = Th(\{m\})$.

This example shows that our translation is not perfect, but it is half the way. Note that the minimal variant faces the same problems (definability and others), so the problems are probably at least not totally due to our perhaps insufficient translation.

We turn to other rules.

$$(\Lambda\cap) If\, A, B \in \Lambda(X), \text{ then there is } C \subseteq A \cap B, C \in \Lambda(X),$$

which seems a minimal requirement for an appropriate limit. It holds in transitive structures by Fact 5.5.1 (p. 148) (2).

The central logical condition for minimal smooth structures is

$$(CUM) T \subseteq T' \subseteq \overline{\overline{T}} \Rightarrow \overline{\overline{T}} = \overline{\overline{T'}}.$$

It would again be wrong – using the limit – to translate this only partly by: If $T \subseteq T' \subseteq \overline{\overline{T}}$, then for all $A \in \Lambda(M(T))$ there is $B \in \Lambda(M(T'))$ such that $A \subseteq B$ – and vice versa. Now, smoothness is in itself a wrong condition for limit structures, as it speaks about minimal elements, which we will not necessarily have. This cannot guide us. But when we consider a more modest version of cumulativity, we see what to do.

$$(CUMfin) \text{ If } T \mid\!\sim \phi, \text{ then } \overline{\overline{T}} = \overline{\overline{T \cup \{\phi\}}}.$$

This translates into algebraic limit conditions as follows – where $Y = M(T)$ and $X = M(T \cup \{\phi\})$:

$(\Lambda CUMfin)$: Let $X \subseteq Y$. If there is $B \in \Lambda(Y)$ such that $B \subseteq X$, then $\left(\forall A \in \Lambda(X)\, \exists B' \in \Lambda(Y), B' \subseteq A \text{ and } \forall B' \in \Lambda(Y)\, \exists A \in \Lambda(X), A \subseteq B'\right)$.

Note that, in this version, we do not have the "ideal" limit on the left of the implication, but one fixed approximation $B \in \Lambda(Y)$. We can now prove that $(\Lambda CUMfin)$ holds in transitive structures: The first part holds by Fact 5.5.1 (p. 148) (2), the second, as $B \cap B' \in \Lambda(Y)$ by Fact 5.5.1 (p. 148) (1). This is true without additional properties of the structure, which might at first sight seems surprising. But note that the initial segments play a similar role as the set of minimal elements: an initial segment has to minimize the other elements, just as the set of minimal elements in the smooth case does.

The central algebraic property of minimal preferential structures is

$$(\mu PR) X \subseteq Y \Rightarrow \mu(Y) \cap X \subseteq \mu(X).$$

This translates naturally and directly to,

$(\Lambda PR) X \subseteq Y \Rightarrow \forall A \in \Lambda(X)\, \exists B \in \Lambda(Y), B \cap X \subseteq A.$

(ΛPR) holds in transitive structures : $Y - X \in \Lambda(Y - X)$,

so the result holds by Fact 5.5.1 (p. 148) (3).

The central algebraic condition of ranked minimal structures is

$$(\mu =) X \subseteq Y, \mu(Y) \cap X \neq \emptyset \Rightarrow \mu(Y) \cap X = \mu(X).$$

We saw above how to translate this condition to $(\Lambda =)$, we also saw that $(\Lambda =)$ holds in ranked structures.

We will see in Corollary 5.5.1 (p. 154) that the following logical version holds in ranked structures:

$$T \not\vdash \neg\gamma implies \overline{\overline{T}} = \overline{\overline{T \cup \{\gamma\}}}$$

We generalize above translation results to a recipe:
Translate

(1) $\mu(X) \subseteq \mu(Y)$ to $\forall B \in \Lambda(Y)\ \exists A \in \Lambda(X),\ A \subseteq B$, and thus
(2) $\mu(Y) \cap X \subseteq \mu(X)$ to $\forall A \in \Lambda(X)\ \exists B \in \Lambda(Y),\ B \cap X \subseteq A$,
(3) $\mu(X) \subseteq Y$ to $\exists A \in \Lambda(X),\ A \subseteq Y$, and thus
(4) $\mu(Y) \cap X \neq \emptyset$ to $\forall B \in \Lambda(Y),\ B \cap X \neq \emptyset$
(5) $X \subseteq \mu(Y)$ to $\forall B \in \Lambda(Y),\ X \subseteq B$,
and quantify expressions separately, thus we repeat:
(6) (μCUM) $\mu(Y) \subseteq X \subseteq Y \Rightarrow \mu(X) = \mu(Y)$ translates to
(7) $(\Lambda CUMfin)$ Let $X \subseteq Y$. If there is $B \in \Lambda(Y)$ such that $B \subseteq X$, then: $\Big(\forall A \in \Lambda(X)\ \exists B' \in \Lambda(Y),\ B' \subseteq A$ and $\forall B' \in \Lambda(Y)\ \exists A \in \Lambda(X),\ A \subseteq B'\Big)$.
(8) $(\mu =)$ $X \subseteq Y,\ \mu(Y) \cap X \neq \emptyset \Rightarrow \mu(Y) \cap X = \mu(X)$ translates to
(9) $(\Lambda =)$ Let $X \subseteq Y$. If $\forall B \in \Lambda(Y),\ B \cap X \neq \emptyset$, then $\Big(\forall A \in \Lambda(X)\ \exists B' \in \Lambda(Y),\ B' \cap X \subseteq A$, and $\forall B' \in \Lambda(Y)\ \exists A \in \Lambda(X),\ A \subseteq B' \cap X\Big)$.

We collect now for easier reference the definitions and some algebraic properties which we saw above to hold:

Definition 5.5.2 $(\Lambda\cap)$ If $A, B \in \Lambda(X)$ then there is $C \subseteq A \cap B,\ C \in \Lambda(X)$,
(ΛPR) $X \subseteq Y \Rightarrow \forall A \in \Lambda(X)\ \exists B \in \Lambda(Y),\ B \cap X \subseteq A$,
$(\Lambda CUMfin)$ Let $X \subseteq Y$. If there is $B \in \Lambda(Y)$ such that $B \subseteq X$, then: $\Big(\forall A \in \Lambda(X)\ \exists B' \in \Lambda(Y),\ B' \subseteq A$ and $\forall B' \in \Lambda(Y)\ \exists A \in \Lambda(X),\ A \subseteq B'\Big)$.
$(\Lambda =)$ Let $X \subseteq Y$. If $\forall B \in \Lambda(Y),\ B \cap X \neq \emptyset$, then
$\Big(\forall A \in \Lambda(X) \exists B' \in \Lambda(Y). B' \cap X \subseteq A$, and $\forall B' \in \Lambda(Y) \exists A \in \Lambda(X). A \subseteq B' \cap X\Big)$.

Fact 5.5.3 In transitive structures the following properties hold:

(1) $(\Lambda\cap)$
(2) (ΛPR)
(3) $(\Lambda CUMfin)$

In ranked structures the following property holds:

(4) $(\Lambda =)$

Proof (1) By Fact 5.5.1 (p. 148), (2).
(2) $Y - X \in \Lambda(Y - X)$, so the result holds by Fact 5.5.1 (p. 148), (3).
(3) By Fact 5.5.1 (p. 148), (1) and (2).
(4) By Fact 5.5.1 (p. 148), (5) and (7). □

To summarize the discussion:

Just as in the minimal case, the algebraic laws may hold, but not the logical ones, due in both cases to definability problems. Thus, we cannot expect a clean proof of correspondence. But we can argue that we did a correct translation, which shows its limitation, too. The part with $\mu(X)$ and $\mu(Y)$ on both sides of $\subseteq$ is obvious, we will have a perfect correspondence. The part with $X \subseteq \mu(Y)$ is obvious, too. The problem is in the part with $\mu(X) \subseteq Y$. As we cannot use the limit, but only its approximation, we are limited here to one (or finitely many) consequences of T, if $X = M(T)$, so we obtain only $T \mid\sim \phi$, if $Y \subseteq M(\phi)$, and if there is $A \in \Lambda(X)$, $A \subseteq Y$.

We consider a limit only appropriate, if it is an algebraic limit which preserves algebraic properties of the minimal version in above translation.

The advantage of such limits is that they allow – with suitable caveats – to show that they preserve the logical properties of the minimal variant, and thus are equivalent to the minimal case (with, of course, perhaps a different relation). Thus, they allow a straightforward trivialization.

5.5.3.2 Logical Properties of the Limit Variant

We begin with some simple logical facts about the limit version.

We abbreviate $\Lambda(T) := \Lambda(M(T))$ etc., assume transitivity.

Fact 5.5.4

(1) $A \in \Lambda(T) \Rightarrow M(\overline{\overline{T}}) \subseteq \widehat{A}$,

(2) $M(\overline{\overline{T}}) = \bigcap\{\widehat{A} : A \in \Lambda(T)\}$,

(2a) $M(\overline{\overline{T'}}) \models \sigma \Rightarrow \exists B \in \Lambda(T').\ \widehat{B} \models \sigma$,

(3) $M(\overline{\overline{T'}}) \cap M(T) \models \sigma \Rightarrow \exists B \in \Lambda(T').\ \widehat{B} \cap M(T) \models \sigma$.

Proof (1) Note that $A \models \phi \Rightarrow T \mid\sim \phi$ by definition, see Definition 5.5.1 (p. 146).

Let $M(\overline{\overline{T}}) \not\subseteq \widehat{A}$, so there is ϕ, $\widehat{A} \models \phi$, so $A \models \phi$, but $M(\overline{\overline{T}}) \not\models \phi$, so $T \not\mid\sim \phi$, *contradiction.*

(2) "$\subseteq$" by (1). "$\supseteq$": Let $x \in \bigcap\{\widehat{A} : A \in \Lambda(T)\} \Rightarrow \forall A \in \Lambda(T), x \models Th(A)$ $\Rightarrow x \models \overline{\overline{T}}$.

(2a) $M(\overline{\overline{T'}}) \models \sigma \Rightarrow T' \mid\sim \sigma \Rightarrow \exists B \in \Lambda(T'), B \models \sigma$. But $B \models \sigma \Rightarrow \widehat{B} \models \sigma$.

(3) $M(\overline{\overline{T'}}) \cap M(T) \models \sigma \Rightarrow \overline{\overline{T'}} \cup T \vdash \sigma \Rightarrow \exists \tau_1 \ldots \tau_n \in \overline{\overline{T'}}$ such that $T \cup \{\tau_1, \ldots, \tau_n\} \vdash \sigma$, so $\exists B \in \Lambda(T'), Th(B) \cup T \vdash \sigma$. So $M(Th(B)) \cap M(T) \models \sigma \Rightarrow \widehat{B} \cap M(T) \models \sigma$. □

We saw in Example 5.5.2 (p. 150) and its discussion the problems which might arise in the limit version, even if the algebraic behaviour is correct.

This analysis leads us to consider the following facts:

Fact 5.5.5 (1) Let $\forall B \in \Lambda(T')\exists A \in \Lambda(T).A \subseteq B \cap M(T)$, then $\overline{\overline{\overline{T'}} \cup T} \subseteq \overline{\overline{T}}$.

Let, in addition, $\{B \in \Lambda(T') : \overbrace{B} \cap \overbrace{M(T)} = \overbrace{B \cap M(T)}\}$ be cofinal in $\Lambda(T')$. Then

(2) $Con(\overline{\overline{T'}}, T)$ implies $\forall A \in \Lambda(T'), A \cap M(T) \neq \emptyset$.

(3) $\forall A \in \Lambda(T)\ \exists B \in \Lambda(T'), B \cap M(T) \subseteq A$ implies $\overline{\overline{T}} \subseteq \overline{\overline{\overline{T'}} \cup T}$.

Note that $M(T) = \overbrace{M(T)}$, so we could have also written $\overbrace{B} \cap M(T) = \overbrace{B \cap M(T)}$, but the above way of writing stresses more the essential condition $\overbrace{X} \cap \overbrace{Y} = \overbrace{X \cap Y}$.

Proof (1) Let $\overline{\overline{T'}} \cup T \vdash \sigma$, so $\exists B \in \Lambda(T'), \overbrace{B} \cap M(T) \models \sigma$ by Fact 5.5.4 (p. 153), (3) above (using compactness). Thus $\exists A \in \Lambda(T), A \subseteq B \cap M(T) \models \sigma$ by prerequisite, so $\sigma \in \overline{\overline{T}}$.

(2) Let $Con(\overline{\overline{T'}}, T)$, so $M(\overline{\overline{T'}}) \cap M(T) \neq \emptyset$. $M(\overline{\overline{T'}}) = \bigcap\{\overbrace{A} : A \in \Lambda(T')\}$ by Fact 5.5.4 (p. 153) (2), so $\forall A \in \Lambda(T'), \overbrace{A} \cap M(T) \neq \emptyset$. As cofinally often $\overbrace{A} \cap M(T) = \overbrace{A \cap M(T)}$, $\forall A \in \Lambda(T'), \overbrace{A \cap M(T)} \neq \emptyset$, so $\forall A \in \Lambda(T'), A \cap M(T) \neq \emptyset$ by $\overbrace{\emptyset} = \emptyset$.

(3) Let $\sigma \in \overline{\overline{T}}$, so $T \mid\!\sim \sigma$, so $\exists A \in \Lambda(T), A \models \sigma$, so $\exists B \in \Lambda(T'), B \cap M(T) \subseteq A$ by prerequisite, so $\exists B \in \Lambda(T').\big(B \cap M(T) \subseteq A$ and $\overbrace{B} \cap \overbrace{M(T)} = \overbrace{B \cap M(T)}\big)$. So for such B $\overbrace{B} \cap \overbrace{M(T)} = \overbrace{B \cap M(T)} \subseteq \overbrace{A} \models \sigma$. By Fact 5.5.4 (p. 153) (1) $M(\overline{\overline{T'}}) \subseteq \overbrace{B}$, so $M(\overline{\overline{T'}}) \cap M(T) \models \sigma$, so $\overline{\overline{T'}} \cup T \vdash \sigma$. □

We obtain now as easy corollaries of a more general situation the following properties shown in [Sch04] by direct proofs. Thus, we have the trivialization results shown there.

Corollary 5.5.1 *Let the structure be transitive.*

(1) *Let* $\{B \in \Lambda(T') : \overbrace{B} \cap \overbrace{M(T)} = \overbrace{B \cap M(T)}\}$ *be cofinal in* $\Lambda(T')$, *then* $(PR)\ T \vdash T' \Rightarrow \overline{\overline{T}} \subseteq \overline{\overline{\overline{T'}} \cup T}$ *holds.*

(2) $\overline{\overline{\phi \wedge \phi'}} \subseteq \overline{\overline{\overline{\phi}} \cup \{\phi'\}}$ *holds.*

If the structure is ranked, then also:

(3) *Let* $\{B \in \Lambda(T') : \overbrace{B} \cap \overbrace{M(T)} = \overbrace{B \cap M(T)}\}$ *be cofinal in* $\Lambda(T')$, *then* $(\mid\!\sim=)\ T \vdash T', Con(\overline{\overline{T'}}, T) \Rightarrow \overline{\overline{T}} = \overline{\overline{\overline{T'}} \cup T}$ *holds.*

(4) $T \not\mid\!\sim \neg\gamma \Rightarrow \overline{\overline{T}} = \overline{\overline{\overline{T}} \cup \{\gamma\}}$ *holds.*

Proof (1) $\forall A \in \Lambda(M(T))\ \exists B \in \Lambda(M(T')), B \cap M(T) \subseteq A$ by Fact 5.5.3 (p. 152), (2). So the result follows from Fact 5.5.5 (p. 154) (3).

(2) Set $T' := \{\phi\}$, $T := \{\phi, \phi'\}$. Then for $B \in \Lambda(T')$ $\widehat{B} \cap M(T) = \widehat{B} \cap M(\phi') = \widehat{B \cap M(\phi')}$ by Fact 2.1.1 (p. 32) $(Cl \cap +)$, so the result follows by (1).

(3) Let $Con(\overline{\overline{T'}}, T)$, then by Fact 5.5.5 (p. 154), (2) $\forall A \in \Lambda(T'), A \cap M(T) \neq \emptyset$, so by Fact 5.5.3 (p. 152), (4) $\forall B \in \Lambda(T') \exists A \in \Lambda(T), A \subseteq B \cap M(T)$, so $\overline{\overline{T'}} \cup T \subseteq \overline{\overline{T}}$ by Fact 5.5.5 (p. 154) (1).

The other direction follows from (1).

(4) Set $T := T' \cup \{\gamma\}$. Then for $B \in \Lambda(T')$ $\widehat{B} \cap M(T) = \widehat{B} \cap M(\gamma) = \widehat{B \cap M(\gamma)}$ again by Fact 2.1.1 (p. 32) $(Cl \cap +)$, so the result follows from (3). □

We summarize for easier reference here our main positive logical results on the limit variant of general preferential structures where each model occurs in one copy only, Proposition 3.4.7 and Proposition 3.10.19 from [Sch04]:

Proposition 5.5.1 *Let the relation be transitive. Then*

(1) *Every instance of the the limit version, where the definable closed minimizing sets are cofinal in the closed minimizing sets, is equivalent to an instance of the minimal version.*

(2) *If we consider only formulas on the left of* $\mid\!\sim$, *the resulting logic of the limit version can also be generated by the minimal version of a (perhaps different) preferential structure. Moreover, the structure can be chosen smooth.*

Proposition 5.5.2 *When considering just formulas, in the ranked case without copies,* Λ *is equivalent to* μ *– so* Λ *is trivialized in this case. More precisely*:

Let a logic $\phi \mid\!\sim \psi$ *be given by the limit variant without copies, i.e. by Definition 5.5.1* (p. 146). *Then there is a ranked structure, which gives exactly the same logic, but interpreted in the minimal variant.*

(As Example 3.10.2 in [Sch04] *has shown, this is NOT necessarily true if we consider full theories* T *and* $T \mid\!\sim \psi$.)

This shows that there is an important difference between considering full theories and considering just formulas (on the left of $\mid\!\sim$). If we consider full theories, we can "grab" single models, and thus determine the full order. As long as we restrict ourselves to formulas, we are much more shortsighted, and see only a blurred picture. In particular, we can make sequences of models to converge to some model, but put this model elsewhere. Such suitable manipulations will pass unobserved by formulas. The example also shows that there are structures whose limit version for theories is unequal to any minimal structure.

(The negative results for the general not definability preserving minimal case apply also to the general limit case – see Sect. 5.2.3 in [Sch04] for details.)

Chapter 6
Higher Preferential Structures

6.1 Introduction

Definition 6.1.1 An IBR is called a *generalized preferential structure* iff the origins of all arrows are points. We will usually write x, y, etc. for points, and α, β, etc. for arrows.

Definition 6.1.2 Consider a generalized preferential structure $\mathcal{X}$.

(1) *Level n arrow*:
Definition by upward induction.
If $\alpha : x \to y$, x, y are points, then α is a level 1 arrow.
If $\alpha : x \to \beta$, x is a point, β a level n arrow, then α is a level $n+1$ arrow. ($o(\alpha)$ is the origin, $d(\alpha)$ is the destination of α.)
$\lambda(\alpha)$ will denote the level of α.

(2) *Level n structure*:
$\mathcal{X}$ is a level n structure iff all arrows in $\mathcal{X}$ are atmost level n arrows.
We consider here only structures of some arbitrary but finite level n.

(3) We define for an arrow α by induction $O(\alpha)$ and $D(\alpha)$.
If $\lambda(\alpha) = 1$, then $O(\alpha) := \{o(\alpha)\}$, $D(\alpha) := \{d(\alpha)\}$.
If $\alpha : x \to \beta$, then $D(\alpha) := D(\beta)$, and $O(\alpha) := \{x\} \cup O(\beta)$.

Thus, for example, if $\alpha : x \to y$, $\beta : z \to \alpha$, then $O(\beta) := \{x, z\}$, $D(\beta) = \{y\}$.

Comment 6.1.1 A counterargument to α is NOT an argument for $\neg\alpha$ (this is asking for too much), but just showing one case where $\neg\alpha$ holds. In preferential structures, an argument for α is a set of level 1 arrows, eliminating $\neg\alpha$-models. A counterargument is one level 2 arrow attacking one such level 1 arrow.

Of course, when we have copies, we may need many successful attacks, on all copies, to achieve the goal. As we may have copies of level 1 arrows, we may need many level 2 arrows to destroy them all.

D.M. Gabbay, K. Schlechta, *Logical Tools for Handling Change in Agent-Based Systems*, Cognitive Technologies, DOI 10.1007/978-3-642-04407-6_6,

We will not consider here diagrams with arbitrarily high levels. One reason is that diagrams like the following will have an unclear meaning:

Example 6.1.1
$\langle\alpha, 1\rangle : x \to y,$
$\langle\alpha, n+1\rangle : x \to \langle\alpha, n\rangle \; (n \in \omega).$
Is $y \in \mu(X)$?

Definition 6.1.3 Let $\mathcal{X}$ be a generalized preferential structure of (finite) level n. We define (by downward induction):

(1) $Valid\ X - to - Y\ arrow$:
Let $X, Y \subseteq \mathbf{P}(\mathcal{X})$.
$\alpha \in \mathbf{A}(\mathcal{X})$ is a $valid\ X - to - Y\ arrow$ iff
(1.1) $O(\alpha) \subseteq X,\ D(\alpha) \subseteq Y,$
(1.2) $\forall \beta : x' \to \alpha.\ (x' \in X \Rightarrow \exists \gamma : x'' \to \beta.$ (γ is a valid $X - to - Y$ arrow).)
We will also say that α is a $valid$ $arrow$ in X, or just $valid$ in X, iff α is a valid $X - to - X$ arrow.

(2) $Valid\ X \Rightarrow Y\ arrow$:
Let $X \subseteq Y \subseteq \mathbf{P}(\mathcal{X})$.
$\alpha \in \mathbf{A}(\mathcal{X})$ is a $valid\ X \Rightarrow Y\ arrow$ iff
(2.1) $o(\alpha) \in X,\ O(\alpha) \subseteq Y,\ D(\alpha) \subseteq Y,$
(2.2) $\forall \beta : x' \to \alpha.\ (x' \in Y \Rightarrow \exists \gamma : x'' \to \beta.$ (γ is a valid $X \Rightarrow Y$ arrow).)

(Note that in particular $o(\gamma) \in X$, and that $o(\beta)$ need not be in X, but can be in the bigger Y.)

Example 6.1.2

(1) Consider the arrow $\beta := \langle\beta', l'\rangle$ in Diagram 6.1.1 (p. 160). $D(\beta) = \{\langle x, i\rangle\}$, $O(\beta) = \{\langle z', m'\rangle, \langle y, j\rangle\}$, and the only arrow attacking β originates outside X, so β is a valid $X - to - \mu(X)$ arrow.
(2) Consider the arrows $\langle\alpha', k'\rangle$ and $\langle\gamma', n'\rangle$ in Diagram 6.1.2 (p. 161). Both are valid $\mu(X) \Rightarrow X$ arrows.

Fact 6.1.1

(1) If α is a valid $X \Rightarrow Y$ arrow, then α is a valid $Y - to - Y$ arrow.
(2) If $X \subseteq X' \subseteq Y' \subseteq Y \subseteq \mathbf{P}(\mathcal{X})$ and $\alpha \in \mathbf{A}(\mathcal{X})$ is a valid $X \Rightarrow Y$ arrow, and $O(\alpha) \subseteq Y'$, $D(\alpha) \subseteq Y'$, then α is a valid $X' \Rightarrow Y'$ arrow.

Proof Let α be a valid $X \Rightarrow Y$ arrow. We show (1) and (2) together by downward induction (both are trivial).

By prerequisite $o(\alpha) \in X \subseteq X'$, $O(\alpha) \subseteq Y' \subseteq Y$, $D(\alpha) \subseteq Y' \subseteq Y$.

Case 1: $\lambda(\alpha) = n$. So α is a valid $X' \Rightarrow Y'$ arrow and a valid $Y - to - Y$ arrow.
Case 2: $\lambda(\alpha) = n - 1$. So there is no $\beta : x' \to \alpha,\ y \in Y$, so α is a valid $Y - to - Y$ arrow. By $Y' \subseteq Y$, α is a valid $X' \Rightarrow Y'$ arrow.

Case 3: Let the result be shown down to m, $n > m > 1$, let $\lambda(\alpha) = m - 1$. So $\forall \beta : x' \to \alpha$ ($x' \in Y \Rightarrow \exists \gamma : x'' \to \beta$ ($x'' \in X$ and γ is a valid $X \Rightarrow Y$ arrow)). By induction hypothesis γ is a valid $Y - to - Y$ arrow and a valid $X' \Rightarrow Y'$ arrow. So α is a valid $Y - to - Y$ arrow, and by $Y' \subseteq Y$, α is a valid $X' \Rightarrow Y'$ arrow. □

Definition 6.1.4 Let $\mathcal{X}$ be a generalized preferential structure of level n, $X \subseteq \boldsymbol{P}(\mathcal{X})$.
$\mu(X) := \{x \in X : \exists \langle x, i \rangle . \neg \exists$ valid $X - to - X$ arrow $\alpha : x' \to \langle x, i \rangle\}$.

Comment 6.1.2 The purpose of smoothness is to guarantee cumulativity. Smoothness achieves cumulativity by mirroring all information present in X also in $\mu(X)$. Closer inspection shows that smoothness does more than necessary. This is visible when there are copies (or, equivalently, non injective labelling functions). Suppose we have two copies of $x \in X$, $\langle x, i \rangle$ and $\langle x, i' \rangle$, and there is $y \in X$, $\alpha : \langle y, j \rangle \to \langle x, i \rangle$, but there is no $\alpha' : \langle y', j' \rangle \to \langle x, i' \rangle$, $y' \in X$. Then $\alpha : \langle y, j \rangle \to \langle x, i \rangle$ is irrelevant, as $x \in \mu(X)$ anyhow. So mirroring $\alpha : \langle y, j \rangle \to \langle x, i \rangle$ in $\mu(X)$ is not necessary, i.e. it is not necessary to have some $\alpha' : \langle y', j' \rangle \to \langle x, i \rangle$, $y' \in \mu(X)$.

On the other hand, Example 6.1.4 (p. 162) shows that, if we want smooth structures to correspond to the property (μCUM), we need at least some valid arrows from $\mu(X)$ also for higher level arrows. This "some" is made precise (essentially) in Definition 6.1.5 (p. 159).

From a more philosophical point of view, when we see the (inverted) arrows of preferential structures as attacks on non-minimal elements, then we should see smooth structures as always having attacks also from valid (minimal) elements. So, in general structures, also attacks from non valid elements are valid; in smooth structures we always also have attacks from valid elements.
The analogue to usual smooth structures, on level 2, is then that any successfully attacked level 1 arrow is also attacked from a minimal point.

Definition 6.1.5 Let $\mathcal{X}$ be a generalized preferential structure.
$X \sqsubseteq X'$ iff

(1) $X \subseteq X' \subseteq \boldsymbol{P}(\mathcal{X})$,
(2) $\forall x \in X' - X\ \forall \langle x, i \rangle\ \exists \alpha : x' \to \langle x, i \rangle$($\alpha$ is a valid $X \Rightarrow X'$ arrow),
(3) $\forall x \in X\ \exists \langle x, i \rangle$,
($\forall \alpha : x' \to \langle x, i \rangle (x' \in X' \Rightarrow \exists \beta : x'' \to \alpha$. ($\beta$ is a valid $X \Rightarrow X'$ arrow)).)

Note that (3) is not simply the negation of (2):
Consider a level 1 structure. Thus all level 1 arrows are valid, but the source of the arrows must not be neglected.

(2) reads now: $\forall x \in X' - X\ \forall \langle x, i \rangle\ \exists \alpha : x' \to \langle x, i \rangle.,\ x' \in X$;
(3) reads: $\forall x \in X\ \exists \langle x, i \rangle\ \neg \exists \alpha : x' \to \langle x, i \rangle.,\ x' \in X'$.

This is intended: Intuitively, $X = \mu(X')$, and minimal elements must not be attacked at all, but non minimals must be attacked from X – which is a modified version of smoothness.

Remark 6.1.1 We note the special case of Definition 6.1.5 (p. 159) for level 3 structures, as it will be used later. We also write it immediately for the intended case $\mu(X) \sqsubseteq X$, and explicitly with copies.
$x \in \mu(X)$ iff

(1) $\exists\langle x,i\rangle\forall\langle\alpha,k\rangle : \langle y,j\rangle \to \langle x,i\rangle$
$(y \in X \Rightarrow \exists\langle\beta',l'\rangle : \langle z',m'\rangle \to \langle\alpha,k\rangle.$
$(z' \in \mu(X) \wedge \neg\exists\langle\gamma',n'\rangle : \langle u',p'\rangle \to \langle\beta',l'\rangle.u' \in X)).$
See Diagram 6.1.1 (p. 160).

$x \in X - \mu(X)$ iff
(2) $\forall\langle x,i\rangle\exists\langle\alpha',k'\rangle : \langle y',j'\rangle \to \langle x,i\rangle$
$(y' \in \mu(X) \wedge$

(*a*) $\neg\exists\langle\beta',l'\rangle : \langle z',m'\rangle \to \langle\alpha',k'\rangle.z' \in X$
or
(*b*) $\forall\langle\beta',l'\rangle : \langle z',m'\rangle \to \langle\alpha',k'\rangle$
$(z' \in X \Rightarrow \exists\langle\gamma',n'\rangle : \langle u',p'\rangle \to \langle\beta',l'\rangle.u' \in \mu(X))).$

See Diagram 6.1.2 (p. 161).

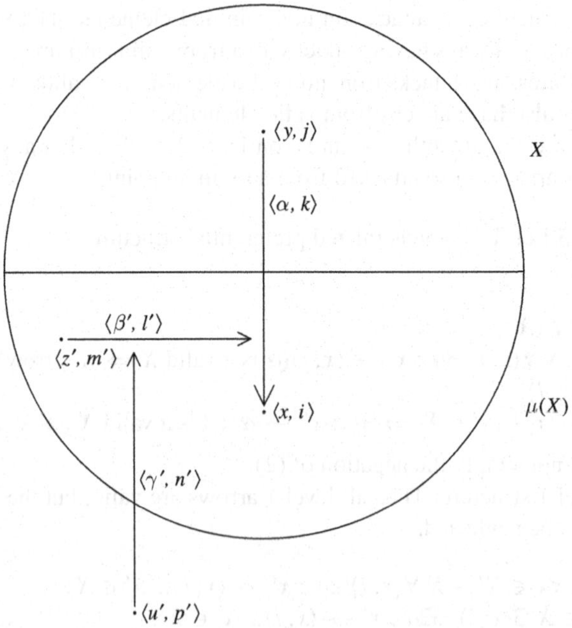

Diagram 6.1.1 Case 3-1-2

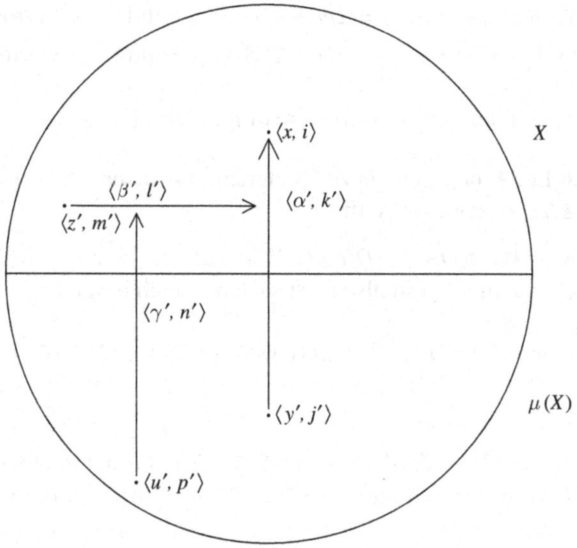

Diagram 6.1.2 Case 3-2

Fact 6.1.2

(1) If $X \sqsubseteq X'$, then $X = \mu(X')$,
(2) $X \sqsubseteq X'$, $X \subseteq X'' \subseteq X' \Rightarrow X \sqsubseteq X''$. (This corresponds to (μCUM) .)
(3) $X \sqsubseteq X'$, $X \subseteq Y'$, $Y \sqsubseteq Y'$, $Y \subseteq X' \Rightarrow X = Y$. (This corresponds to $(\mu \subseteq\supseteq)$.)

Proof

(1) Trivial by Fact 6.1.1 (p. 158) (1).
(2) We have to show

(a) $\forall x \in X'' - X\ \forall \langle x, i\rangle\ \exists \alpha : x' \to \langle x, i\rangle(\alpha$ is a valid $X \Rightarrow X''$ arrow) and
(b) $\forall x \in X\ \exists \langle x, i\rangle\ (\forall \alpha : x' \to \langle x, i\rangle(x' \in X'' \Rightarrow \exists \beta : x'' \to \alpha.(\beta$ is a valid $X \Rightarrow X''$ arrow))).

Both follow from the corresponding condition for $X \Rightarrow X'$, the restriction of the universal quantifier, and Fact 6.1.1 (p. 158) (2).
(3) Let $x \in X-$Y.

(a) By $x \in X \sqsubseteq X'$, $\exists \langle x, i\rangle$ s.t. $(\forall \alpha : x' \to \langle x, i\rangle(x' \in X' \Rightarrow \exists \beta : x'' \to \alpha.$ (β is a valid $X \Rightarrow X'$ arrow))).
(b) By $x \notin Y \sqsubseteq \exists \alpha_1 : x' \to \langle x, i\rangle\ \alpha_1$ is a valid $Y \Rightarrow Y'$ arrow, in particular $x' \in Y \subseteq X'$. Moreover, $\lambda(\alpha_1) = 1$.

So by (a) $\exists \beta_2 : x'' \to \alpha_1.(\beta_2$ is a valid $X \Rightarrow X'$ arrow), in particular $x'' \in X \subseteq Y'$, moreover, $\lambda(\beta_2) = 2$.
It follows by induction from the definition of valid $A \Rightarrow B$ arrows that

$$\forall n \exists \alpha_{2m+1}, \lambda(\alpha_{2m+1}) = 2m+1, \alpha_{2m+1} \text{a valid} Y \Rightarrow Y' \text{arrow and}$$
$$\forall n \exists \beta_{2m+2}, \lambda(\beta_{2m+2}) = 2m+2, \beta_{2m+2} \text{a valid} X \Rightarrow X' \text{arrow,}$$

which is impossible, as $\mathcal{X}$ is a structure of finite level. □

Definition 6.1.6 Let $\mathcal{X}$ be a generalized preferential structure, $X \subseteq \boldsymbol{P}(\mathcal{X})$.
$\mathcal{X}$ is called *totally smooth* for X iff

(1) $\forall \alpha : x \to y \in \boldsymbol{A}(\mathcal{X})(O(\alpha) \cup D(\alpha) \subseteq X \Rightarrow \exists \alpha' : x' \to y. x' \in \mu(X))$;
(2) if α is valid, then there must also exist such α' which is valid.
(y a point or an arrow).
If $\mathcal{Y} \subseteq \boldsymbol{P}(\mathcal{X})$, then $\mathcal{X}$ is called *$\mathcal{Y}$-totally smooth* iff for all $X \in \mathcal{Y}$, $\mathcal{X}$ is totally smooth for X.

Example 6.1.3
$X := \{\alpha : a \to b, \alpha' : b \to c, \alpha'' : a \to c, \beta : b \to \alpha'\}$ is not totally smooth,
$X := \{\alpha : a \to b, \alpha' : b \to c, \alpha'' : a \to c, \beta : b \to \alpha', \beta' : a \to \alpha'\}$ is totally smooth.

Example 6.1.4 Consider $\alpha' : a \to b, \alpha'' : b \to c, \alpha : a \to c, \beta : a \to \alpha$.
Then $\mu(\{a, b, c\}) = \{a\}$, $\mu(\{a, c\}) = \{a, c\}$. Thus, (μCUM) does not hold in this structure. Note that there is no valid arrow from $\mu(\{a, b, c\})$ to c.

Definition 6.1.7 Let $\mathcal{X}$ be a generalized preferential structure, $X \subseteq \boldsymbol{P}(\mathcal{X})$.
$\mathcal{X}$ is called *essentially smooth* for X iff $\mu(X) \sqsubseteq X$. If $\mathcal{Y} \subseteq \boldsymbol{P}(\mathcal{X})$, then $\mathcal{X}$ is called *$\mathcal{Y}$-essentially smooth* iff for all $X \in \mathcal{Y}$, $\mu(X) \sqsubseteq X$.

Example 6.1.5 It is easy to see that we can distinguish total and essential smoothness in richer structures, as the following example shows:
We add an accessibility relation R, and consider only those models which are accessible.

Let, e.g. $a \to b \to \langle c, 0\rangle, \langle c, 1\rangle$, without transitivity. Thus, only c has two copies. This structure is essentially smooth, but of course not totally so.

Let now mRa, mRb, $mR\langle c, 0\rangle$, $mR\langle c, 1\rangle$, $m'Ra$, $m'Rb$, $m'R\langle c, 0\rangle$.

Thus, seen from m, $\mu(\{a, b, c\}) = \{a, c\}$, but seen from m', $\mu(\{a, b, c\}) = \{a\}$, but $\mu(\{a, c\}) = \{a, c\}$, contradicting (CUM).□

6.2 The General Case

The idea to solve the representation problem illustrated by Example 2.3.2 (p. 43) is to use the points c and d as bases for counterarguments against $\alpha : b \to a$ – as is possible in IBRS. We do this now. We will obtain a representation for logics weaker than P by generalized preferential structures.

We will now prove a representation theorem, but will make it more general than for preferential structures only. For this purpose, we will introduce some definitions first.

Definition 6.2.1 Let $\eta, \rho : \mathcal{Y} \to \mathcal{P}(U)$.

(1) If $\mathcal{X}$ is a simple structure:
$\mathcal{X}$ is called an *attacking structure* relative to η representing ρ iff $\rho(X) = \{x \in \eta(X) :$ there is no valid $X - to - \eta(X)$ arrow $\alpha : x' \to x\}$ for all $X \in \mathcal{Y}$.

(2) If $\mathcal{X}$ is a structure with copies:
$\mathcal{X}$ is called an *attacking structure* relative to η representing ρ iff $\rho(X) = \{x \in \eta(X) :$ there is $\langle x, i\rangle$ and no valid $X-to-\eta(X)$ arrow $\alpha : \langle x', i'\rangle \to \langle x, i\rangle\}$ for all $X \in \mathcal{Y}$.

Obviously, in those cases $\rho(X) \subseteq \eta(X)$ for all $X \in \mathcal{Y}$. Thus, $\mathcal{X}$ is a preferential structure iff η is the identity. See Diagram 6.2.1 (p. 163).

(Note that it does not seem very useful to generalize the notion of smoothness from preferential structures to general attacking structures, as, in the general case, the minimizing set X and the result $\rho(X)$ may be disjoint.)

The following result is the first positive representation result here, and shows that we can obtain (almost) anything with level 2 structures.

Proposition 6.2.1 *Let $\eta, \rho : \mathcal{Y} \to \mathcal{P}(U)$. Then there is an attacking level 2 structure relative to η representing ρ iff*

(1) $\rho(X) \subseteq \eta(X)$ for all $X \in \mathcal{Y}$ and
(2) $\rho(\emptyset) = \eta(\emptyset)$ if $\emptyset \in \mathcal{Y}$.

(2) *is, of course, void for preferential structures.*

Proof

(A) The construction
We make a two-stage construction.

(A.1) Stage 1:
In stage one, consider (almost as usual)
$\mathcal{U} := \langle \mathcal{X}, \{\alpha_i : i \in I\}\rangle$, where
$\mathcal{X} := \{\langle x, f\rangle : x \in U,\ f \in \Pi\{X \in \mathcal{Y} : x \in \eta(X) - \rho(X)\}\}$,
$\alpha : x' \to \langle x, f\rangle :\Leftrightarrow x' \in ran(f)$. Attention: $x' \in X$, not $x' \in \rho(X)$!

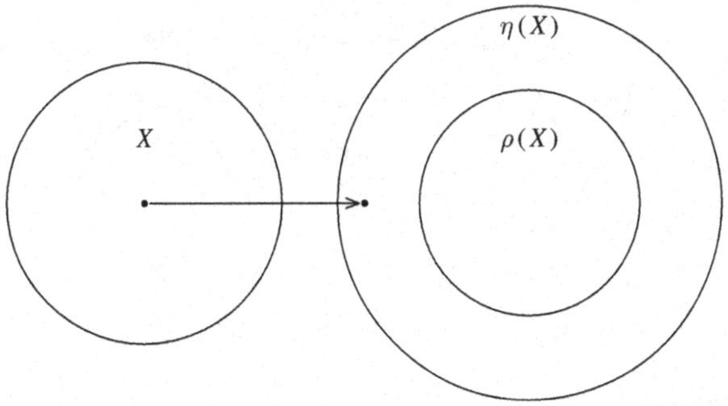

Diagram 6.2.1 Attacking structure

(A.2) Stage 2:
Let $\mathcal{X}'$ be the set of all $\langle x, f, X\rangle$ s.t. $\langle x, f\rangle \in \mathcal{X}$ and

(a) either X is some dummy value, say $*$
or
(b) all of the following (1)–(4) hold:

(1) $X \in \mathcal{Y}$,
(2) $x \in \rho(X)$,
(3) there is $X' \subseteq X$, $x \in \eta(X') - \rho(X')$, $X' \in \mathcal{Y}$, (thus $ran(f) \cap X \neq \emptyset$ by definition), and
(4) $\forall X'' \in \mathcal{Y}.(X \subseteq X'', x \in \eta(X'') - \rho(X'') \Rightarrow (ran(f) \cap X'') - X \neq \emptyset)$.
(Thus, f chooses in (4) for X'' also outside X. If there is no such X'', (4) is void, and only (1)–(3) need to hold, i.e. we may take any f with $\langle x, f\rangle \in \mathcal{X}$.)

See Diagram 6.2.2 (p. 164).
Note: If (1)–(3) are satisfied for x and X, then we will find f s.t. $\langle x, f\rangle \in \mathcal{X}$, and $\langle x, f, X\rangle$ satisfies (1)–(4): As $X \subset X''$ for X'' as in (4), we find f which chooses for such X'' outside of X.

So for any $\langle x, f\rangle \in \mathcal{X}$, there is $\langle x, f, *\rangle$, and maybe also some $\langle x, f, X\rangle$ in $\mathcal{X}'$.
Let again for any x', $\langle x, f, X\rangle \in \mathcal{X}'$
$\alpha : x' \to \langle x, f, X\rangle :\Leftrightarrow x' \in ran(f)$

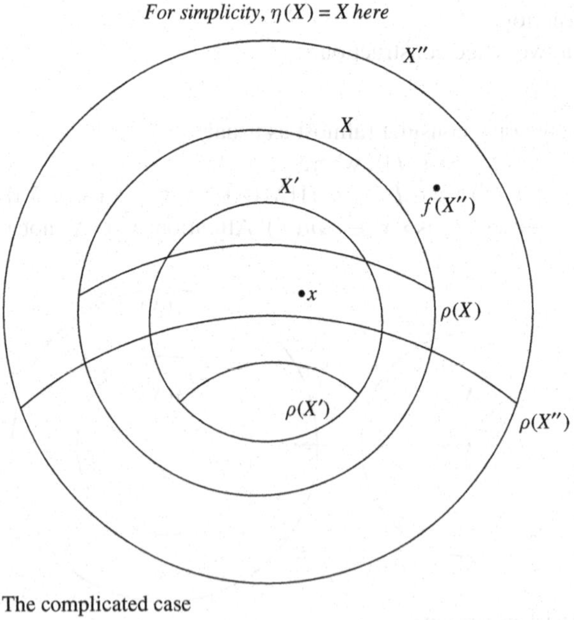

Diagram 6.2.2 The complicated case

(A.3) Adding arrows:
Consider x' and $\langle x, f, X\rangle$.
If $X = *$, or $x' \notin X$, we do nothing, i.e. leave a simple arrow $\alpha : x' \to \langle x, f, X\rangle \Leftrightarrow x' \in ran(f)$.
If $X \in \mathcal{Y}$, $x' \in X$, and $x' \in ran(f)$, we make X many copies of the attacking arrow and have then: $\langle \alpha, x''\rangle : x' \to \langle x, f, X\rangle$ for all $x'' \in X$.
In addition, we add attacks on the $\langle \alpha, x''\rangle : \langle \beta, x''\rangle : x'' \to \langle \alpha, x''\rangle$ for all $x'' \in X$.
The full structure $\mathcal{Z}$ is thus:
$\mathcal{X}'$ is the set of elements.
If $x' \in ran(f)$, and $X = *$ or $x' \notin X$ then $\alpha : x' \to \langle x, f, X\rangle$.
If $x' \in ran(f)$, and $X \neq *$ and $x' \in X$ then
(a) $\langle \alpha, x''\rangle : x' \to \langle x, f, X\rangle$ for all $x'' \in X$,
(b) $\langle \beta, x''\rangle : x'' \to \langle \alpha, x''\rangle$ for all $x'' \in X$.

See Diagram 6.2.3 (p. 165).

(B) Representation:
We have to show that this structure represents ρ relative to η.
Let $y \in \eta(Y)$, $Y \in \mathcal{Y}$.

Case 1: $y \in \rho(Y)$.
We have to show that there is $\langle y, g, Y''\rangle$ s.t. there is no valid $\alpha : y' \to \langle y, g, Y''\rangle$, $y' \in Y$. In Case 1.1 below, Y'' will be $*$, in Case 1.2, Y'' will be Y, g will be chosen suitably.

Case 1.1: There is no $Y' \subseteq Y$, $y \in \eta(Y') - \rho(Y')$, $Y' \in \mathcal{Y}$.
So for all Y' with $y \in \eta(Y') - \rho(Y')$ $Y' - Y \neq \emptyset$. Let $g \in \Pi\{Y' - Y : y \in \eta(Y') - \rho(Y')\}$. Then $ran(g) \cap Y = \emptyset$, and $\langle y, g, *\rangle$ is not attacked from Y. ($\langle y, g\rangle$ was already not attacked in $\mathcal{X}$.)

Case 1.2: There is $Y' \subseteq Y$, $y \in \eta(Y') - \rho(Y')$, $Y' \in \mathcal{Y}$.
Let now $\langle y, g, Y\rangle \in \mathcal{X}'$, s.t. $g(Y'') \notin Y$, if $Y \subseteq Y''$, $y \in \eta(Y'') - \rho(Y'')$, $Y'' \in \mathcal{Y}$. As noted above, such g and thus $\langle y, g, Y\rangle$ exist. Fix $\langle y, g, Y\rangle$.
Consider any $y' \in ran(g)$. If $y' \notin Y$, y' does not attack $\langle y, g, Y\rangle$ in Y.
Suppose $y' \in Y$. We had made Y many copies $\langle \alpha, y''\rangle$, $y'' \in Y$ with

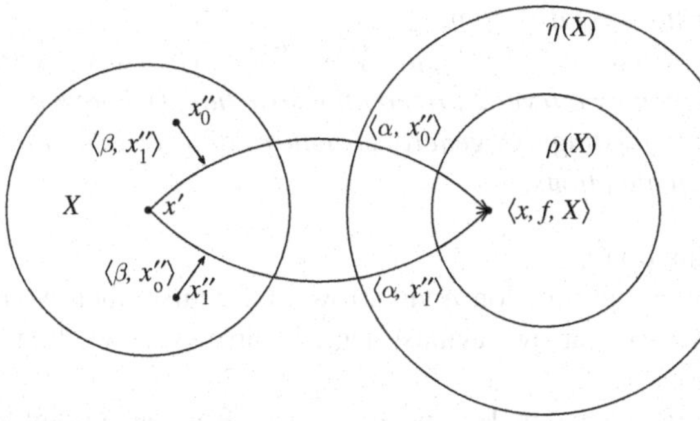

Diagram 6.2.3 Attacking structure

$\langle \alpha, y'' \rangle : y' \to \langle y, g, Y \rangle$ and had added the level 2 arrows $\langle \beta, y'' \rangle : y'' \to \langle \alpha, y'' \rangle$ for $y'' \in Y$. So all copies $\langle \alpha, y'' \rangle$ are destroyed in Y. This was done for all $y' \in Y$, $y' \in ran(g)$, so $\langle y, g, Y \rangle$ is now not (validly) attacked in Y anymore.

Case 2: $y \in \eta(Y) - \rho(Y)$.

Let $\langle y, g, Y' \rangle$ (where Y' can be $*$) be any copy of y, we have to show that there is $z \in Y$, $\alpha : z \to \langle y, g, Y' \rangle$, or some $\langle \alpha, z' \rangle : z \to \langle y, g, Y' \rangle$, $z' \in Y'$, which is not destroyed by some level 2 arrow $\langle \beta, z' \rangle : z' \to \langle \alpha, z' \rangle$, $z' \in Y$.

As $y \in \eta(Y) - \rho(Y)$, $ran(g) \cap Y \neq \emptyset$, so there is $z \in ran(g) \cap Y$. Fix such z. (We will modify the choice of z only in Case 2.2.2 below.)

Case 2.1: $Y' = *$.

As $z \in ran(g)$, $\alpha : z \to \langle y, g, * \rangle$. (There are no level 2 arrows introduced for this copy.)

Case 2.2: $Y' \neq *$.

So $\langle y, g, Y' \rangle$ satisfies the conditions (1)–(4) of (b) at the beginning of the proof.

If $z \notin Y'$, we are done, as $\alpha : z \to \langle y, g, Y' \rangle$, and there are no level 2 arrows introduced in this case. If $z \in Y'$, we had made Y' many copies $\langle \alpha, z' \rangle$, $\langle \alpha, z' \rangle : z \to \langle y, g, Y' \rangle$, one for each $z' \in Y'$. Each $\langle \alpha, z' \rangle$ was destroyed by $\langle \beta, z' \rangle : z' \to \langle \alpha, z' \rangle$, $z' \in Y'$.

Case 2.2.1: $Y' \not\subseteq Y$.

Let $z'' \in Y'$–Y, then $\langle \alpha, z'' \rangle : z \to \langle y, g, Y' \rangle$ is destroyed only by $\langle \beta, z'' \rangle : z'' \to \langle \alpha, z'' \rangle$ in Y', but not in Y, as $z'' \notin Y$, so $\langle y, g, Y' \rangle$ is attacked by $\langle \alpha, z'' \rangle : z \to \langle y, g, Y' \rangle$, valid in Y.

Case 2.2.2: $Y' \subset Y$ ($Y = Y'$ is impossible, as $y \in \rho(Y')$, $y \notin \rho(Y)$).

Then there was by definition (condition (b) (4)) some $z' \in (ran(g) \cap Y) - Y'$ and $\alpha : z' \to \langle y, g, Y' \rangle$ is valid, as $z' \notin Y'$. (In this case, there are no copies of α and no level 2 arrows.) □

Corollary 6.2.1

(1) *We cannot distinguish general structures of level* 2 *from those of higher levels by their* ρ*-functions relative to* η.

(2) *Let* U *be the universe,* $\mathcal{Y} \subseteq \mathcal{P}(U)$, $\mu : \mathcal{Y} \to \mathcal{P}(U)$. *Then any* μ *satisfying* ($\mu \subseteq$) *can be represented by a level* 2 *preferential structure. (Choose* $\eta =$ identity.) *Again, we cannot distinguish general structures of level* 2 *from those of higher levels by their* μ*-functions.*

A remark on the function η:

We can also obtain the function η via arrows. Of course, then we need positive arrows (not only negative arrows against negative arrows, as we first need to have something positive).

If η is the identity, we can make a positive arrow from each point to itself. Otherwise, we can connect every point to every point by a positive arrow, and then choose those we really want in η by a choice function obtained from arrows just as we obtained ρ from arrows.

6.3 Discussion of the Totally Smooth Case

Fact 6.3.1 Let $X, Y \in \mathcal{Y}$, and $\mathcal{X}$ a level n structure. Let $\langle\alpha, k\rangle : \langle x, i\rangle \to \langle y, j\rangle$, where $\langle y, j\rangle$ may itself be (a copy of) an arrow.

(1) Let $n > 1$, $X \subseteq Y$, $\langle\alpha, k\rangle \in X$ a level $n-1$ arrow in $\mathcal{X}\lceil X$. If $\langle\alpha, k\rangle$ is valid in $\mathcal{X}\lceil Y$, then it is valid in $\mathcal{X}\lceil X$.
(2) Let $\mathcal{X}$ be totally smooth, $\mu(X) \subseteq Y$, $\mu(Y) \subseteq X$, $\langle\alpha, k\rangle \in \mathcal{X}\lceil X \cap Y$, then $\langle\alpha, k\rangle$ is valid in $\mathcal{X}\lceil X$ iff it is valid in $\mathcal{X}\lceil Y$.

Note that we will also sometimes write X for $\mathcal{X}\lceil X$, when the context is clear.

Proof

(1) If $\langle\alpha, k\rangle$ is not valid in $\mathcal{X}\lceil X$, then there must be a level n arrow $\langle\beta, r\rangle : \langle z, s\rangle \to \langle\alpha, k\rangle$ in $\mathcal{X}\lceil X \subseteq \mathcal{X}\lceil Y$. $\langle\beta, r\rangle$ must be valid in $\mathcal{X}\lceil X$ and $\mathcal{X}\lceil Y$, as there are no level $n+1$ arrows. So $\langle\alpha, k\rangle$ is not valid in $\mathcal{X}\lceil Y$, *contradiction*.
(2) By downward induction. Case $n : \langle\alpha, k\rangle \in \mathcal{X}\lceil X \cap Y$, so it is valid in both as there are no level $n+1$ arrows. Case $m \to m-1$: Let $\langle\alpha, k\rangle \in \mathcal{X}\lceil X \cap Y$ be a level $m-1$ arrow valid in $\mathcal{X}\lceil X$, but not in $\mathcal{X}\lceil Y$. So there must be a level m arrow $\langle\beta, r\rangle : \langle z, s\rangle \to \langle\alpha, k\rangle$ valid in $\mathcal{X}\lceil Y$. By total smoothness, we may assume $z \in \mu(Y) \subseteq X$, so $\langle\beta, r\rangle \in \mathcal{X}\lceil X$ is valid by induction hypothesis. So $\langle\alpha, k\rangle$ is not valid in $\mathcal{X}\lceil X$, *contradiction*. □

Corollary 6.3.1 *Let $X, Y \in \mathcal{Y}$, $\mathcal{X}$ a totally smooth level n structure, $\mu(X) \subseteq Y$, $\mu(Y) \subseteq X$. Then $\mu(X) = \mu(Y)$.*

Proof Let $x \in \mu(X) - \mu(Y)$. Then by $\mu(X) \subseteq Y$, $x \in Y$, so there must be for all $\langle x, i\rangle \in \mathcal{X}$ an arrow $\langle\alpha, k\rangle : \langle y, j\rangle \to \langle x, i\rangle$ valid in $\mathcal{X}\lceil Y$; without loss of generality $y \in \mu(Y) \subseteq X$ by total smoothness. So by Fact 6.3.1 (p. 167), (2), $\langle\alpha, k\rangle$ is valid in $\mathcal{X}\lceil X$. This holds for all $\langle x, i\rangle$, so $x \notin \mu(X)$, a *contradiction*. □

Fact 6.3.2 There are situations satisfying $(\mu \subseteq) + (\mu CUM) + (\cap)$ which cannot be represented by level 2 totally smooth preferential structures.
The proof is given in the following example.

Example 6.3.1 Let $Y := \{x, y, y'\}$, $X := \{x, y\}$, $X' := \{x, y'\}$. Let $\mathcal{Y} := \mathcal{P}(Y)$. Let $\mu(Y) := \{y, y'\}$, $\mu(X) := \mu(X') := \{x\}$, and $\mu(Z) := Z$ for all other sets.
Obviously, this satisfies $(\cap)$, $(\mu \subseteq)$, and (μCUM).
Suppose $\mathcal{X}$ is a totally smooth level 2 structure representing μ.
So $\mu(X) = \mu(X') \subseteq Y - \mu(Y)$, $\mu(Y) \subseteq X \cup X'$. Let $\langle x, i\rangle$ be minimal in $\mathcal{X}\lceil X$.

As $\langle x, i\rangle$ cannot be minimal in $\mathcal{X}\lceil Y$, there must be $\alpha : \langle z, j\rangle \rightarrow \langle x, i\rangle$ valid in $\mathcal{X}\lceil Y$.

Case 1: $z \in X'$.

So $\alpha \in \mathcal{X}\lceil X'$. If α is valid in $\mathcal{X}\lceil X'$, there must be $\alpha' : \langle x', i'\rangle \rightarrow \langle x, i\rangle$, $x' \in \mu(X')$, valid in $\mathcal{X}\lceil X'$, and thus in $\mathcal{X}\lceil X$, by $\mu(X) = \mu(X')$ and Fact 6.3.1 (p. 167) (2). This is impossible, so there must be $\beta : \langle x', i'\rangle \rightarrow \alpha$, $x' \in \mu(X')$, valid in $\mathcal{X}\lceil X'$. As β is in $\mathcal{X}\lceil Y$ and $\mathcal{X}$ a level ≤ 2 structure, β is valid in $\mathcal{X}\lceil Y$, so α is not valid in $\mathcal{X}\lceil Y$, *contradiction*.

Case 2: $z \in X$.

α cannot be valid in $\mathcal{X}\lceil X$, so there must be $\beta : \langle x', i'\rangle \rightarrow \alpha$, $x' \in \mu(X)$, valid in $\mathcal{X}\lceil X$. Again, as β is in $\mathcal{X}\lceil Y$ and $\mathcal{X}$ a level ≤ 2 structure, β is valid in $\mathcal{X}\lceil Y$, so α is not valid in $\mathcal{X}\lceil Y$, *contradiction*.

It is unknown to the authors whether an analogue is true for essential smoothness, i.e. whether there are examples of such μ-function which need at least level 3 essentially smooth structures for representation. Proposition 6.4.1 (p. 171) below shows that such structures suffice, but we do not know whether level 3 is necessary.

Fact 6.3.3 Above Example 6.3.1 (p. 167) can be solved by a totally smooth level 3 structure:
Let $\alpha_1 : x \rightarrow y$, $\alpha_2 : x \rightarrow y'$, $\alpha_3 : y \rightarrow x$, $\beta_1 : y \rightarrow \alpha_2$, $\beta_2 : y' \rightarrow \alpha_1$, $\beta_3 : y \rightarrow \alpha_3$, $\beta_4 : x \rightarrow \alpha_3$, $\gamma_1 : y' \rightarrow \beta_3$, $\gamma_2 : y' \rightarrow \beta_4$.
See Diagram 6.3.1 (p. 168).

The subdiagram generated by X contains $\alpha_1, \alpha_3, \beta_3, \beta_4$. $\alpha_1, \beta_3, \beta_4$ are valid, so $\mu(X) = \{x\}$.

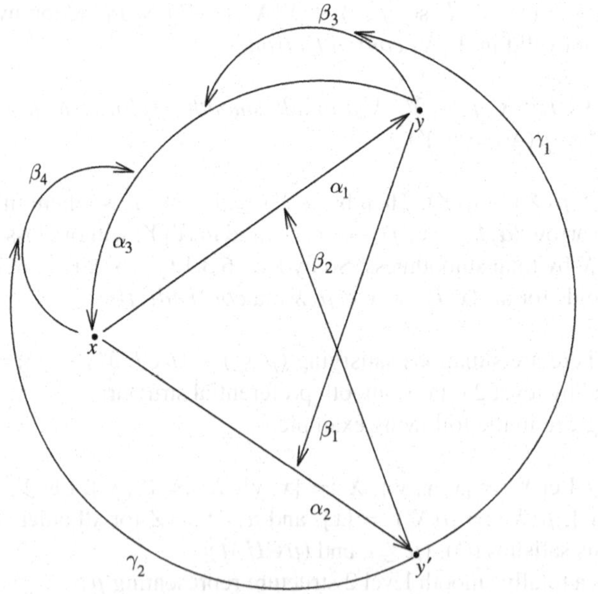

Diagram 6.3.1 Solution by smooth level 3 structure

The subdiagram generated by X' contains α_2. α_2 is valid, so $\mu(X') = \{x\}$.
In the full diagram, $\alpha_3, \beta_1, \beta_2, \gamma_1, \gamma_2$ are valid, so $\mu(Y) = \{y, y'\}$. □

Remark 6.3.1 Example 2.3.1 (p. 43) together with Corollary 6.3.1 (p. 167) show that $(\mu \subseteq)$ and (μCUM) without $(\cap)$ do not guarantee representability by a level n totally smooth structure.

6.4 The Essentially Smooth Case

Definition 6.4.1 Let $\mu : \mathcal{Y} \to \mathcal{P}(U)$ and $\mathcal{X}$ be given, let $\alpha : \langle y, j\rangle \to \langle x, i\rangle \in \mathcal{X}$.
Define

$$
\begin{aligned}
&\boldsymbol{O}(\alpha) := \{Y \in \mathcal{Y} : x \in Y - \mu(Y), y \in \mu(Y)\},\\
&\boldsymbol{D}(\alpha) := \{X \in \mathcal{Y} : x \in \mu(X), y \in X\},\\
&\Pi(\boldsymbol{O}, \alpha) := \Pi\{\mu(Y) : Y \in \boldsymbol{O}(\alpha)\},\\
&\Pi(\boldsymbol{D}, \alpha) := \Pi\{\mu(X) : X \in \boldsymbol{D}(\alpha)\}.
\end{aligned}
$$

Lemma 6.4.1 *Let U be the universe, $\mu : \mathcal{Y} \to \mathcal{P}(U)$. Let μ satisfy $(\mu \subseteq)$+$(\mu \subseteq\supseteq)$. Let $\mathcal{X}$ be a level 1 preferential structure, $\alpha : \langle y, j\rangle \to \langle x, i\rangle$, $\boldsymbol{O}(\alpha) \neq \emptyset$, $\boldsymbol{D}(\alpha) \neq \emptyset$. We can modify $\mathcal{X}$ to a level 3 structure $\mathcal{X}'$ by introducing level 2 and level 3 arrows s.t. no copy of α is valid in any $X \in \boldsymbol{D}(\alpha)$, and in every $Y \in \boldsymbol{O}(\alpha)$ at least one copy of α is valid. (More precisely, we should write $\mathcal{X}' \lceil X$, etc.) Thus, in $\mathcal{X}'$,*

(1) *$\langle x, i\rangle$ will not be minimal in any $Y \in \boldsymbol{O}(\alpha)$,*
(2) *if α is the only arrow minimizing $\langle x, i\rangle$ in $X \in \boldsymbol{D}(\alpha)$, $\langle x, i\rangle$ will now be minimal in X.*
The construction is made independently for all such arrows $\alpha \in \mathcal{X}$.

(This is probably the main technical result of the chapter.)

Proof

(1) The construction:
Make $\Pi(\boldsymbol{D}, \alpha)$ many copies of α : $\{\langle \alpha, f\rangle : f \in \Pi(\boldsymbol{D}, \alpha)\}$, all $\langle \alpha, f\rangle : \langle y, j\rangle \to \langle x, i\rangle$.
Note that $\langle \alpha, f\rangle \in X$ for all $X \in \boldsymbol{D}(\alpha)$ and $\langle \alpha, f\rangle \in Y$ for all $Y \in \boldsymbol{O}(\alpha)$.
Add to the structure $\langle \beta, f, X_r, g\rangle : \langle f(X_r), i_r\rangle \to \langle \alpha, f\rangle$, for any $X_r \in \boldsymbol{D}(\alpha)$, and $g \in \Pi(\boldsymbol{O}, \alpha)$ (and some or all i_r – this does not matter).
For all $Y_s \in \boldsymbol{O}(\alpha)$:
if $\mu(Y_s) \not\subseteq X_r$ and $f(X_r) \in Y_s$, then add to the structure $\langle \gamma, f, X_r, g, Y_s\rangle : \langle g(Y_s), j_s\rangle \to \langle \beta, f, X_r, g\rangle$ (again for all or some j_s),
if $\mu(Y_s) \subseteq X_r$ or $f(X_r) \notin Y_s$, $\langle \gamma, f, X_r, g, Y_s\rangle$ is not added.
See Diagram 6.4.1 (p. 170).
(2) Let $X_r \in \boldsymbol{D}(\alpha)$. We have to show that no $\langle \alpha, f\rangle$ is valid in X_r. Fix f.
$\langle \alpha, f\rangle$ is in X_r, so we have to show that for at least one $g \in \Pi(\boldsymbol{O}, \alpha)$ $\langle \beta, f, X_r, g\rangle$ is valid in X_r, i.e. for this g, no $\langle \gamma, f, X_r, g, Y_s\rangle : \langle g(Y_s), j_s\rangle \to \langle \beta, f, X_r, g\rangle$, $Y_s \in \boldsymbol{O}(\alpha)$ attacks $\langle \beta, f, X_r, g\rangle$ in X_r.

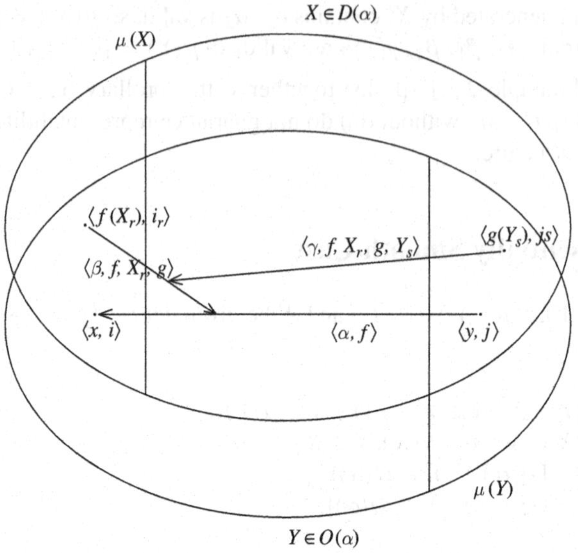

Diagram 6.4.1 The construction

We define g. Take $Y_s \in \boldsymbol{O}(\alpha)$.

Case 1: $\mu(Y_s) \subseteq X_r$ or $f(X_r) \notin Y_s$: choose arbitrary $g(Y_s) \in \mu(Y_s)$.

Case 2: $\mu(Y_s) \not\subseteq X_r$ and $f(X_r) \in Y_s$: Choose $g(Y_s) \in \mu(Y_s) - X_r$.

In Case 1, $\langle \gamma, f, X_r, g, Y_s\rangle$ does not exist, so it cannot attack $\langle \beta, f, X_r, g\rangle$.

In Case 2, $\langle \gamma, f, X_r, g, Y_s\rangle : \langle g(Y_s), j_s\rangle \to \langle \beta, f, X_r, g\rangle$ is not in X_r, as $g(Y_s) \notin X_r$.

Thus, no $\langle \gamma, f, X_r, g, Y_s\rangle : \langle g(Y_s), j_s\rangle \to \langle \beta, f, X_r, g\rangle$, $Y_s \in \boldsymbol{O}(\alpha)$ attacks $\langle \beta, f, X_r, g\rangle$ in X_r.

So $\forall \langle \alpha, f\rangle : \langle y, j\rangle \to \langle x, i\rangle$

$y \in X_r \Rightarrow \exists \langle \beta, f, X_r, g\rangle : \langle f(X_r), i_r\rangle \to \langle \alpha, f\rangle$

$(f(X_r) \in \mu(X_r) \wedge \neg\exists \langle \gamma, f, X_r, g, Y_s\rangle : \langle g(Y_s), j_s\rangle \to \langle \beta, f, X_r, g\rangle . g(Y_s) \in X_r)$.

But $\langle \beta, f, X_r, g\rangle$ was constructed only for $\langle \alpha, f\rangle$, so was $\langle \gamma, f, X_r, g, Y_s\rangle$, and there was no other $\langle \gamma, i\rangle$ attacking $\langle \beta, f, X_r, g\rangle$, so we are done.

(3) Let $Y_s \in \boldsymbol{O}(\alpha)$. We have to show that at least one $\langle \alpha, f\rangle$ is valid in Y_s.

We define $f \in \Pi(\boldsymbol{D}, \alpha)$. Take X_r.

If $\mu(X_r) \not\subseteq Y_s$, choose $f(X_r) \in \mu(X_r) - Y_s$. If $\mu(X_r) \subseteq Y_s$, choose arbitrary $f(X_r) \in \mu(X_r)$.

All attacks on $\langle x, f\rangle$ have the form $\langle \beta, f, X_r, g\rangle : \langle f(X_r), i_r\rangle \to \langle \alpha, f\rangle$, $X_r \in \boldsymbol{D}(\alpha)$, $g \in \Pi(\boldsymbol{O}, \alpha)$. We have to show that they are either not in Y_s, or that they are themselves attacked in Y_s.

Case 1: $\mu(X_r) \not\subseteq Y_s$. Then $f(X_r) \notin Y_s$, so $\langle \beta, f, X_r, g\rangle : \langle f(X_r), i_r\rangle \to \langle \alpha, f\rangle$ is not in Y_s (for no g).

Case 2: $\mu(X_r) \subseteq Y_s$. Then $\mu(Y_s) \not\subseteq X_r$ by $(\mu \subseteq\supseteq)$ and $f(X_r) \in Y_s$, so $\langle \beta, f, X_r, g \rangle : \langle f(X_r), i_r \rangle \to \langle \alpha, f \rangle$ is in Y_s (for all g). Take any $g \in \Pi(\boldsymbol{O}, \alpha)$. As $\mu(Y_s) \not\subseteq X_r$ and $f(X_r) \in Y_s$, $\langle \gamma, f, X_r, g, Y_s \rangle : \langle g(Y_s), j_s \rangle \to \langle \beta, f, X_r, g \rangle$ is defined, and $g(Y_s) \in \mu(Y_s)$, so it is in Y_s (for all g). Thus, $\langle \beta, f, X_r, g \rangle$ is attacked in Y_s.
Thus, for this f, all $\langle \beta, f, X_r, g \rangle$ are either not in Y_s, or attacked in Y_s.
Thus for this f, $\langle \alpha, f \rangle$ is valid in Y_s.
So for this $\langle x, i \rangle$
$\exists \langle \alpha, f \rangle : \langle y, j \rangle \to \langle x, i \rangle . y \in \mu(Y_s) \wedge$
(a) $\neg \exists \langle \beta, f, X_r, g \rangle : \langle f(X_r), i \rangle \to \langle \alpha, f \rangle . f(X_r) \in Y_s$
or
(b) $\forall \langle \beta, f, X_r, g \rangle : \langle f(X_r), i \rangle \to \langle \alpha, f \rangle$
$(f(X_r) \in Y_s \Rightarrow$
$\exists \langle \gamma, f, X_r, g, Y_s \rangle : \langle g(Y_s), j_s \rangle \to \langle \beta, f, X_r, g \rangle . g(Y_s) \in \mu(Y_s))$.

As we made copies of α only, introduced only βs attacking the α-copies, and γs attacking the βs, the construction is independent for different αs.
□

Proposition 6.4.1 *Let U be the universe, $\mu : \mathcal{Y} \to \mathcal{P}(U)$.*
Then any μ satisfying $(\mu \subseteq)$, $(\cap)$, (μCUM) (or, alternatively, $(\mu \subseteq)$ and $(\mu \subseteq\supseteq)$) can be represented by a level 3 essentially smooth structure.

Proof In stage one, consider as usual $\mathcal{U} := \langle \mathcal{X}, \{\alpha_i : i \in I\} \rangle$ where $\mathcal{X} := \{\langle x, f \rangle : x \in U, f \in \Pi\{\mu(X) : X \in \mathcal{Y}, x \in X - \mu(X)\}\}$, and set $\alpha : \langle x', f' \rangle \to \langle x, f \rangle :\Leftrightarrow x' \in ran(f)$.
For stage two:
Any level 1 arrow $\alpha : \langle y, j \rangle \to \langle x, i \rangle$ was introduced in stage one by some $Y \in \mathcal{Y}$ s.t. $y \in \mu(Y)$, $x \in Y - \mu(Y)$. Do the construction of Lemma 6.4.1 (p. 169) for all level 1 arrows of $\mathcal{X}$ in parallel or successively.

We have to show that the resulting structure represents μ and is essentially smooth. (Level 3 is obvious.)

(1) Representation:
Suppose $x \in Y - \mu(Y)$. Then there was in stage 1 for all $\langle x, i \rangle$ some $\alpha : \langle y, j \rangle \to \langle x, i \rangle$, $y \in \mu(Y)$. We examine the y.
If there is no X s.t. $x \in \mu(X)$, $y \in X$, then there were no βs and γs introduced for this $\alpha : \langle y, j \rangle \to \langle x, i \rangle$, so α is valid.
If there is X s.t. $x \in \mu(X)$, $y \in X$, consider $\alpha : \langle y, j \rangle \to \langle x, i \rangle$. So $X \in \boldsymbol{D}(\alpha)$, $Y \in \boldsymbol{O}(\alpha)$, so we did the construction of Lemma 6.4.1 (p. 169), and by its result, $\langle x, i \rangle$ is not minimal in Y.
Thus, in both cases, $\langle x, i \rangle$ is successfully attacked in Y, and no $\langle x, i \rangle$ is a minimal element in Y.
Suppose $x \in \mu(X)$ (we change notation to conform to Lemma 6.4.1 (p. 169)). Fix $\langle x, i \rangle$.
If there is no $\alpha : \langle y, j \rangle \to \langle x, i \rangle$, $y \in X$, then $\langle x, i \rangle$ is minimal in X, and we are done.

If there is α or $\langle \alpha, k \rangle : \langle y, j \rangle \to \langle x, i \rangle$, $y \in X$, then α originated from stage one through some Y s.t. $x \in Y - \mu(Y)$ and $y \in \mu(Y)$. (Note that stage 2 of the construction did not introduce any new level 1 arrows - only copies of existing level 1 arrows.) So $X \in \boldsymbol{D}(\alpha)$, $Y \in \boldsymbol{O}(\alpha)$; hence we did the construction of Lemma 6.4.1 (p. 169), and by its result, $\langle x, i \rangle$ is minimal in X, and we are done again.
In both cases, all $\langle x, i \rangle$ are minimal elements in X.

(2) Essential smoothness. We have to show the conditions of Definition 6.1.5 (p. 159). We will, however, work with the reformulation given in Remark 6.1.1 (p. 160).

Case (1): $x \in \mu(X)$.
Case (1.1): there is $\langle x, i \rangle$ with no $\langle \alpha, f \rangle : \langle y, j \rangle \to \langle x, i \rangle$, $y \in X$. There is nothing to show.
Case (1.2): for all $\langle x, i \rangle$ there is $\langle \alpha, f \rangle : \langle y, j \rangle \to \langle x, i \rangle$, $y \in X$.
α was introduced in stage 1 by some Y s.t. $x \in Y - \mu(Y)$, $y \in X \cap \mu(Y)$, so $X \in \boldsymbol{D}(\alpha)$, $Y \in \boldsymbol{O}(\alpha)$. In the proof of Lemma 6.4.1 (p. 169), at the end of (2), it was shown that

$$\begin{aligned}
&\exists \langle \beta, f, X_r, g \rangle : \langle f(X_r), i_r \rangle \to \langle \alpha, f \rangle \\
&\qquad (f(X_r) \in \mu(X_r) \wedge \\
&\qquad \neg\exists \langle \gamma, f, X_r, g, Y_s \rangle : \langle g(Y_s), j_s \rangle \to \langle \beta, f, X_r, g \rangle . g(Y_s) \in X_r).
\end{aligned}$$

By $f(X_r) \in \mu(X_r)$, condition (1) of Remark 6.1.1 (p. 160) is true.
Case (2): $x \notin \mu(Y)$. Fix $\langle x, i \rangle$. (We change notation back to Y.)
In stage 1, we constructed $\alpha : \langle y, j \rangle \to \langle x, i \rangle$, $y \in \mu(Y)$, so $Y \in \boldsymbol{O}(\alpha)$.
If $\boldsymbol{D}(\alpha) = \emptyset$, then there is no attack on α, and the condition (2) of Remark 6.1.1 (p. 160) is trivially true.
If $\boldsymbol{D}(\alpha) \neq \emptyset$, we did the construction of Lemma 6.4.1 (p. 169), so
$\exists \langle \alpha, f \rangle : \langle y, j \rangle \to \langle x, i \rangle . y \in \mu(Y_s) \wedge$

(a) $\neg\exists \langle \beta, f, X_r, g \rangle : \langle f(X_r), i \rangle \to \langle \alpha, f \rangle . f(X_r) \in Y_s$
or
(b) $\forall \langle \beta, f, X_r, g \rangle : \langle f(X_r), i \rangle \to \langle \alpha, f \rangle$
$(f(X_r) \in Y_s \Rightarrow$
$\exists \langle \gamma, f, X_r, g, Y_s \rangle : \langle g(Y_s), j_s \rangle \to \langle \beta, f, X_r, g \rangle . g(Y_s) \in \mu(Y_s)$.

As the only attacks on $\langle \alpha, f \rangle$ had the form $\langle \beta, f, X_r, g \rangle$, and $g(Y_s) \in \mu(Y_s)$, condition (2) of Remark 6.1.1 (p. 160) is satisfied. □

As said after Example 6.3.1 (p. 167), we do not know if level 3 is necessary for representation. We also do not know whether the same can be achieved with level 3, or higher, totally smooth structures.

6.5 Translation to Logic

We turn to the translation to logics.

Proposition 6.5.1 *Let* $\mid\sim$ *be a logic for* $\mathcal{L}$. *Set* $T^{\mathcal{M}} := Th(\mu_{\mathcal{M}}(M(T)))$, *where* $\mathcal{M}$ *is a generalized preferential structure and* $\mu_{\mathcal{M}}$ *its choice function. Then*

(1) *there is a level* 2 *preferential structure* $\mathcal{M}$ *s.t.* $\overline{\overline{T}} = T^{\mathcal{M}}$ *iff* (LLE), (CCL), (SC) *hold for all* $T, T' \subseteq \mathcal{L}$;

(2) *there is a level* 3 *essentially smooth preferential structure* $\mathcal{M}$ *s.t.* $\overline{\overline{T}} = T^{\mathcal{M}}$ *iff* (LLE), (CCL), (SC), $(\subseteq\supseteq)$ *hold for all* $T, T' \subseteq \mathcal{L}$.

Proof The proof is an immediate consequence of Corollary 6.2.1 (p. 166) (2), Fact 6.1.2 (p. 161), Proposition 6.4.1 (p. 171), and Proposition 2.3.1 (p. 45) (10) and (11).
(More precisely, for (2): Let $\mathcal{M}$ be an essentially smooth structure, then by Definition 6.1.7 (p. 162) for all X, $\mu(X) \sqsubseteq X$. Consider (μCUM). So by Fact 6.1.2 (p. 161) (2) $\mu(X') \subseteq X'' \subseteq X' \Rightarrow \mu(X') \sqsubseteq X''$, so by Fact 6.1.2 (p. 161) (1) $\mu(X') = \mu(X'')$. $(\mu \subseteq\supseteq)$ is analogous, using Fact 6.1.2 (p. 161) (3). □

We leave aside the generalization of preferential structures to attacking structures relative to η, as this can cause problems, without giving real insight. It might well be that $\rho(X) \not\sqsubseteq \eta(X)$, but, still, $\rho(X)$ and $\eta(X)$ might define the same theory – due to definability problems.

Chapter 7
Deontic Logic and Hierarchical Conditionals

7.1 Semantics of Deontic Logic

The material in this section is taken from [GS08f].

7.1.1 Introductory Remarks

We see some relation of "better" as central for obligations.

Obligations determine what is "better" and what is "worse"; conversely, given such a relation of "better", we can define obligations. The problems lie, in our opinion, in the fact that an adequate treatment of such a relation is somewhat complicated and leads to many ramifications.

On the other hand, obligations have sufficiently many things in common with facts, so we can in a useful way say that an obligation is satisfied in a situation, and one can also define a notion of derivation for obligations.

Our approach is almost exclusively semantical.

7.1.1.1 Context

The problem with formalization using logic is that the natural movements in the application area being formalized do not exactly correspond to natural movements in the logic being used as a tool of formalization. Put differently, we may be able to express statement A of the application area by a formula ϕ of the logic, but the subtleties of the way A is manipulated in the application area cannot be matched in the formal logic used. This gives rise to paradoxes. To resolve the paradoxes one needs to improve the logic. So the progress in the formalization programme depends on the state of development of logic itself. Recent serious advances in logical tools by the authors of this paper enable us to offer some better formalizations possibilities for the notion of obligation. This is what we offer in this chapter.

Historically, articles on deontic logic include collections of problems, see e.g. [MDW94], semantical approaches, see e.g. [Han69], and others, like [CJ02]. Our basic idea is to see obligations as tightly connected to some relation of "better".

D.M. Gabbay, K. Schlechta, *Logical Tools for Handling Change in Agent-Based Systems*, Cognitive Technologies, DOI 10.1007/978-3-642-04407-6_7,

An immediate consequence is that negation, which inverses such a relation, behaves differently in the case of obligations and of classical logic. ("And" and "or" seem to show analogue behaviour in both logics.) The relation of "better" has to be treated with some caution; however, and we introduce and investigate local and global properties about "better" of obligations. Most of these properties coincide in sufficiently nice situations, while in others, they are different.

We do not come to a final conclusion which properties obligations should or should not have. Perhaps, this will be answered in future, perhaps there is no universal answer. We provide a list of ideas which seem reasonable to us.

Throughout, we work in a finite (propositional) setting.

7.1.1.2 Central Idea

We see a tight connection between obligations and a relation of "morally" better between situations. Obligations are there to guide us for "better" actions, and, conversely, given some relation of "better", we can define obligations.

The problems lie in the fact that a simple approach via quality is not satisfactory. We examine a number of somewhat more subtle ideas, some of them also using a notion of distance.

7.1.1.3 A Common Property of Facts and Obligations

We are fully aware that an obligation has a conceptually different status than a fact. The latter is true or false, an obligation has been set by some instance as a standard for behaviour, or whatever.

Still, we will say that an obligation holds in a situation, or that a situation satisfies an obligation. If the letter is in the mail box, the obligation to post it is satisfied, but if it is in the trash bin, the obligation is not satisfied. In some set of worlds, the obligation is satisfied, in the complement of this set, it is not satisfied. Thus, obligations behave in this respect like facts, and we put for this aspect all distinctions between facts and obligations aside.

Thus, we will treat obligations most of the time as subsets of the set of all models, but also sometimes as formulas. As we work mostly in a finite propositional setting, both are interchangeable.

We are *not* concerned here with actions to fulfill obligations, developments or so, just simply situations which satisfy or not obligations. This is perhaps also the right place to point out that one has to distinguish between facts that *hold* in "good" situations (they will be closed under arbitrary right weakening) and obligations which describe what *should* be, they will not be closed under arbitrary right weakening. This article is only about the latter.

7.1.1.4 Derivations of Obligations

Again, we are aware that "deriving" obligations is different from "deriving" facts. Derivation of facts is supposed to conclude from truth to truth, deriving obligations while will be about concluding what can also be considered an obligation, given some set of "basic" obligations. The parallel is sufficiently strong to justify the double use of the word "derive".

Very roughly, we will say that conjunctions (or intersections) and disjunctions (unions) of obligations lead in a reasonable way to derived obligations, but negations do not. We take the Ross paradox (see below) very seriously. It was, as a matter of fact, our starting point to avoid it in a reasonable notion of derivation.

We mention two simple postulates that derived obligations should probably satisfy.

(1) Every original obligation should also be a derived obligation, corresponding to $\alpha, \beta \mid\sim \alpha$.
(2) A derived obligation should not be trivial, i.e. neither empty nor U, the universe we work in.

The last property is not very important from an algebraic point of view, and easily satisfiable, so we will not give it too much attention.

7.1.1.5 Orderings and Obligations

There is, in our opinion, a deep connection between obligations and orderings (and, in addition, distances), which works both ways.

First, given a set of obligations, we can say that one situation is "better" than a second situation, if the first satisfies "more" obligations than the second does. "More" can be measured by the set of obligations satisfied, and then by the subset/superset relation, or by the number of obligations. Both are variants of the Hamming distance. "Distance" between two situations can be measured by the set or number of obligations in which they differ (i.e. one situation satisfies them, the other not). In both cases, we will call the variants the set or the counting variant.

This is also the deeper reason why we have to be careful with negation. Negation inverses such orderings, if ϕ is better than $\neg\phi$, then $\neg\phi$ is worse than $\neg\neg\phi = \phi$. But in some reasonable sense $\wedge$ and $\vee$ preserve the ordering, thus they are compatible with obligations.

Conversely, given a relation (of quality), we might for instance require that obligations are closed under improvement. More subtle requirements might work with distances. The relations of quality and distance can be given abstractly (as the notion of size used for "soft" obligations), or as above by a starting set of obligations. We will also define important auxiliary concepts on such abstract relations.

7.1.1.6 Derivation Revisited

A set of "basic" obligations generates an ordering and a distance between situations, and ordering and distance can be used to define properties obligations should have. It is thus natural to define obligations derived from the basic set as those sets of situations which satisfy the desirable properties of obligations defined via the order and distance generated by the basic set of obligations. Our derivation is thus a two-step procedure: First, generate the order and distance, which define suitable sets of situations.

We will call properties which are defined without using distances global properties (like closure under improving quality), properties involving distance (like being a neighbourhood) local properties.

7.1.1.7 Relativization

An important problem is relativization. Suppose $\mathcal{O}$ is a set of obligations for all possible situations, e.g. O is the obligation to post the letter and O' is the obligation to water the flowers. Ideally, we will do both. Suppose we consider now a subset, where we cannot do both (e.g. for lack of time). What are our obligations in this subset? Are they just the restrictions to the subset? Conversely, if O is an obligation for a subset of all situations, is then some O' with $O \subseteq O'$ an obligation for the set of all situations?

In more complicated requirements, it might be reasonable, e.g. to choose the ideal situations still in the big set, even if they are not in the subset to be considered, but use an approximation inside the subset. Thus, relativizations present a non trivial problem with many possible solutions, and it seems doubtful whether a universally adequate solution can be found.

7.1.1.8 Numerous Possibilities

Seeing the possibilities presented so far (set or counting order, set or counting distance, various relativizations), we can already guess that there are numerous possible reasonable approaches to what an obligation is or should be. Consequently, it seems quite impossible to pursue all these combinations in detail. Thus, we concentrate mostly on one combination, and leave it to the reader to fill in details for the others, if (s)he is so interested.

We will also treat the defeasible case here. Perhaps somewhat surprisingly, this is straightforward, and largely due to the fact that there is one natural definition of "big" sets for the product set, given that "big" sets are defined for the components. So there are no new possibilities to deal with here.

The multitude of possibilities is somewhat disappointing. It may, of course, be due to an incapacity of the present authors to find *the* right notion. But it may also be a genuine problem of ambiguous intuitions, and thus generate conflicting views and misunderstandings on the one side, and loopholes for not quite honest argumentation in practical juridical reasoning on the other hand.

7.1.1.9 Overview

We will work in a finite propositional setting, so there is a trivial and 1-1 correspondence between formulas and model sets. Thus, we can just work with model sets – which implies, of course, that obligations will be robust under logical reformulation. So we will formulate most results only for sets.

- In Sect. 7.1.2 (p. 179), we give the basic definitions, together with some simple results about those definitions.
- Sect. 7.1.3 (p. 186) will present a more philosophical discussion, with more examples, and we will argue that our definitions are relevant for our purpose.
- As said already, there seems to be a multitude of possible and reasonable definitions of what an obligation can or should be, so we limit our formal investigation to a few cases, this is given in Sect. 7.1.4 (p. 193).
- In Sect. 7.1.5 (p. 197), we give a tentative definition of an obligation.

(The concept of neighbourhood semantics is not new, and was already introduced by D. Scott, [Sco70], and R.Montague, [Mon70]. Further investigations showed that it was also used by O. Pacheco, [Pac07], precisely to avoid unwanted weakening for obligations. We came the other way and started with the concept of independent strengthening, see Definition 7.1.12 (p. 186), and introduced the abstract concept of neighbourhood semantics only at the end. This is one of the reasons we also have different descriptions which turned out to be equivalent: we came from elsewhere.)

7.1.2 Basic Definitions

We give here all definitions needed for our treatment of obligations. The reader may continue immediately to Sect. 7.1.3 (p. 186), and come back to the present section whenever necessary. Intuitively, U is supposed to be the set of models of some propositional language $\mathcal{L}$, but we will stay purely algebraic whenever possible. $U' \subseteq U$ is some subset.

For ease of writing, we will here and later sometimes work with propositional variables, and also identify models with the formula describing them, still this is just shorthand for arbitrary sets and elements. pq will stand for $p \wedge q$, etc. If a set $\mathcal{O}$ of obligations is given, these will be just arbitrary subsets of the universe U. We will also say that $\mathcal{O}$ is over U.

Before we deepen the discussion of more conceptual aspects, we give some basic definitions (for which we claim no originality). We will need them quite early, and think it is better to put them together here, not to be read immediately, but so the reader can leaf back and find them easily.

We work here with a notion of size (for the defeasible case), a notion d of distance, and a quality relation $\preceq$. The latter two can be given abstractly, but may also be defined from a set of (basic) obligations.

We use these notions to describe properties that obligations should, in our opinion, have. A careful analysis will show later interdependencies between the different properties.

7.1.2.1 Product Size

For each $U' \subseteq U$, we suppose an abstract notion of size to be given. We may assume this notion to be a filter or an ideal. Coherence properties between the filters/ideals of different U', U'' will be left open. The reader may assume them to be the conditions of the system P of preferential logic, see Sect. 2.1 (p. 31).

Given such notions of size on U' and U'', we will also need a notion of size on $U' \times U''$. We take the evident solution:

Definition 7.1.1 Let a notion of "big subset" be defined by a principal filter for all $X \subseteq U$ and all $X' \subseteq U'$. Thus, for all $X \subseteq U$ there exists a fixed principal filter $\mathcal{F}(X) \subseteq \mathcal{P}(X)$, and likewise for all $X' \subseteq U'$. (This is the situation in the case of preferential structures, where $\mathcal{F}(X)$ is generated by $\mu(X)$, the set of minimal elements of X.)

Define now $\mathcal{F}(X \times X')$ as generated by $\{A \times A' : A \in \mathcal{F}(X),\ A' \in \mathcal{F}(X')\}$, i.e. if A is the smallest element of $\mathcal{F}(X)$, A' the smallest element of $\mathcal{F}(X')$, then $\mathcal{F}(X \times X') := \{B \subseteq X \times X' : A \times A' \subseteq B\}$.

Fact 7.1.1 If $\mathcal{F}(X)$ and $\mathcal{F}(X')$ are generated by preferential structures $\preceq_X$, $\preceq_{X'}$, then $\mathcal{F}(X \times X')$ is generated by the product structure defined by
$\langle x, x' \rangle \preceq_{X \times X'} \langle y, y' \rangle :\Leftrightarrow x \preceq_X y$ and $x' \preceq_{X'} y'$.

Proof We will omit the indices of the orderings when this causes no confusion.

Let $A \in \mathcal{F}(X)$, $A' \in \mathcal{F}(X')$, i.e. A minimizes X, A' minimizes X'. Let $\langle x, x' \rangle \in X \times X'$, then there are $a \in A$, $a' \in A'$ with $a \preceq x$, $a' \preceq x'$, so $\langle a, a' \rangle \preceq \langle x, x' \rangle$.

Conversely, suppose $U \subseteq X \times X'$, U minimizes $X \times X'$. But there is no $A \times A' \subseteq U$ s.t. $A \in \mathcal{F}(X)$, $A' \in \mathcal{F}(X')$. Assume $A = \mu(X)$, $A' = \mu(X')$, so there is $\langle a, a' \rangle \in \mu(X) \times \mu(X')$, $\langle a, a' \rangle \notin U$. But only $\langle a, a' \rangle \preceq \langle a, a' \rangle$, and U does not minimize $X \times X'$, *contradiction*. □

Note that a natural modification of our definition,

There is $A \in \mathcal{F}(X)$ s.t. for all $a \in A$ there is a (maybe varying) $A'_a \in \mathcal{F}(X')$, and $U := \{\langle a, a' \rangle : a \in A,\ a' \in A'_a\}$ as generating sets, will result in the same definition, as our filters are principal, and thus stable under arbitrary intersections.

7.1.2.2 Distance

We consider a set of sequences Σ, for $x \in \Sigma\ x : I \to S$, I a finite index set, S some set. Often, S will be $\{0, 1\}$, $x(i) = 1$ will mean that $x \in i$, when $I \subseteq \mathcal{P}(U)$ and $x \in U$. For abbreviation, we will call this (unsystematically, often context will tell) the $\in$-case . Often, I will be written $\mathcal{O}$, intuitively, $O \in \mathcal{O}$ is then an obligation, and $x(O) = 1$ means $x \in O$, or x "satisfies" the obligation O.

Definition 7.1.2 In the $\in$-case, set $\mathcal{O}(x) := \{O \in \mathcal{O} : x \in O\}$.

Recall Definition 2.1.6 (p. 33).

Remark 7.1.1 If $\sigma(i)$ are equivalence classes, one has to be careful not to confound the distance between the classes and the resulting distance between elements of the classes, as two different elements in the same class have distance 0. So in Fact 2.1.2 (p. 33), (2.1) only one direction holds.

Definition 7.1.3

(1) We can define for any distance d with some minimal requirements a notion of "between".
If the codomain of d has an ordering $\leq$, but no addition, we define

$$\langle x, y, z\rangle_d :\Leftrightarrow d(x, y) \leq d(x, z) \text{ and } d(y, z) \leq d(x, z).$$

If the codomain has a commutative addition, we define

$$\langle x, y, z\rangle_d :\Leftrightarrow d(x, z) = d(x, y) + d(y, z) \text{ – in } d_s + \text{ will be replaced by } \cup,$$

i.e.

$$\langle x, y, z\rangle_s :\Leftrightarrow d(x, z) = d(x, y) \cup d(y, z).$$

For above two Hamming distances, we will write $\langle x, y, z\rangle_s$ and $\langle x, y, z\rangle_c$.
(2) We further define
$[x, z]_d := \{y \in X : \langle x, y, x\rangle_d\}$ – where X is the set we work in.
We will write $[x, z]_s$ and $[x, z]_c$ when appropriate.
(3) For $x \in U$, $X \subseteq U$ set $x \parallel_d X := \{x' \in X : \neg\exists x'' \neq x' \in X.d(x, x') \geq d(x, x'')\}$.
Note that, if $X \neq \emptyset$, then $x \parallel X \neq \emptyset$.
We omit the index when this does not cause confusion. Again, when adequate, we write $\parallel_s$ and $\parallel_c$.

For problems with characterizing "between" see [Sch04].

Fact 7.1.2

(0) $\langle x, y, z\rangle_d \Leftrightarrow \langle z, y, x\rangle_d$.
Consider the situation of a set of sequences Σ.
Let $A := A_{\sigma,\sigma''} := \{\sigma' : \forall i \in I(\sigma(i) = \sigma''(i) \rightarrow \sigma'(i) = \sigma(i) = \sigma''(i))\}$. Then
(1) If $\sigma' \in A$, then $d_s(\sigma, \sigma'') = d_s(\sigma, \sigma') \cup d_s(\sigma', \sigma'')$, so $\langle \sigma, \sigma', \sigma''\rangle_s$.
(2) If $\sigma' \in A$ and S consists of two elements (as in classical two-valued logic), then $d_s(\sigma, \sigma')$ and $d_s(\sigma', \sigma'')$ are disjoint.
(3) $[\sigma, \sigma'']_s = A$.
(4) If, in addition, S consists of two elements, then $[\sigma, \sigma'']_c = A$.

Proof

(0) Trivial.
(1) "$\subseteq$" follows from Fact 2.1.2 (p. 33), (2.3).
Conversely, if e.g. $i \in d_s(\sigma, \sigma')$, then by prerequisite $i \in d_s(\sigma, \sigma'')$.
(2) Let $i \in d_s(\sigma, \sigma') \cap d_s(\sigma', \sigma'')$, then $\sigma(i) \neq \sigma'(i)$ and $\sigma'(i) \neq \sigma''(i)$, but then by $card(S) = 2$ $\sigma(i) = \sigma''(i)$, but $\sigma' \in A$, *contradiction*.

We turn to (3) and (4):

If $\sigma' \notin A$, then there is i' s.t. $\sigma(i') = \sigma''(i') \neq \sigma'(i')$. On the other hand, for all i s.t. $\sigma(i) \neq \sigma''(i)$ $i \in d_s(\sigma, \sigma') \cup d_s(\sigma', \sigma'')$. Thus:

(3) By (1) $\sigma' \in A \Rightarrow \langle \sigma, \sigma', \sigma'' \rangle_s$. Suppose $\sigma' \notin A$, so there is i' s.t. $i' \in d_s(\sigma, \sigma') - d_s(\sigma, \sigma'')$, so $\langle \sigma, \sigma', \sigma'' \rangle_s$ cannot be.

(4) By (1) and (2) $\sigma' \in A \Rightarrow \langle \sigma, \sigma', \sigma'' \rangle_c$. Conversely, if $\sigma' \notin A$, then $card(d_s(\sigma, \sigma')) + card(d_s(\sigma', \sigma'')) \geq card(d_s(\sigma, \sigma'')) + 2$. □

Definition 7.1.4 Given a finite propositional language $\mathcal{L}$ defined by the set $v(\mathcal{L})$ of propositional variables, let $\mathcal{L}_\wedge$ be the set of all consistent conjunctions of elements from $v(\mathcal{L})$ or their negations. Thus, $p \wedge \neg q \in \mathcal{L}_\wedge$ if $p, q \in v(\mathcal{L})$, but $p \vee q$, $\neg(p \wedge q) \notin \mathcal{L}_\wedge$. Finally, let $\mathcal{L}_{\vee\wedge}$ be the set of all (finite) disjunctions of formulas from $\mathcal{L}_\wedge$. (As we will later not consider all formulas from $\mathcal{L}_\wedge$, this will be a real restriction.)

Given a set of models M for a finite language $\mathcal{L}$, define $\phi_M := \bigwedge\{p \in v(\mathcal{L}) : \forall m \in M, m(p) = v\} \wedge \bigwedge\{\neg p : p \in v(\mathcal{L}), \forall m \in M, m(p) = f\} \in \mathcal{L}_\wedge$. (If there are no such p, set $\phi_M := TRUE$.)

This is the strongest $\phi \in \mathcal{L}_\wedge$ which holds in M.

Fact 7.1.3 If x, y are models, then $[x, y] = M(\phi_{\{x,y\}})$.

Proof

$m \in [x, y] \Leftrightarrow \forall p\ (x \models p, y \models p \Rightarrow m \models p$ and $x \not\models p, y \not\models p \Rightarrow m \not\models p)$,
$m \models \phi_{\{x,y\}} \Leftrightarrow m \models \bigwedge\{p : x(p) = y(p) = v\} \wedge \bigwedge\{\neg p : x(p) = y(p) = f\}$. □

7.1.2.3 Quality and Closure

Definition 7.1.5 Given any relation $\preceq$ (of quality), we say that $X \subseteq U$ is (downward) closed (with respect to $\preceq$) iff $\forall x \in X \forall y \in U\ (y \preceq x \Rightarrow y \in X)$.

(Warning, we follow the preferential tradition, "smaller" will mean "better".)

We think that being closed is a desirable property for obligations: what is at least as good as one element in the obligation should be "in", too.

Fact 7.1.4 Let $\preceq$ be given.

(1) Let $D \subseteq U' \subseteq U''$; if D is closed in U'', then D is also closed in U'.
(2) Let $D \subseteq U' \subseteq U''$; if D is closed in U', U' closed in U'', then D is closed in U''.
(3) Let $D_i \subseteq U'$ be closed for all $i \in I$, then so are $\bigcup\{D_i : i \in I\}$ and $\bigcap\{D_i : i \in I\}$.

Proof

(1) Trivial.
(2) Let $x \in D \subseteq U'$, $x' \preceq x$, $x' \in U''$, then $x' \in U'$ by closure of U'', so $x' \in D$ by closure of U'.
(3) Trivial. □

We may have an abstract relation $\preceq$ of quality on the domain, but we may also define it from the structure of the sequences, as we will do now.

Definition 7.1.6 Consider the case of sequences.

Given a relation $\preceq$ (of quality) on the codomain, we extend this to sequences in Σ :

$$x \sim y :\Leftrightarrow \forall i \in I\ (x(i) \sim y(i)),$$
$$x \preceq y :\Leftrightarrow \forall i \in I\ (x(i) \preceq y(i)),$$
$$x \prec y :\Leftrightarrow \forall i \in I\ (x(i) \preceq y(i)) \text{ and } \exists i \in I (x(i) \prec y(i)).$$

In the $\in$-case, we will consider $x \in i$ better than $x \notin i$. As we have only two values, true/false, it is easy to count the positive and negative cases (in more complicated situations, we might be able to multiply), so we have an analogue of the two Hamming distances, which we might call the Hamming quality relations.

Let $\mathcal{O}$ be given now.

(Recall that we follow the preferential tradition, "smaller" will mean "better".)

$$x \sim_s y :\Leftrightarrow \mathcal{O}(x) = \mathcal{O}(y),$$
$$x \preceq_s y :\Leftrightarrow \mathcal{O}(y) \subseteq \mathcal{O}(x),$$
$$x \prec_s y :\Leftrightarrow \mathcal{O}(y) \subset \mathcal{O}(x),$$
$$x \sim_c y :\Leftrightarrow card(\mathcal{O}(x)) = card(\mathcal{O}(y)),$$
$$x \preceq_c y :\Leftrightarrow card(\mathcal{O}(y)) \leq card(\mathcal{O}(x)),$$
$$x \prec_c y :\Leftrightarrow card(\mathcal{O}(y)) < card(\mathcal{O}(x)).$$

The requirement of closure causes a problem for the counting approach: Given, e.g. two obligations $O,\ O'$, then any two elements in just one obligation have the same quality, so if one is in, the other should be too. But this prevents now any of the original obligations to have the desirable property of closure. In the counting case, we will obtain a ranked structure, where elements satisfy 0, 1, 2, etc. obligations, and we are unable to differentiate inside those layers. Moreover, the set variant seems to be closer to logic, where we do not count the propositional variables which hold in a model, but consider them individually. For these reasons, we will not pursue the counting approach as systematically as the set approach. One should, however, keep in mind that the counting variant gives a ranking relation of quality, as all qualities are comparable, and the set variant does not. A ranking seems to be appreciated sometimes in the literature, though we are not really sure why.

Of particular interest is the combination of d_s and $\preceq_s$ (d_c and $\preceq_c$), respectively – where by $\preceq_s$ we also mean $\prec_s$ and $\sim_s$, etc. We turn to this now.

Fact 7.1.5 We work in the $\in$-case.

(1) $x \preceq_s y \Rightarrow d_s(x, y) = \mathcal{O}(x) - \mathcal{O}(y)$
Let $a \prec_s b \prec_s c$. Then
(2) $d_s(a, b)$ and $d_s(b, c)$ are not comparable,
(3) $d_s(a, c) = d_s(a, b) \cup d_s(b, c)$, and thus $b \in [a, c]_s$.
This does not hold in the counting variant, as Example 7.1.1 (p. 184) shows.
(4) Let $x \prec_s y$ and $x' \prec_s y$ with x, x' $\prec_s$-incomparable. Then $d_s(x, y)$ and $d_s(x', y)$ are incomparable.

(This does not hold in the counting variant, as then all distances are comparable.)

(5) If $x \prec_s z$, then for all $y \in [x, z]_s$ $x \preceq_s y \preceq_s z$.

Proof

(1) Trivial.
(2) We have $\mathcal{O}(c) \subset \mathcal{O}(b) \subset \mathcal{O}(a)$, so the results follows from (1).
(3) By definition of d_s and (1).
(4) x and x' are $\preceq_s$-incomparable, so there are $O \in \mathcal{O}(x) - \mathcal{O}(x')$, $O' \in \mathcal{O}(x') - \mathcal{O}(x)$.
As $x, x' \prec_s y$, $O, O' \notin \mathcal{O}(y)$, so $O \in d_s(x, y) - d_s(x', y)$, $O' \in d_s(x', y) - d_s(x, y)$.
(5) $x \prec_s z \Rightarrow \mathcal{O}(z) \subset \mathcal{O}(x)$, $d_s(x, z) = \mathcal{O}(x) - \mathcal{O}(z)$. By prerequisite $d_s(x, z) = d_s(x, y) \cup d_s(y, z)$. Suppose $x \not\preceq_s y$. Then there is $i \in \mathcal{O}(y) - \mathcal{O}(x) \subseteq d_s(x, y)$, so $i \notin \mathcal{O}(x) - \mathcal{O}(z) = d_s(x, z)$, *contradiction*.
Suppose $y \not\preceq_s z$. Then there is $i \in \mathcal{O}(z) - \mathcal{O}(y) \subseteq d_s(y, z)$, so $i \notin \mathcal{O}(x) - \mathcal{O}(z) = d_s(x, z)$, *contradiction*. □

Example 7.1.1 In this and similar examples, we will use the model notation. Some propositional variables p, q, etc. are given, and models are described by $p\neg qr$, etc. Moreover, the propositional variables are the obligations, so in this example we have the obligations p, q, r.

Consider $x := \neg p\neg qr$, $y := pq\neg r$, $z := \neg p\neg q\neg r$. Then $y \prec_c x \prec_c z$, $d_c(x, y) = 3$, $d_c(x, z) = 1$, $d_c(z, y) = 2$, so $x \notin [y, z]_c$.

Definition 7.1.7 Given a quality relation $\prec$ between elements, and a distance d, we extend the quality relation to sets and define:

(1) $x \prec Y :\Leftrightarrow \forall y \in (x \parallel Y), x \prec y$. (The closest elements – i.e. there are no closer ones – of Y, seen from x, are less good than x.)
Analogously $X \prec y :\Leftrightarrow \forall x \in (y \parallel X), x \prec y$
(2) $X \prec_l Y :\Leftrightarrow \forall x \in X, x \prec Y$ and $\forall y \in Y, X \prec y$ (X is locally better than Y).
When necessary, we will write $\prec_{l,s}$ or $\prec_{l,c}$ to distinguish the set from the counting variant.
For the next definition, we use the notion of size: $\nabla\phi$ iff for almost all ϕ holds, i.e. the set of exceptions is small.
(3) $X \ll_l Y :\Leftrightarrow \nabla x \in X, x \prec Y$ and $\nabla y \in Y, X \prec y$.
We will likewise write $\ll_{l,s}$ etc.
This definition is supposed to capture quality difference under minimal change, the "ceteris paribus " idea: $X \prec_l \mathbf{C}X$ should hold for an obligation X. Minimal change is coded by $\parallel$, and "ceteris paribus" by minimal change.

Fact 7.1.6 If $X \prec_l \mathbf{C}X$ and $x \in U$ an optimal point (there is no better one), then $x \in X$.

Proof If not, then take $x' \in X$ closest to x, this must be better than x, contradiction.

Fact 7.1.7 Take the set version.
If $X \prec_{l,s} CX$, then X is downward $\prec_s$-closed.

Proof Suppose $X \prec_{l,s} CX$, but X is not downward closed.

Case 1: There are $x \in X$, $y \notin X$, $y \sim_s x$. Then $y \in x \parallel_s CX$, but $x \not\prec y$, *contradiction*.

Case 2: There are $x \in X$, $y \notin X$, $y \prec_s x$. By $X \prec_{l,s} CX$, the elements in X closest to y must be better than y. Thus, there is $x' \prec_s y$, $x' \in X$, with minimal distance from y. But then $x' \prec_s y \prec_s x$, so $d_s(x', y)$ and $d_s(y, x)$ are incomparable by Fact 7.1.5 (p. 183), so x is among those with minimal distance from y, so $X \prec_{l,s} CX$ does not hold. □

Example 7.1.2 We work with the set variant.

This example shows that $\preceq_s$-closed does not imply $X \prec_{l,s} CX$, even if X contains the best elements.

Let $\mathcal{O} := \{p, q, r, s\}$, $U' := \{x := p\neg q\neg r\neg s,\ y := \neg pq\neg r\neg s,\ x' := pqrs\}$, $X := \{x, x'\}$. x' is the best element of U', so X contains the best elements, and X is downward closed in U', as x and y are not comparable. $d_s(x, y) = \{p, q\}$, $d_s(x', y) = \{p, r, s\}$, so the distances from y are not comparable, so x is among the closest elements in X, seen from y, but $x \not\prec_s y$.

The lack of comparability is essential here, as the following fact shows.

We have, however, for the counting variant:

Fact 7.1.8 Consider the counting variant. Then
If X is downward closed, then $X \prec_{l,c} CX$.

Proof Take any $x \in X$, $y \notin X$. We have $y \preceq_c x$ or $x \prec_c y$, as any two elements are $\preceq_c$-comparable. $y \preceq_c x$ contradicts closure, so $x \prec_c y$, and $X \prec_{l,c} CX$ holds trivially. □

7.1.2.4 Neighbourhood

Definition 7.1.8 Given a distance d, we define:

(1) Let $X \subseteq Y \subseteq U'$, then Y is a neighbourhood of X in U' iff
$\forall y \in Y\ \forall x \in X$ (x is closest to y among all x' with $x' \in X \Rightarrow [x, y] \cap U' \subseteq Y$).
(Closest means that there are no closer ones.)
When we also have a quality relation $\prec$, we define:

(2) Let $X \subseteq Y \subseteq U'$, then Y is an improving neighbourhood of X in U' iff
$\forall y \in Y\ \forall x$(($x$ is closest to y among all x' with $x' \in X$ and $x' \preceq y$) $\Rightarrow [x, y] \cap U' \subseteq Y$).
When necessary, we will have to say for (3) and (4) which variant, i.e. set or counting, we mean.

Fact 7.1.9

(1) If $X \subseteq X' \subseteq \Sigma$, and $d(x, y) = 0 \Rightarrow x = y$, then X and X' are Hamming neighbourhoods of X in X'.
(2) If $X \subseteq Y_j \subseteq X' \subseteq \Sigma$ for $j \in J$, and all Y_j are Hamming neighbourhoods of X in X', then so are $\bigcup\{Y_j : j \in J\}$ and $\bigcap\{Y_j : j \in J\}$.

Proof

(1) is trivial (we need here that $d(x, y) = 0 \Rightarrow x = y$).
(2) Trivial. □

7.1.2.5 Unions of Intersections and Other Definitions

Definition 7.1.9 Let $\mathcal{O}$ over U be given.

$X \subseteq U'$ is (ui) (for union of intersections) iff there is a family $\mathcal{O}_i \subseteq \mathcal{O}$, $i \in I$ s.t.
$X = (\bigcup\{\bigcap \mathcal{O}_i : i \in I\}) \cap U'$.

Unfortunately, this definition is not very useful for simple relativization.

Definition 7.1.10 Let $\mathcal{O}$ be over U. Let $\mathcal{O}' \subseteq \mathcal{O}$. Define for $m \in U$ and $\delta : \mathcal{O}' \to 2 = \{0, 1\}$

$$m \models \delta :\Leftrightarrow \forall O \in \mathcal{O}' \, (m \in O \Leftrightarrow \delta(O) = 1).$$

Definition 7.1.11 Let $\mathcal{O}$ be over U.
$\mathcal{O}$ is independent iff $\forall \delta : \mathcal{O} \to 2 \; \exists m \in U$ s.t. $m \models \delta$.
Obviously, independence does not inherit downward to subsets of U.

Definition 7.1.12 This definition is only intended for the set variant.
Let $\mathcal{O}$ be over U.

$$\begin{aligned}\mathcal{D}(\mathcal{O}) := \{X \subseteq U' : \forall \mathcal{O}' \subseteq \mathcal{O} \forall \delta : \mathcal{O}' \to 2 \\ ((\exists m, m' \in U, m, m' \models \delta, m \in X, m' \notin X) \Rightarrow \\ (\exists m'' \in X, m'' \models \delta \wedge m'' \prec_s m'))\}.\end{aligned}$$

This property expresses that we can satisfy obligations independently: If we respect O, we can, in addition, respect O', and if we are hopeless kleptomaniacs, we may still not be a murderer. If $X \in \mathcal{D}(\mathcal{O})$, we can go from $U - X$ into X by improving on all $O \in \mathcal{O}$, which we have not fixed by δ, if δ is not too rigid.

7.1.3 Philosophical Discussion of Obligations

We take now a closer look at obligations, in particular at the ramifications of the treatment of the relation "better". Some aspects of obligations will also need a notion of distance, we call them local properties of obligations.

7.1.3.1 A Fundamental Difference Between Facts and Obligations: Asymmetry and Negation

There is an important difference between facts and obligations. A situation which satisfies an obligation is in some sense "good", while a situation which does not is in some sense "bad". This is not true of facts. Being "round" is a priori not better than "having corners" or vice versa. But given the obligation to post the letter, the letter in the mail box is "good", and the letter in the trash bin is "bad". Consequently, negation has to play a different role for obligations and facts.

This is a fundamental property, which can also be found in orders, planning (we move towards the goal or not), reasoning with utility (is ϕ or $\neg\phi$ more useful?), and probably others, like perhaps the black raven paradox.

We also think that the Ross paradox (see below) is a true paradox and should be avoided. A closer look shows that this paradox involves arbitrary weakening, in particular by the "negation" of an obligation. This was a starting point of our analysis.

"Good" and "bad" cannot mean that any situation satisfying obligation O is better than any situation not satisfying O, as the following example shows.

Example 7.1.3 If we have three independent and equally strong obligations, O, O', O'', then a situation satisfying O but neither O' nor O'' will not be better than one satisfying O' and O'', but not O.

We have to introduce some kind of "ceteris paribus". All other things being equal, a situation satisfying O is better than a situation not satisfying O, see Sect. 7.1.3.3 (p. 188).

Example 7.1.4 The original version of the Ross paradox reads: If we have the obligation to post the letter, then we have the obligation to post or burn the letter. Implicit here is the background knowledge that burning the letter implies not to post it, and is even worse than not posting it.

We prefer a modified version, which works with two independent obligations: We have the obligation to post the letter, and we have the obligation to water the plants. We conclude by unrestricted weakening that we have the obligation to post the letter or *not* to water the plants. This is obvious nonsense.

It is not the "or" itself which is the problem. For instance, in case of an accident, to call an ambulance or to help the victims by giving first aid is a perfectly reasonable obligation. It is the negation of the obligation to water the plants which is the problem. More generally, it must not be that the system of suitable sets is closed under arbitrary supersets, otherwise we have closure under arbitrary right weakening, and thus the Ross paradox. Notions like "big subset" or "small exception sets" from the semantics of nonmonotonic logics are closed under supersets, so they are not suitable.

7.1.3.2 "And" and "Or" for Obligations

"Not" behaves differently for facts and for obligations. If O and O' are obligations, can $O \wedge O'$ be considered an obligation? We think, yes. "Ceteris paribus", satisfying

O and O' together, is better than not to do so. If O the obligation is to post the letter, O' to water the plants, then doing both is good, and better than doing none, or only one. Is $O \vee O'$ an obligation? Again, we think, yes. Satisfying one (or even both, a non exclusive or) is better than doing nothing. We might not have enough time to do both, so we do our best, and water the plants or post the letter. Thus, if α and β are obligations, then so will be $\alpha \wedge \beta$ and $\alpha \vee \beta$, but not anything involving $\neg\alpha$ or $\neg\beta$. (In a non trivial manner, leaving aside tautologies and contradictions which have to be considered separately.) To summarize: "and" and "or" preserve the asymmetry, "not" does not; therefore we can combine obligations using "and" and "or", but not "not". Thus, a reasonable notion of derivation of obligations will work with $\wedge$ and $\vee$, but not with $\neg$.

We should not close under inverse $\wedge$, i.e. if $\phi \wedge \phi'$ is an obligation, we should not conclude that ϕ and ϕ' separately are obligations, as the following example shows.

Example 7.1.5 Let p stand for: post letter, w: water plants, s: strangle grandmother.

Consider now $\phi \wedge \phi'$, where $\phi = p \vee (\neg p \wedge \neg w)$, $\phi' = p \vee (\neg p \wedge w \wedge s)$. $\phi \wedge \phi'$ is equivalent to p – though it is perhaps a bizarre way to express the obligation to post the letter. ϕ leaves us the possibility not to water the plants and ϕ' to strangle the grandmother, and neither seems good obligations.

Remark 7.1.2 This is particularly important in the case of soft obligations, as we see now, when we try to apply the rules of preferential reasoning to obligations.

One of the rules of preferential reasoning is the (OR) rule:

$$\phi \mid\!\sim \psi, \phi' \mid\!\sim \psi \Rightarrow \phi \vee \phi' \mid\!\sim \psi.$$

Suppose we have $\phi \mid\!\sim \psi' \wedge \psi''$ and $\phi' \mid\!\sim \psi'$. We might be tempted to split $\psi' \wedge \psi''$ – as ψ' is a "legal" obligation, and argue: $\phi \mid\!\sim \psi' \wedge \psi''$, so $\phi \mid\!\sim \psi'$; moreover, $\phi' \mid\!\sim \psi'$, so $\phi \vee \phi' \mid\!\sim \psi'$. The following example shows that this is not always justified.

Example 7.1.6 Consider the following obligations for a physician:

Let ϕ' imply that the patient has no heart disease, and if ϕ' holds, we should give drug A or (not drug A, but drug B), abbreviated $A \vee (\neg A \wedge B)$. (B is considered dangerous for people with heart problems.)

Let ϕ imply that the patient has heart problems. Here, the obligation is $(A \vee (\neg A \wedge B)) \wedge (A \vee (\neg A \wedge \neg B))$, equivalent to A.

The false conclusion would then be $\phi' \mid\!\sim A \vee (\neg A \wedge B)$ and $\phi \mid\!\sim A \vee (\neg A \wedge B)$, so $\phi \vee \phi' \mid\!\sim A \vee (\neg A \wedge B)$, so in both situations we should either give A or B, but B is dangerous in "one half" of the situations.

We captured this idea about "and" and "or" in Definition 7.1.9 (p. 186).

7.1.3.3 Ceteris Paribus – A Local Property

Basically, the set of points "in" an obligation has to be better than the set of "exterior" points. As above, Example 7.1.3 (p. 187) with three obligations shows demanding that any element inside is better than any element outside is too strong. We use instead the "ceteris paribus" idea.

"All other things being equal" seems to play a crucial role in understanding obligations. Before we try to analyse it, we look for other concepts which have something to do with it.

The Stalnaker/Lewis semantics for counterfactual conditionals (see [Sta68], [Lew73]) also works with some kind of "ceteris paribus". "If it were to rain, *I* would use an umbrella" means something like: "If it were to rain, and there were not a very strong wind" (there is no such wind now), "if *I* had an umbrella" (I have one now), etc., i.e. if things were mostly as they are now, with the exception that now it does not rain, and in the situation *I* speak about it rains, then *I* will use an umbrella.

But also theory revision in the AGM sense contains – at least as objective – this idea: Change things as little as possible to incorporate some new information in a consistent way.

When looking at the "ceteris paribus" in obligations, a natural interpretation is to read it as "all other obligations being unchanged" (i.e. satisfied or not as before). This is then just a Hamming distance considering the obligations (but not other information). Then, in particular, if $\mathcal{O}$ is a family of obligations, and if x and x' are in the same subset $\mathcal{O}' \subseteq \mathcal{O}$ of obligations, then an obligation derived from $\mathcal{O}$ should not separate them. More precisely, if $x \in O \in \mathcal{O} \Leftrightarrow x' \in O \in \mathcal{O}$, and D is a derived obligation, then $x \in D \Leftrightarrow x' \in D$.

Example 7.1.7 If the only obligation is not to kill, then it should not be derivable not to kill and to eat spaghetti.

Often, this is impossible, as obligations are not independent. In this case, and also in other situations, we can push "ceteris paribus" into an abstract distance d (as in the Stalnaker/Lewis semantics), which we postulate as given, and say that satisfying an obligation makes things better when going from "outside" the obligation to the d-closest situation "inside". Conversely, whatever the analysis of "ceteris paribus", and given a quality order on the situations, we can now define an obligation as a formula where (perhaps among other criteria) "ceteris paribus" improves the situation when we go from "outside" the formula to "inside".

A simpler way to capture "ceteris paribus" is to connect it directly to obligations, see Definition 7.1.12 (p. 186). This is probably too much tied to independence (see below), and thus too rigid.

7.1.3.4 Hamming Neighbourhoods

A combination concept is a Hamming neighbourhood:

X is called a Hamming neighbourhood of the best cases iff for any $x \in X$ and y a best case with minimal distance from x, all elements between x and y are in X.

For this, we need a notion of distance (also to define "between"). This was made precise in Definition 2.1.6 (p. 33) and Definition 7.1.8 (p. 185).

7.1.3.5 Global and Mixed Global/Local Properties of Obligations

We now look at some global properties (or mixtures of global and local) which seem desirable for obligations:

(1) Downward closure: Consider the following example:

Example 7.1.8 Let $U' := \{x, x', y, y'\}$ with $x' := pqrs$, $y' := pqr\neg s$, $x := \neg p\neg qr\neg s$, $y := \neg p\neg q\neg r\neg s$.
Consider $X := \{x, x'\}$.
The counting version:
Then x' has quality 4 (the best), y' has quality 3, x has 1, y has 0.

$$d_c(x', y') = 1, d_c(x, y) = 1, d_c(x, y') = 2.$$

Then above "ceteris paribus" criterion is satisfied, as y' and x do not "see" each other, so $X \prec_{l,c} CX$.

But X is not downward closed, below $x \in X$ is a better element $y' \notin X$.

This seems an argument against X being an obligation.

The set version:

We still have $x' \prec_s y' \prec_s x \prec_s y$. As shown in Fact 7.1.5 (p. 183), $d_s(x, y)$ (and also $d_s(x', y')$) and $d_s(x, y')$ are not comparable, so our argument collapses.

As a matter of fact, we have the result that the "ceteris paribus" criterion entails downward closure in the set variant, see Fact 7.1.7 (p. 185).

Note that a sufficiently rich domain (put elements between y' and x) will make this local condition (for $\prec$) a global one, so we have here a domain problem. Domain problems are discussed, e.g. in [Sch04] and [GS08a].

(2) Best states

It seems also reasonable to postulate that obligations contain all best states. In particular, obligations then have to be consistent – under the condition that best states exist. We are aware that this point can be debated, there is, of course, an easy technical way out: We take, when necessary, unions of obligations to cover the set of ideal cases. So obligations will be certain "neighbourhoods" of the "best" situations.

We think that some such notion of neighbourhood is a good candidate for a semantics:

- A system of neighbourhoods is not necessarily closed under supersets.
- Obligations express something like an approximation to the ideal case where all obligations (if possible, or, as many as possible) are satisfied, so we try to be close to the ideal. If we satisfy an obligation, we are (relatively) close, and stay so as long as the obligation is satisfied.
- The notion of neighbourhood expresses the idea of being close, and containing everything which is sufficiently close. Behind "containing everything which is sufficiently close" is the idea of being in some sense convex. Thus, "convex" or "between" is another basic notion to be investigated. See here also the discussion of "between" in [Sch04].

7.1.3.6 Soft Obligations

"Soft" obligations are obligations which have exceptions. Normally, one is obliged to do O, but there are cases where one is not obliged. This is like soft rules, as "Birds fly" (but penguins do not), where exceptions are not explicitly mentioned.

The semantic notions of size are very useful here too. We will content ourselves that soft obligations satisfy the postulates of usual obligations everywhere except on a small set of cases. For instance, a soft obligation O should be downward closed "almost" everywhere, i.e. for a small subset of pairs $\langle a, b\rangle$ in $U \times U$ we accept that $a \prec b$, $b \in O$, $a \notin O$. We transplanted a suitable and cautious notion of size from the components to the product in Definition 7.1.1 (p. 180).

When we look at the requirement to contain the best cases, we might have to soften this too. We will admit that a small set of the ideal cases might be excluded. Small can be relative to all cases, or only to all ideal cases.

Soft obligations generate an ordering which takes care of exceptions, like the normality ordering of birds will take care of penguins: Within the set of penguins, non-flying animals are the normal ones. Based on this ordering, we define "derived soft obligations", they may have (a small set of) exceptions with respect to this ordering.

7.1.3.7 Overview of Different Types of Obligations

(1) Hard obligations: They hold without exceptions, as in the Ten Commandments. You should not kill.

- (1.1) In the simplest case, they apply everywhere and can be combined arbitrarily, i.e. for any $\mathcal{O}' \subseteq \mathcal{O}$ there is a model where all $O \in \mathcal{O}'$ hold, and no $O' \in \mathcal{O} - \mathcal{O}'$.
- (1.2) In a more complicated case, not all combinations are possible. This is the same as considering just an arbitrary subset of U with the same set $\mathcal{O}$ of obligations. This case is very similar to the case of conditional obligations (which might not be defined outside a subset of U), and we treat them together.

A good example is the considerate assassin:

Example 7.1.9 Normally, one should not offer a cigarette to someone, out of respect for his health. But the considerate assassin might do so nonetheless, on the cynical reasoning that the victim's health is going to suffer anyway:

(1) One should not kill, $\neg k$.
(2) One should not offer cigarettes, $\neg o$.
(3) The assassin should offer his victim a cigarette before killing him, if k, then o.

Here, globally, $\neg k$ and $\neg o$ is best, but among k-worlds, o is better than $\neg o$. The model ranking is $\neg k \wedge \neg o \prec \neg k \wedge o \prec k \wedge o \prec k \wedge \neg o$.

Recall that an obligation for the whole set need not be an obligation for a subset any more, as it need not contain all best states. In this case, we may have to take a union with other obligations.

(2) Soft obligations: Many obligations have exceptions. Consider the following example:

Example 7.1.10 You are in a library. Of course, you should not pour water on a book. But if the book has caught fire, you should pour water on it to prevent worse damage. In stenographic style, these obligations read: "Do not pour water on books." "If a book is on fire, do pour water on it." It is like "birds fly", but "penguins do not fly". "Soft" or nonmonotonic obligations are obligations which have exceptions that are not formulated in the original obligation, but added as exceptions.

We could have formulated the library obligation also without exceptions: "When you are in a library, and the book is not on fire, do not pour water on it." "When you are in a library, and the book is on fire, pour water on it." This formulation avoids exceptions. Conditional obligations behave like restricted quantifiers: They apply in a subset of all possible cases.
We treat now the considerate assassin case as an obligation (not to offer) with exceptions. Consider the full set U, and consider the obligation $\neg o$. This is not downward closed, as $k \wedge o$ is better than $k \wedge \neg o$. Downward closure will only hold for "most" cases, but not for all.

(3) Contrary-to-duty obligations: Contrary-to-duty obligations are about different degrees of fulfillment. If you should ideally not have any fence, but are not willing or able to fulfill this obligation (e.g. you have a dog which might stray), then you should at least paint it white to make it less conspicuous. This is also a conditional obligation. Conditional, as it specifies what has to be done if there is a fence. The new aspect in contrary-to-duty obligations is the different degree of fulfillment.

We will not treat contrary-to-duty obligations here, as they do not seem to have any import on our basic ideas and solutions.

(4) A still more complicated case is when the language of obligations is not uniform, i.e. there are subsets $V \subseteq U$ where obligations are defined, which are not defined in $U - V$.

We will not pursue this case here.

7.1.3.8 Summary of the Philosophical Remarks

(1) It seems justifiable to say that an obligation is satisfied or holds in a certain situation.
(2) Obligations are fundamentally asymmetrical, thus negation has to be treated with care. "Or" and "and" behave as for facts.
(3) Satisfying obligations improves the situation with respect to some given grading – ceteris paribus.

(4) "Ceteris paribus" can be defined by minimal change with respect to other obligations, or by an abstract distance.
(5) Conversely, given a grading and some distance, we can define an obligation locally as describing an improvement with respect to this grading when going from "outside" to the closest point "inside" the obligation.
(6) Obligations should also have global properties: they should be downward (i.e. under increasing quality) closed and cover the set of ideal cases.
(7) The properties of "soft" obligations, i.e. with exceptions, have to be modified appropriately. Soft obligations generate an ordering, which in turn may generate other obligations, where exceptions to the ordering are permitted.
(8) Quality and distance can be defined from an existing set of obligations in the set or the counting variant. Their behaviour is quite different.
(9) We distinguished various cases of obligations, soft and hard, with and without all possibilities, etc.

Finally, we should emphasize that the notions of distance, quality, and size are in principle independent, even if they may be based on a common substructure.

7.1.4 Examination of the Various Cases

We will concentrate here on the set version of hard obligations.

7.1.4.1 Hard Obligations for the Set Approach

Introduction

We work here in the set version, the $\in$-case, and examine mostly the set version only.

We will assume a set $\mathcal{O}$ of obligations to be given. We define the relation $\prec := \prec_{\mathcal{O}}$ as described in Definition 2.1.6 (p. 33), and the distance d is the Hamming distance based on $\mathcal{O}$.

The Not Necessarily Independent Case

Example 7.1.11 Work in the set variant. We show that X $\preceq_s$-closed does not necessarily imply that X contains all $\preceq_s$-best elements.

Let $\mathcal{O} := \{p, q\}$, $U' := \{p\neg q, \neg pq\}$, then all elements of U' have best quality in U', $X := \{p\neg q\}$ is closed, but does not contain all best elements.

Example 7.1.12 Work in the set variant. We show that X $\preceq_s$-closed does not necessarily imply that X is a neighbourhood of the best elements, even if X contains them.

Consider $x := pq\neg rstu$, $x' := \neg pqrs\neg t\neg u$, $x'' := p\neg qr\neg s\neg t\neg u$, $y := p\neg q\neg r\neg s\neg t\neg u$, $z := pq\neg r\neg s\neg t\neg u$. In $U := \{x, x', x'', y, z\}$, the $\prec_s$-best elements are x, x', x'', which are contained in $X := \{x, x', x'', z\}$. $d_s(z, x) = \{s, t, u\}$,

$d_s(z, x') = \{p, r, s\}$, $d_s(z, x'') = \{q, r\}$, so x'' is one of the best elements closest to z. $d(z, y) = \{q\}$, $d(y, x'') = \{r\}$, so $[z, x''] = \{z, y, x''\}$, $y \notin X$, but X is downward closed.

Fact 7.1.10 Work in the set variant.
Let $X \neq \emptyset$, X $\preceq_s$-closed. Then

(1) X does not necessarily contain all best elements.
Assume now that X contains, in addition, all best elements. Then
(2) $X \prec_{l,s} CX$ does not necessarily hold.
(3) X is (ui).
(4) $X \in \mathcal{D}(\mathcal{O})$ does not necessarily hold.
(5) X is not necessarily a neighbourhood of the best elements.
(6) X is an improving neighbourhood of the best elements.

Proof

(1) See Example 7.1.11 (p. 193).
(2) See Example 7.1.2 (p. 185).
(3) If there is $m \in X$, $m \notin O$ for all $O \in \mathcal{O}$, then by closure $X = U$, take $\mathcal{O}_i := \emptyset$.
For $m \in X$ let $\mathcal{O}_m := \{O \in \mathcal{O} : m \in O\}$. Let $X' := \bigcup\{\bigcap \mathcal{O}_m : m \in X\}$.
$X \subseteq X'$: Trivial, as $m \in X \rightarrow m \in \bigcap \mathcal{O}_m \subseteq X'$.
$X' \subseteq X$: Let $m' \in \bigcap \mathcal{O}_m$, for some $m \in X$. It suffices to show that $m' \preceq_s m$.
$m' \in \bigcap \mathcal{O}_m = \bigcap\{O \in \mathcal{O} : m \in O\}$, so for all $O \in \mathcal{O}$ $(m \in O \rightarrow m' \in O)$.
(4) Consider Example 7.1.2 (p. 185), let $dom(\delta) = \{r, s\}$, $\delta(r) = \delta(s) = 0$. Then $x, y \models \delta$, but $x' \not\models \delta$ and $x \in X$, $y \notin X$, but there is no $z \in X$, $z \models \delta$ and $z \prec y$, so $X \notin \mathcal{D}(\mathcal{O})$.
(5) See Example 7.1.12 (p. 193).
(6) By Fact 7.1.5 (p. 183), (5). □

Fact 7.1.11 Work in the set variant.

(1.1) $X \prec_{l,s} CX$ implies that X is $\preceq_s$-closed.
(1.2) $X \prec_{l,s} CX \Rightarrow X$ contains all best elements.
(2.1) X is $(ui) \Rightarrow X$ is $\preceq_s$-closed.
(2.2) X is (ui) does not necessarily imply that X contains all $\preceq_s$-best elements.
(3.1) $X \in \mathcal{D}(\mathcal{O}) \Rightarrow X$ is $\preceq_s$-closed.
(3.2) $X \in \mathcal{D}(\mathcal{O})$ implies that X contains all $\preceq_s$-best elements.
(4.1) X is an improving neighbourhood of the $\preceq_s$-best elements $\Rightarrow X$ is $\preceq_s$-closed.
(4.2) X is an improving neighbourhood of the best elements $\Rightarrow X$ contains all best elements.

Proof

(1.1) By Fact 7.1.7 (p. 185).
(1.2) By Fact 7.1.6 (p. 184).

(2.1) Let $O \in \mathcal{O}$, then O is downward closed (no $y \notin O$ can be better than $x \in O$). The rest follows from Fact 7.1.4 (p. 182) (3).

(2.2) Consider Example 7.1.11 (p. 193), p is (ui) (formed in $U!$), but $p \cap X$ does not contain $\neg pq$.

(3.1) Let $X \in \mathcal{D}(\mathcal{O})$, but let X not be closed. Thus, there are $m \in X$, $m' \preceq_s m$, $m' \notin X$.
Case 1: Suppose $m' \sim m$. Let $\delta_m : \mathcal{O} \to 2$, $\delta_m(O) = 1$ iff $m \in O$. Then $m, m' \models \delta_m$, and there cannot be any $m'' \models \delta_m$, $m'' \prec_s m'$, so $X \notin \mathcal{D}(\mathcal{O})$.
Case 2: $m' \prec_s m$. Let $\mathcal{O}' := \{O \in \mathcal{O} : m \in O \not\Leftrightarrow m' \in O\}$, $dom(\delta) = \mathcal{O}'$, $\delta(O) := 1$ iff $m \in O$ for $O \in \mathcal{O}'$. Then $m, m' \models \delta$. If there is $O \in \mathcal{O}$ s.t. $m' \notin O$, then by $m' \preceq_s m$ $m \notin O$, so $O \in \mathcal{O}'$. Thus for all $O \notin dom(\delta)$, $m' \in O$. But then there is no $m'' \models \delta$, $m'' \prec_s m'$, as m' is already optimal among the n with $n \models \delta$.

(3.2) Suppose $X \in \mathcal{D}(\mathcal{O})$, $x' \in U - X$ is a best element. Take $\delta := \emptyset$, $x \in X$. Then there must be $x'' \prec x'$, $x'' \in X$, but this is impossible as x' was best.

(4.1) By Fact 7.1.5 (p. 183), (4) all minimal elements have incomparable distance. But if $z \preceq y$, $y \in X$, then either z is minimal or it is above a minimal element, with minimal distance from y, so $z \in X$ by Fact 7.1.5 (p. 183), (3).

(4.2) Trivial. □

The Independent Case

Assume now the system to be independent, i.e. all combinations of $\mathcal{O}$ are present.

Note that there is now only one minimal element; then the notions of Hamming neighbourhood of the best elements and improving Hamming neighbourhood of the best elements coincide.

Fact 7.1.12 Work in the set variant.
Let $X \neq \emptyset$, X $\preceq_s$-closed. Then

(1) X contains the best element.
(2) $X \prec_{l,s} CX$.
(3) X is (ui).
(4) $X \in \mathcal{D}(\mathcal{O})$.
(5) X is a (improving) Hamming neighbourhood of the best elements.

Proof

(1) Trivial.
(2) Fix $x \in X$, let y be closest to x, $y \notin X$. Suppose $x \not\prec y$, then there must be $O \in \mathcal{O}$ s.t. $y \in O$, $x \notin O$. Choose y' s.t. y' is like y, only $y' \notin O$. If $y' \in X$, then by closure $y \in X$, so $y' \notin X$. But y' is closer to x than y is, *contradiction*.
Fix $y \in U - X$. Let x be closest to y, $x \in X$. Suppose $x \not\prec y$, then there is $O \in \mathcal{O}$ s.t. $y \in O$, $x \notin O$. Choose x' s.t. x' is like x, only $x' \in O$. By closure of X, $x' \in X$, but x' is closer to y than x is, *contradiction*.
(3) By Fact 7.1.10 (p. 194), (3).

(4) Let X be closed, and $\mathcal{O}' \subseteq \mathcal{O}$, $\delta : \mathcal{O}' \to 2$, $m, m' \models \delta$, $m \in X$, $m' \notin X$. Let m'' be s.t. $m'' \models \delta$, and for all $O \in \mathcal{O} - dom(\delta)$, $m'' \in O$. This exists by independence. Then $m'' \preceq_s m'$, but also $m'' \preceq_s m$, so $m'' \in X$. Suppose $m'' \sim m'$, then $m' \preceq_s m''$, so $m' \in X$, contradiction, so $m'' \prec_s m'$.
(5) Trivial by (1), the remark preceding this fact, and Fact 7.1.10 (p. 194), (6). □

Fact 7.1.13 Work in the set variant.

(1) $X \prec_{l,s} Cx \Rightarrow X$ is $\preceq_s$-closed.
(2) X is $(ui) \Rightarrow X$ is $\preceq_s$-closed.
(3) $X \in \mathcal{D}(\mathcal{O}) \Rightarrow X$ is $\preceq_s$-closed.
(4) X is a (improving) neighbourhood of the best elements $\Rightarrow X$ is $\preceq_s$-closed.

Proof

(1) Suppose there are $x \in X$, $y \in U - X$, $y \prec x$. Choose them with minimal distance. If $card(d_s(x, y)) > 1$, then there is z, $y \prec_s z \prec_s x$, $z \in X$ or $z \in U - X$, contradicting minimality. So $card(d_s(x, y)) = 1$. So y is among the closest elements of $U - X$ seen from x, but then by prerequisite $x \prec y$, *contradiction*.
(2) By Fact 7.1.11 (p. 194), (2.1).
(3) By Fact 7.1.11 (p. 194), (3.1).
(4) There is just one best element z, so if $x \in X$, then $[x, z]$ contains all y, $y \prec x$ by Fact 7.1.5 (p. 183), (3). □

The $\mathcal{D}(\mathcal{O})$ condition seems to be adequate only for the independent situation, so we stop considering it now.

Fact 7.1.14 Let $X_i \subseteq U$, $i \in I$, be a family of sets; we note the following about closure under unions and intersections:

(1) If the X_i are downward closed, then so are their unions and intersections.
(2) If the X_i are (ui), then so are their unions and intersections.

Proof Trivial. □

We do not know whether $\prec_{l,s}$ is preserved under unions and intersections; it does not seem to be an easy problem.

Fact 7.1.15

(1) Being downward closed is preserved while going to subsets.
(2) Containing the best elements is not preserved (and thus neither the neighbourhood property).
(3) The $\mathcal{D}(\mathcal{O})$ property is not preserved.
(4) $\preceq_{l,s}$ is not preserved.

Proof (4) Consider Example 7.1.8 (p. 190), and eliminate y from U', then the closest to x not in X is y', which is better. □

7.1.4.2 Remarks on the Counting Case

Remark 7.1.3 In the counting variant all qualities are comparable. So if X is closed, it will contain all minimal elements.

Example 7.1.13 We measure distance by counting.

Consider $a := \neg p \neg q \neg r \neg s$, $b := \neg p \neg q \neg r s$, $c := \neg p \neg q r \neg s$, $d := p q r \neg s$. Let $U := \{a, b, c, d\}$, $X := \{a, c, d\}$. d is the best element, $[a, d] = \{a, d, c\}$, so X is an improving Hamming neighbourhood, but $b \prec a$, so $X \not\prec_{l,c} CX$.

Fact 7.1.16 We measure distances by counting.

$X \prec_{l,c} CX$ does not necessarily imply that X is an improving Hamming neighbourhood of the best elements.

Proof Consider Example 7.1.8 (p. 190). There $X \prec_{l,c} CX$. x' is the best element, and $y' \in [x', x]$, but $y' \notin X$. □

7.1.5 What Is An Obligation?

The reader will probably not expect a final definition. All we can do is to give a tentative definition, which, in all probability, will not be satisfactory in all cases.

Definition 7.1.13 We decide for the set relation and distance.

(1) Hard obligation:
A hard obligation has the following properties:

(1.1) It contains all ideal cases in the set considered.
(1.2) It is closed under increasing quality, Definition 7.1.5 (p. 182).
(1.3) It is an improving neighbourhood of the ideal cases (this also implies (1.1)), Definition 7.1.8 (p. 185).
We are less committed to:
(1.4) It is ceteris paribus improving, Definition 7.1.7 (p. 184).
An obligation O is a derived obligation of a system $\mathcal{O}$ of obligations iff it is a hard obligation based on the set variant of the order and distance generated by $\mathcal{O}$.

(2) Soft obligations:
A set is a soft obligation iff it satisfies the soft versions of above postulates. The notion of size has to be given, and is transferred to products as described in Definition 7.1.1 (p. 180). More precisely, strict universal quantifiers are transformed into their soft variant "almost all", and the other operators are left as they are. Of course, one might also want to use a mixture of soft and hard conditions, e.g. we might want to have all ideal cases, but renounce on closure for a small set of pairs $\langle x, x' \rangle$.

An obligation O is derived from $\mathcal{O}$ iff it is a soft obligation based on the set variant of the order and distance generated by the translation of $\mathcal{O}$ into their hard versions. (i.e. Exceptions will be made explicit.)

Fact 7.1.17 Let $O \in \mathcal{O}$, then $\mathcal{O} \mid\sim O$ in the independent set case.

Proof We check (1.1)–(1.3) of Definition 7.1.13 (p. 197).

(1.1) holds by independence.
(1.2) If $x \in O$, $x' \notin O$, then $x' \not\preceq_s x$.
(1.3) By Fact 7.1.10 (p. 194), (6).
Note that (1.4) will also hold by Fact 7.1.12 (p. 195) (2). □

Corollary 7.1.1 *Every derived obligation is a classical consequence of the original set of obligations in the independent set case.*

Proof This follows from Fact 7.1.12 (p. 195), (3), and Fact 7.1.17 (p. 198). □

Example 7.1.14 The Ross paradox is not a derived obligation.

Proof Suppose we have the alphabet p, q and the obligations $\{p, q\}$, let $R := p \vee \neg q$. This is not closed, as $\neg p \wedge q \prec \neg p \wedge \neg q \in R$. □

7.1.6 Conclusion

Obligations differ from facts in the behaviour of negation, but not of conjunction and disjunction. The Ross paradox originates, in our opinion, from the differences in negation. Central to the treatment of obligations seems to be a relation of "better", which can generate obligations, but also be generated by obligations. The connection between obligations and this relation of "better" seems to be somewhat complicated and leads to a number of ramifications. A tentative definition of a derivation of obligations is given.

7.2 A Comment on Work by Aqvist

7.2.1 Introduction

The article [Aqv00] discusses three systems, which are presented now in outlines. (When necessary, we will give details later in this section.)

(1) The systems Hm, where $m \in \omega$. The (Kripke style) semantics has for each $i \in \omega$ a subset opt_i of the model set, s.t. the opt_i form a disjoint cover of the model set, all opt_i for all $i \leq m$ are noempty, and all other opt_i are empty. The opt_i interpret new frame constants Q_i. The opt_i describe intuitively levels of perfection, where opt_1 is the best and opt_m the worst level.

(2) The dyadic deontic logics Gm, $m \in \omega$. The semantics is again given by a cover opt_i as above, and in addition, a function best, which assigns to each formula ϕ the "best" models of ϕ, i.e. those which are in the best opt_i set. The language has the Q_i operators and a new binary operator $O(\phi/\psi)$ (and its dual $P(./.)$), which expresses that in the best ψ-models ϕ holds. Note that there is no explicit "best" operator in the language.
(3) The dyadic deontic logic G. The semantics does not contain the opt_i any more, but still the "best" operator as in case (2), which now corresponds to a ranking of the models (sufficient axioms are given). The language does not have the Q_i anymore, but contains the O (and P) operator, which is interpreted in the natural way: $O(\phi/\psi)$ holds iff the best ψ-models are ϕ-models. Note again that there is no explicit "best" operator in the language.

Thus, it corresponds to putting the $\mid\!\sim$-relation of ranked models in the object language.

In particular, there is no finiteness restriction any more, as in cases (1) and (2) – and here lies the main difference.

Aqvist gives a theorem (Theorem 6) which shows (among other things) that If G is a G-sentence, provable in G, then it is also provable in Gm.

The converse is left open, and Aqvist thought to have found proof using another result of his, but there was a loophole, as he had found out.

We close this hole here.

7.2.2 There Are (At Least) Two Solutions

(1) We take a detour via a language which contains an operator β to be interpreted by "best".
(2) We work with the original language.

As a matter of fact, Aqvist's paper contains already an almost complete solution of type (1), just the translation part is lacking. We give a slightly different proof (which will be self-contained), which works with the original (i.e. possibly infinite) model set, but reduces the quantity of levels to a finite number. Thus, the basic idea is the same, our technique is more specific to the problem at hand, thus less versatile, but also less "brutal".

Yet, we can also work with the original language, even if the operator $O(\phi/\psi)$ will not allow to describe "best" as the operator β can. We can approximate "best" sufficiently well to suit our purposes. (Note also that O allows, using all formulas, to approximate "best" from above: $best(\phi) = \bigcap\{M(\psi) : O(\psi/\phi)\}$.)

We may describe the difference between solution (1) and (2) as follows: Solution (1) will preserve the exact "best" value of some – but not necessarily all, and this cannot be really improved – sets. Solution (2) will not even allow this, but still, we stay sufficiently close to the original "best" value. So the formula at hand, and its subformulas, will preserve their truth values.

In both cases, the basic idea will be the same: We have a G-model for ϕ, and now construct a Gm-model (i.e. with finitely many levels) for ϕ. ϕ is a finite entity, containing only finitely many propositional variables, say $p_1, \ldots, p_n$. Then we look at all set of the type $\pm p_1 \wedge \ldots \wedge \pm p_n$, where $\pm$ is nothing or the negation symbol. This is basically how fine our structure will have to be. If we work with β, we have to get better, as $\beta(p_i)$ will, in general, be something new, so will $\beta(p_i \wedge \neg\beta(p_i))$, etc., thus we have to take the nesting of β's in ϕ into account, too. (This will be done below.) If we work directly with the $O(\alpha/\beta)$ operator, then we need not go so far down as $O(\alpha/\beta)$ will always evaluate to (universal) true or false. If, e.g. ϕ contains $O(\alpha/\beta)$ and $O(\alpha'/\beta)$, then we will try to define the "best" elements of $M(\beta)$ as $M(\beta \wedge \alpha \wedge \alpha')$, etc. We have to be careful to make a ranking, i.e. $O(true)$ will give the lowest layer, and if $\psi \vdash \psi'$, $O(\phi/\psi)$, $O(\neg\phi/\psi)$, then the rank of ψ' will be strictly smaller than the rank of ψ, etc. This is all straightforward, but a bit tedious.

As said, we will take the first approach, which seems a bit "cleaner".

We repeat now the definitions of [Aqv00] only as far as necessary to understand the subsequent pages. In particular, we will not introduce the axiom systems, as we will work on the semantic side only.

All systems are propositional.

Definition 7.2.1 The systems Hm, $m \in \omega$. The language:

A set $Prop$ of propositional variables, True, the usual Boolean operators, N (universal necessity), a set $\{Q_i : i \in \omega\}$ of systematic frame constants (zero place connectives like True) (and the duals False, M).

The semantics:

$M = \langle W, V, \{opt_i : i \in \omega\}, m\rangle$, where W is a set of possible worlds, V a valuation as usual, each opt_i is a subset of W.

Validity:

Let $x \in W$.
$M, x \models p$ iff $x \in V(p)$ for $x \in W$, $p \in Prop$,
$M, x \models True$,
$M, x \models N\phi$ iff for all $y \in W$, $M, y \models \phi$,
$M, x \models Q_i$ iff $x \in opt_i$.

Conditions on the opt_i:

(a) $opt_i \cap opt_j = \emptyset$ if $i \neq j$,
(b) $opt_1 \cup \ldots \cup opt_m = W$,
(c) $opt_i \neq \emptyset$ for $i \leq m$,
(d) $opt_i = \emptyset$ for $i > m$.

Definition 7.2.2 The systems Gm, $m \in \omega$.

The language:

It is just like that for the systems Hm, with, in addition, a new binary connective $O(\phi/\psi)$, with the meaning that the "best" ψ-worlds satisfy ϕ (and its dual).

The semantics:

It is also just like the one for Hm, with, in addition, a function B (for "best"), assigning to each formula ϕ a set of worlds, the best ϕ-worlds, thus $M = \langle W, V, \{opt_i : i \in \omega\}, B, m\rangle$.

Validity:

$M, x \models O(\phi, \psi)$ iff $B(\psi) \subseteq M(\phi)$.

Conditions:

We have to connect B to the opt_i. This is done by the condition $(\gamma 0)$ in the obvious way:

$(\gamma 0)$ $x \in B(\phi)$ iff $M, x \models \phi$ and for each $y \in W$, if $M, y \models \phi$, then x is at least as good as y – i.e. there is no strictly lower opt_i-level where ϕ holds.

Definition 7.2.3 The system G.

The language:

It is just like that for the systems Gm, but without the Q_i.

The semantics:

It is now just $M = \langle W, V, B\rangle$.

Validity:

As for Gm.

Conditions:

We have no opt_i now, which are replaced by suitable conditions on B, which make B the choice function of a ranked structure:

(σ_0) $M(\phi) = M(\phi') \rightarrow B(\phi) = B(\phi')$,
(σ_1) $B(\phi) \subseteq M(\phi)$,
(σ_2) $B(\phi) \cap M(\psi) \subseteq B(\phi \wedge \psi)$,
(σ_3) $M(\phi) \neq \emptyset \rightarrow B(\phi) \neq \emptyset$,
(σ_4) $B(\phi) \cap M(\psi) \neq \emptyset \rightarrow B(\phi \wedge \psi) \subseteq B(\phi) \cap M(\psi)$.

7.2.3 Outline

We first show that the language (and logic) G may necessitate infinitely many levels (whereas the languages Gm ($m \in \omega$) admit only finitely many ones). Thus, when trying to construct a Gm-model for a G-formula ϕ, we have to construct from a possibly "infinite" function B via finitely many opt_i levels a new function B' with only finitely many levels, which is sufficient for the formula ϕ under consideration. Crucial for the argument is that ϕ is a finite formula, and we need only a limited degree of discernation for any finite formula. Most of the argument is standard reasoning about ranked structures, as it is common for sufficiently strong preferential structures.

We will first reformulate the problem slightly using directly an operator β, which will be interpreted by the semantic function B, and which results in a slightly richer language, as the operators O and P can be expressed using B, but not necessarily conversely. Thus, we show slightly more than what is sufficient to solve Aqvist's problem.

We make this official:

Definition 7.2.4 The language G' is like the language G, without $O(./.)$ and $P(./.)$, but with a new unary operatorβ, which is interpreted by the semantic function B.

Remark 7.2.1

(1) $O(\phi/\psi)$ can thus be expressed by $N(\beta(\psi) \rightarrow \phi)$ – N universal necessity. In particular, and this is the important aspect of β, $\beta(\phi)$ is now (usually) a non-trivial set of models, whereas O and P result always in $\emptyset$ or the set of all models.

(2) Note that by $(\sigma 3)$, if $M(\phi) \neq \emptyset$, then $M(\beta(\phi)) \neq \emptyset$, but it may very well be that $M(\beta(\phi)) = M(\phi)$ – all ϕ-models may be equally good. ($M(\phi)$ is the set of ϕ-models.)

We now use an infinite theory to force B to have infinitely many levels.

Example 7.2.1 We construct ω many levels for the models of the language $\{p_i : i \in \omega\}$, going downward:

Top level: $M(p_0)$;
Second level: $M(\neg p_0 \wedge p_1)$;
Third level: $M(\neg p_0 \wedge \neg p_1 \wedge p_2)$, etc.
We can express via β, and even via $O(./.)$, that these levels are all distinct, e.g. by $\beta(p_0 \vee (\neg p_0 \wedge p_1)) \models \neg p_0$, or by $O(\neg p_0 / p_0 \vee (\neg p_0 \wedge p_1))$, etc. So we will necessarily have ω many non empty opt_i-levels, for any B which satisfies the condition connecting the opt_i and B, condition (γ_0).

We now work in the fixed G-model Γ (with the β operator.) Let in the sequel all X, Y, X_i be non empty (to avoid trivial cases) model sets. By prerequisite, B satisfies $(\sigma 0)$–$(\sigma 4)$.

Definition 7.2.5

(1) $X \prec Y$ iff $B(X \cup Y) \cap Y = \emptyset$,
(2) $X \sim Y$ iff $B(X \cup Y) \cap X \neq \emptyset$ and $B(X \cup Y) \cap Y \neq \emptyset$, i.e. iff $X \not\prec Y$ and $Y \not\prec X$.

Remark 7.2.2

(1) $\prec$ and $\sim$ behave nicely:

(1.1) $\prec$ and $\sim$ are transitive, $\sim$ is reflexive and symmetric, and for no X, $X \prec X$.
(1.2) $\prec$ and $\sim$ cooperate: $X \prec Y \sim Z \rightarrow X \prec Z$ and $X \succ Y \sim Z \rightarrow X \succ Z$
Thus, $\prec$ and $\sim$ give a ranking.
(2.1) $B(X \cup Y) = B(X)$ if $X \prec Y$
(2.2) $B(X \cup Y) = B(X) \cup B(Y)$ if $X \sim Y$

(3) $B(X_1 \cup \ldots \cup X_n) = \bigcup\{B(X_i) : \neg\exists X_j (X_j \prec X_i)\}$ $1 \leq i, j \leq n$, – i.e. $B(X_1 \cup \ldots \cup X_n)$ is the union of the $B(X_i)$ with minimal $\prec$-rank.

We come to the main idea and its formalization.

Let a fixed ϕ be given. Let ϕ contain the propositional variables $p_1, \ldots, p_m$ and the operator β to a depth of nesting n.

Definition 7.2.6 (1) Define by induction:
Elementary sets of degree 0: all intersections of type $M(\pm p_1) \cap M(\pm p_2) \cap \ldots \cap M(\pm p_m)$, where $\pm p_j$ is either p_j or $\neg p_j$.
Elementary sets of degree $i+1$: If s is an elementary set of degree i, then $B(s)$ and $s - B(s)$ are elementary sets of degree $i+1$.
(2) Unions of degree i are either $\emptyset$ or (arbitrary) unions of elementary sets of degree i.
This definition goes up to $i = n$, though we could continue for all $i \in \omega$, but this is not necessary.
(3) $\mathcal{E}$ is the set of all elementary sets of degree n.

Remark 7.2.3

(1) For any degree i, the elementary sets of degree i form a disjoint cover of $M_{\mathcal{L}}$ – the set of all classical models of the language.
(2) The elementary sets of degree $i+1$ form a refinement of the elementary sets of degree i.
(3) The set of unions of degree i is closed under union, intersection, set difference, i.e. if X, Y are unions of degree i, so are $X \cup Y$, $X - Y$, etc.
(4) If X is a union of degree i, $B(X)$ and $X - B(X)$ are unions of degree $i+1$.
(5) A union of degree i is also a union of degree j if $j > i$.

We construct now the new Gm-structure. We first define the opt_i levels, and from these levels, the new function B'. Thus, (γ_0) will hold automatically. As we know from above example, we may forcibly loose some discerning power, so a word where it is lost and where it goes may be adequate. The construction will not look inside $B(s)$ and $s - B(s)$ for $s \in \mathcal{E}$. For $B(s)$, this is not necessary, as we know that all elements inside $B(s)$ are on the same level ($B(s) = B(B(s))$). Inside $s - B(s)$, the original function B may well be able to discern still (even infinitely many) different levels, but our formula ϕ does not permit us to look down at these details – $s - B(s)$ is treated as an atom, one chunk without any details inside.

Definition 7.2.7 We define the rank of $X \in \mathcal{E}$, and the opt_i and B'. This is the central definition.
Let $X \neq \emptyset$, $X \in \mathcal{E}$.

(1) Set $rank(X) = 0$, iff there is no $Y \prec X$, $Y \in \mathcal{E}$.
Set $rank(X) = i+1$, iff $X \in \mathcal{E} - \{Z : rank(z) \leq i\}$ and there is no $Y \prec X$, $Y \in \mathcal{E} - \{Z : rank(z) \leq i\}$.
So, $rank(X)$ is the $\prec$-level of X.
(2) Set $opt_i := \bigcup\{X \in \mathcal{E} : rank(X) = i\}$, and $opt_i = \emptyset$ iff there is no X s.t. $rank(X) = i$.
(3) $B'(A) := A \cap opt_i$, where i is the smallest j s.t. $A \cap opt_j \neq \emptyset$, for $A \neq \emptyset$.

Remark 7.2.4

(1) The opt_i forms again a disjoint cover of $M_{\mathcal{L}}$, all opt_i are $\neq \emptyset$ up to some k, and $\emptyset$ beyond.

(2) (γ_0) will now hold by definition.
(3) B' is not necessarily B, but sufficiently close.

Lemma 7.2.1 *(Main result)*
For any union X of degree $k < n$, $B(X) = B'(X)$.

Proof Write X as a union of degree $n-1$, and let $\mathcal{X} := \{X' \subseteq X : X'$ is of degree $n-1\}$. Note that the construction of $\prec$ / $\sim$ /opt splits all $X' \in \mathcal{X}$ into $B(X')$ and $X' - B(X')$, both of degree n, and that both parts are always present, there will not be any isolated $X' - B(X')$ without its counterpart $B(X')$.

Let $X' \in \mathcal{X}$. Then $(X' - B(X')) \cap B'(X) = \emptyset$, as the opt-level of $B(X')$ is better than the opt-level of $X' - B(X')$. Obviously, also $(X' - B(X')) \cap B(X) = \emptyset$. Thus, $B(X)$ and $B'(X)$ are the union of certain $B(X')$, $X' \in \mathcal{X}$. Suppose $B(X') \subseteq B(X)$ for some $X' \in \mathcal{X}$. Then for no $X'' \in \mathcal{X}$, $B(B(X') \cup B(X'')) \cap B(X') = \emptyset$, so $B(X')$ has minimal opt-level in X, and $B(X') \subseteq B'(X)$. Conversely, let $B(X') \not\subseteq B(X)$. Then there is $X'' \in \mathcal{X}$ s.t. $B(B(X') \cup B(X'')) \cap B(X') = \emptyset$, so $B(X')$ has not minimal opt-level in X, and $B(X') \cap B'(X) = \emptyset$. □

Corollary 7.2.1 *If ϕ' is built up from the ingredients of ϕ, then $[\phi'] = [\phi']'$ – where $[\phi']$ ($[\phi']'$) is the set of models where ϕ' holds in the original (new) structure.*

Proof Let Φ be the set of formulas which are built up from the ingredients of ϕ, i.e. using (some of) the propositional variables of ϕ, and up to the nesting depth of β of ϕ.

Case 0: Let $p \in \Phi$ be a propositional variable. Then $[p] = [p]'$, as we did not change the classical model. Moreover, $[p]$ is a union of degree 0.

Case 1: Let $\phi', \phi'' \in \Phi$, and let $[\phi'] = [\phi']'$, $[\phi''] = [\phi'']'$ be unions of degree k' and k'', respectively, and let $k := max(k', k'') \leq n$. Then both are unions of degree k, and so are $[\phi' \wedge \phi'']$, $[\neg\phi']$ etc., and $[\phi' \wedge \phi''] = [\phi' \wedge \phi'']'$, as $\wedge$ is interpreted by intersection, etc. Let now $k < n$. We have to show that $[\beta(\phi')] = [\beta(\phi')]'$. But $[\beta(\phi')] = B([\phi']) = B'([\phi']) = B'([\phi']') = [\beta(\phi')]'$ by above Lemma and induction hypothesis. □

Example 7.2.2 We interpret $\phi = \beta(p)$. The elementary sets of degree 0 are $M(p)$, $M(\neg p)$, those of degree 1 are $M(\beta(p))$, $M(p \wedge \neg\beta(p))$, $M(\beta(\neg p))$, $M(\neg p \wedge \neg\beta(\neg p))$. Suppose the construction of $\prec$, $\sim$ and the opt-levels results in (omitting the Ms for clarity)

$p \wedge \neg\beta(p)$, $\neg p \wedge \neg\beta(\neg p)$ on the worst nonempty level;
$\beta(p)$;
$\beta(\neg p)$ on the best level.

When we calculate now $B'(M(p))$ via opt, we decompose p in its components $\beta(p)$ and $p - \beta(p)$, and see that $\beta(p)$ is on a better level than $p - \beta(p)$, so $\beta'(M(p)) = \beta(M(p))$, as it should be.

We finally have the following.

7.2.4 $Gm \vdash A$ Implies $G \vdash A$ (Outline)

We show that $\forall m\ (Gm \vdash A)$ implies $G \vdash A$ (A a G-formula).

As Aqvist has given equivalent semantics for both systems, we can argue semantically. We turn the problem round: If $G \not\vdash A$, then there is m s.t. $Gm \not\vdash A$. Or, in other words, if there is a G-model and a point in it, where A does not hold, then we find an analogue Gm-model, or, still differently, if ϕ is any G-formula, and there is a G-model Γ and a point x in Γ, where ϕ holds, then we find m and Gm-model Δ, and a point y in Δ where ϕ holds.

By prerequisite, ϕ contains some propositional variables, perhaps the (absolute) quantifiers M and N, usual connectives, and the binary operators O and P. Note that the function "best" intervenes only in the interpretation of O and P. Moreover, the axioms σ_i express that best defines a ranking, e.g. in the sense of ranked models in preferential reasoning. In addition, σ_3 is a limit condition (which essentially excludes unbounded descending chains).

Let ϕ contain n symbols. We thus use "best" for at most n different ψ, where $O(\phi'/\psi)$ (or $P(\phi'/\psi)$) is a subformula of ϕ. We introduce now more structure into the G-model. We make $m := n + 1$ layers in Gs universe, where the first n layers are those mentioned in ϕ. More precisely, we put the best ψ-models for each ψ mentioned as above in its layer – respecting relations between layers when needed (this is possible, as the σ_i are sufficiently strong), and put all other ψ-models somewhere above. The fact that we have one supplementary layer (which, of course, we put on top) guarantees that we can do so. The opt_i will be the layers.

We then have a Gm-model (if we take a little care, so nothing gets empty prematurely), and ϕ will hold in our new structure.

7.3 Hierarchical Conditionals

The material in this section is taken from [GS08d].

7.3.1 Introduction

7.3.1.1 Description of the Problem

We often see a hierarchy of situations, e.g.:

(1) it is better to prevent an accident than to help the victims,
(2) it is better to prove a difficult theorem than to prove an easy lemma,
(3) it is best not to steal, but if we have stolen, we should return the stolen object to its legal owner, etc.

On the other hand, it is sometimes impossible to achieve the best objective.

We might have seen the accident happen from far away, so we were unable to interfere in time to prevent it, but we can still run to the scene and help the victims.

We might have seen friends last night and had a drink too many, so today's headaches will not allow us to do serious work, but we can still prove a little lemma.

We might have needed a hammer to smash the windows of a car involved in an accident, so we stole it from a building site, but will return it afterwards.

We see in all cases:

- a hierarchy of situations
- not all situations are possible or accessible for an agent.

In addition, we often have implicitly a "normality" relation:

Normally, we should help the victims, but there might be situations where not: This would expose ourselves to a very big danger, or this would involve neglecting another, even more important task (we are supervisor in a nuclear power plant ...), etc.

Thus, in all "normal" situations where an accident seems imminent, we should try to prevent it. If this is impossible, in all "normal" situations, we should help the victims, etc.

We combine these three ideas

(1) normality
(2) hierarchy
(3) accessibility

in the present section.

Note that it might be well possible to give each situation a numerical value and decide by this value what is right to do – but humans do not seem to think this way, and we want to formalize human common sense reasoning.

Before we begin the formal part, we elaborate above situations with more examples.

- We might have the overall intention to advance computer science. So we apply for the job of head of department of computer science at Stanford, and promise every tenured scientist his own laptop.

  Unfortunately, we do not get the job, but become head of computer science department at the local community college. The college does not have research as priority, but we can still do our best to achieve our overall intention, by, say buying good books for the library, or buy computers for those still active in research, etc.

  So, it is reasonable to say that, even if we failed in the best possible situation – it was not accessible to us – we still succeeded in another situation, so we achieved the overall goal.

- The converse is also possible, where better solutions become possible, as is illustrated by the following example.

  The daughter and her husband say to have the overall intention to start a family life with a house of their own, and children.

  Suppose the mother now asks her daughter: You have been married now for two years, how come you are not pregnant?

  Daughter – we cannot afford a baby now, we had to take a huge mortgage to buy our house and we both have to work.

  Mother – *I* shall pay off your mortgage. Get on with it!

  In this case, what was formerly inaccessible, is now accessible, and if the daughter was serious about her intentions – the mother can begin to look for baby carriages.

  Note that we do not distinguish here how the situations change, whether by our own doing, or by someone else's doing, or by some events not controlled by anyone.

- Consider the following hierarchy of obligations making fences as unobtrusive as possible, involving contrary to duty obligations.

  (1) You should have no fence (main duty).
  (2) If this is impossible (e.g. you have a dog which might invade neighbours' property), it should be less than 3 feet high (contrary to duty, but second best choice).
  (3) If this is impossible too (e.g. your dog might jump over it), it should be white (even more contrary to duty, but still better than nothing).
  (4) If all is impossible, you should get the neighbours' consent (etc.).

7.3.1.2 Outline of the Solution

The last example can be modelled as follows ($\mu(x)$ is the minimal models of x) :

Layer 1: $\mu(True)$: all best models have no fence.
Layer 2: $\mu(fence)$: all best models with a fence are less than 3 ft. high.
Layer 3: $\mu(fence$ and more than 3 ft. high$)$: all best models with a tall fence have a white fence.
Layer 4s: $\mu(fence$ and nonwhite and $\geq$ 3 ft$)$: in all best models with a nonwhite fence taller than 3 feet, you have permission.
Layer 5: all the rest.

This will be modelled by a corresponding $\mathcal{A}$-structure.

In summary:

(1) We have a hierarchy of situations, where one group (e.g. preventing accidents) is strictly better than another group (e.g. helping victims).
(2) Within each group, preferences are not so clear (first help person A, or person B, first call ambulance, etc.?).

(3) We have a subset of situations which are attainable, this can be modelled by an accessibility relation which tells us which situations are possible or can be reached.

We combine all three ideas, consider what we call $\mathcal{A}$-ranked structures, structures which are organized in levels A_1, A_2, A_3, etc., where all elements of A_1 are better than any element of A_2 – this is basically rankedness – and where inside each A_i we have an arbitrary relation of preference. Thus, an $\mathcal{A}$-ranked structure is between a simple preferential structure and a fully ranked structure. See Diagram 7.3.1 (p. 208).

Remark: It is not at all necessary that the rankedness relation between the different layers and the relation inside the layers express the same concept. For instance, rankedness may express deontic preference, whereas the inside relation expresses normality or some usualness.

In addition, we have an accessibility relation R, which tells us which situations are reachable. It is perhaps easiest to motivate the precise choice of modelling by layered (or contrary to duty) obligations.

For any point t, let $R(t) := \{s : tRs\}$, the set of R-reachable points from t. Given a preferential structure $\mathcal{X} := \langle X, \prec \rangle$, we can relativize $\mathcal{X}$ by considering only those points in X, which are reachable from t.

Let $X' \subseteq X$, and $\mu(X')$ the minimal points of X, we will now consider $\mu(X') \cap R(t)$ – attention, not: $\mu(X' \cap R(t))$! This choice is motivated by the following: norms are universal and do not depend on one's situation t. If $\mathcal{X}$ describes a simple obligation, then we are obliged to Y iff $\mu(X') \cap R(t) \neq \emptyset$, and $\mu(X') \cap R(t) \subseteq Y$. The first clause excludes obligations to the unattainable. We can write this as follows, supposing that X' is the set of models of ϕ' and Y is the set of models of ψ:

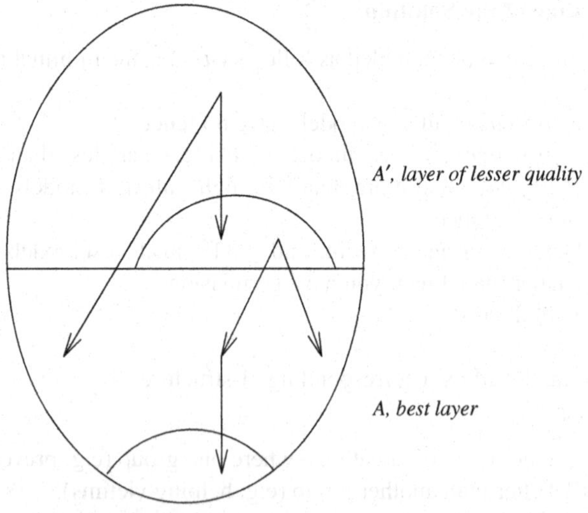

Each layer behaves inside like any preferential structure.
Amongst each other, layers behave like ranked structures.

Diagram 7.3.1 $\mathcal{A}$-ranked structure

$$m \models \phi' > \psi.$$

Thus, we put the usual consequence relation $\mid\sim$ into the object language as $>$, and relativize to the attainable (from m).

If an $\mathcal{A}$-ranked structure has two or more layers, then we are, if possible, obliged to fulfill the lower obligation, e.g. prevent an accident, but if this is impossible, we are obliged to fulfill the upper obligation, e.g. help the victims, etc.

See Diagram 7.3.2 (p. 209).

Let now, for simplicity, $\boldsymbol{B}$ be a subset of the union of all layers A, and let $\boldsymbol{B}$ be the set of models of β. This can be done, as the individual subset can be found by considering $A \cap \boldsymbol{B}$, and call the whole structure $\langle \mathcal{A}, \boldsymbol{B} \rangle$.

Then we say that m satisfies $\langle \mathcal{A}, \boldsymbol{B} \rangle$ iff in the lowest layer A, where $\mu(A) \cap R(m) \neq \emptyset$ and $\mu(A) \cap R(m) \subseteq \boldsymbol{B}$.

When we want a terminology closer to usual conditionals, we may write, e.g. $(A_1 > B_1; A_2 > B_2; \ldots.)$ expressing that the best is A_1, and then B_1 should hold, the second best is A_2, then B_2 should hold, etc. (The B_i are just $A_i \cap \boldsymbol{B}$.) See Diagram 7.3.3 (p. 215).

7.3.2 Formal Modelling and Summary of Results

We started with an investigation of "best fulfillment" of abstract requirements and contrary to duty obligations. See also [Gab08a] and [Gab08b].

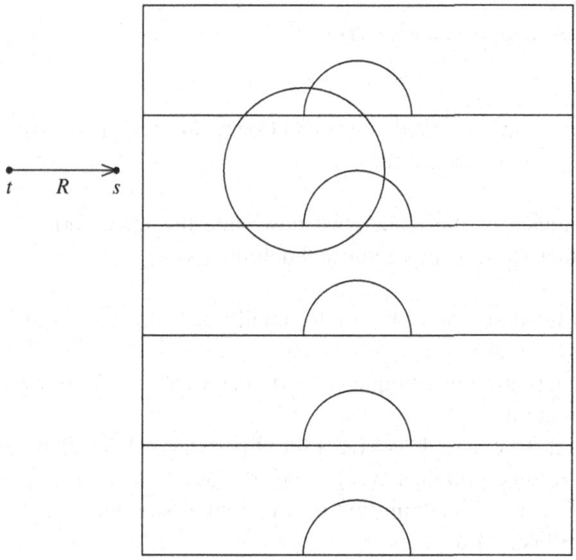

The overall structure is visible from t
Only the inside of the circle is visible from s
Half-circles are the sets of minimal elements of layers

Diagram 7.3.2 A-ranked structure and accessibility

It soon became evident that semi-ranked preferential structures give a natural semantics to contrary to duty obligations, just as simple preferential structures give a natural semantics to simple obligations – the latter goes back to Hansson [Han69].

A semi-ranked – or $\mathcal{A}$-ranked preferential structure, as we will call them later, as they are based on a system of sets $\mathcal{A}$ – has a finite number of layers, which amongst them are totally ordered by a ranking, but the internal ordering is just any (binary) relation. It thus has stronger properties than a simple preferential structure, but not as strong ones as a (totally) ranked structure.

The idea is to put the (cases of the) strongest obligation at the bottom and the weaker ones more towards the top. Then, fulfillment of a strong obligation makes the whole obligation automatically satisfied, and the weaker ones are forgotten.

Beyond giving a natural semantics to contrary to duty obligations, semi-ranked structures seem very useful for other questions of knowledge representation. For instance, any blackbird might seem a more normal bird than any penguin, but we might not be so sure within each set of birds. Thus, this generalization of preferential semantics seems very natural and welcome.

The second point of this paper is to make some, but not necessarily all, situations accessible to each point of departure. Thus, if we imagine agent a to be at point p, some fulfillments of the obligation, which are reachable to agent a' from point p' might just be impossible to reach for him. Thus, we introduce a second relation, of accessibility in the intuitive sense, denoting situations which can be reached. If this relation is transitive, then we have restrictions on the set of reachable situations: if p is accessible from p' and p can access situation s, then so can p', but not necessarily the other way round.

On the formal side, we characterize

(1) $\mathcal{A}$-ranked structures ,
(2) satisfaction of an $\mathcal{A}$-ranked conditional once an accessibility relation between the points p, p', etc. is given.

For the convenience of the reader, we now state the main formal results of this chapter – together with the more unusual definitions.

On (1):

Let $\boldsymbol{A}$ be a fixed set, and $\mathcal{A}$ a finite, totally ordered (by $<$) disjoint cover by nonempty subsets of $\boldsymbol{A}$.

For $x \in \boldsymbol{A}$, let $rg(x)$ the unique $A \in \mathcal{A}$ such that $x \in A$, so $rg(x) < rg(y)$ is defined in the natural way.

A preferential structure $\langle \mathcal{X}, \prec \rangle$ ($\mathcal{X}$ a set of pairs $\langle x, i \rangle$) is called $\mathcal{A}$-ranked iff for all x, x' $rg(x) < rg(x')$ implies $\langle x, i \rangle \prec \langle x', i' \rangle$ for all $\langle x, i \rangle, \langle x', i' \rangle \in \mathcal{X}$. See Definition 4.1.1 (p. 78) for the definition of preferential structures, and Diagram 7.3.1 (p. 208) for an illustration.

We then have:

Let $\mid\!\sim$ be a logic for $\mathcal{L}$. Set $T^{\mathcal{M}} := Th(\mu_{\mathcal{M}}(M(T)))$, and $\overline{\overline{T}} := \{\phi : T \mid\!\sim \phi\}$, where $\mathcal{M}$ is a preferential structure.

(1) Then there is a (transitive) definability preserving classical preferential model $\mathcal{M}$ s.t. $\overline{\overline{T}} = T^{\mathcal{M}}$ iff
(LLE), (CCL), (SC), (PR) hold for all $T, T' \subseteq \mathcal{L}$.
(2) The structure can be chosen smooth, iff, in addition
(CUM) holds.
(3) The structure can be chosen $\mathcal{A}$-ranked, iff, in addition

($\mathcal{A}$-min) $T \not\vdash \neg\alpha_i$ and $T \not\vdash \neg\alpha_j$, $i < j$ implies $\overline{\overline{T}} \vdash \neg\alpha_j$ holds.

See Definition 4.1.2 (p. 79) for the logic defined by a preferential structure, Definition 2.3 (p. 35) for the logical conditions, and Definition 4.1.3 (p. 80) for smoothness.

On (2)

Given a transitive accessibility relation R, $R(m) := \{x : mRx\}$.

Given $\mathcal{A}$ as above, let $\boldsymbol{B} \subseteq \boldsymbol{A}$ be the set of "good" points in $\boldsymbol{A}$, and set $\mathcal{C} := \langle \mathcal{A}, \boldsymbol{B} \rangle$.

We define:

(1) $\mu(\mathcal{A}) := \bigcup\{\mu(A_i) : i \in I\}$,
(warning: this is NOT $\mu(\boldsymbol{A})$)
(2) $\mathcal{A}_m := R(m) \cap \boldsymbol{A}$,
(3) $\mu(\mathcal{A}_m) := \bigcup\{\mu(A_i) \cap R(m) : i \in I\}$,
(3a) $\nu(\mathcal{A}_m) := \mu(\mu(\mathcal{A}_m))$,
(thus $\nu(\mathcal{A}_m) = \{a \in \boldsymbol{A} : \exists A \in \mathcal{A}(a \in \mu(A), a \in R(m)$, and
$\neg\exists a'(\exists A' \in \mathcal{A}(a' \in \mu(A'), a' \in R(m), a' \prec a\}$.
(4) $m \models \mathcal{C} :\leftrightarrow \nu(\mathcal{A}_m)) \subseteq \boldsymbol{B}$.

See Diagram 7.3.3 (p. 215).

Then the following hold:

Let $m, m' \in M$, $A, A' \in \mathcal{A}$, $\boldsymbol{A}$ be the set of models of α.

(1) $m \models \Box\neg\alpha$, $mRm' \Rightarrow m' \models \Box\neg\alpha$
(2) mRm', $\nu(\mathcal{A}_m) \cap A \neq \emptyset$, $\nu(\mathcal{A}_{m'}) \cap A' \neq \emptyset$, $\Rightarrow A \leq A'$ (in the ranking)
(3) mRm', $\nu(\mathcal{A}_m) \cap A \neq \emptyset$, $\nu(\mathcal{A}_{m'}) \cap A' \neq \emptyset$, $m \models \mathcal{C}$, $m' \not\models \mathcal{C}$, $\Rightarrow A < A'$

Conversely, these conditions suffice to construct an accessibility relation between M and $\boldsymbol{A}$ satisfying them, so they are sound and complete.

7.3.3 Overview

We next point out some connections with other domains of artificial intelligence and computer science. We then put our work in perspective with a summary of logical and semantical conditions for nonmonotonic and related logics, and present basic defintions for preferential structures. Next, we will give special definitions for our framework. We then start the main formal part and prove representation results for $\mathcal{A}$-ranked structures. First, for the general case, then for the smooth case.

The general case needs more work, as we have to do a (minor) modification of the not $\mathcal{A}$-ranked case. The smooth case is easy, we simply have to append a small construction. Both proofs are given in full detail, in order to make the text self-contained. Finally, we characterize changes due to restricted accessibility.

Definition 7.3.1 We have the usual framework of preferential structures, i.e. either a set with a possibly noninjective labelling function, or, equivalently, a set of possible worlds with copies. The relation of the preferential structure will be fixed, and will not depend on the point m from where we look at it.

Next, we have a set $\boldsymbol{A}$, and a finite, disjoint cover $A_i : i < n$ of $\boldsymbol{A}$, with a relation "of quality" $<$, $\mathcal{A}$ will denote the A_i (and thus $\boldsymbol{A}$), and $<$, i.e. $\mathcal{A} = \langle\{A_i : i \in I\}, <\rangle$.

By Fact 5.2.3 (p. 130), we may assume that all A_i are described by a formula.

Finally, we have $\boldsymbol{B} \subseteq \boldsymbol{A}$, the subset of "good" elements of $\boldsymbol{A}$ – which we also assume to be described by a formula.

In addition, we have a binary relation of accessibility, R, which we assume transitive – modal operators will be defined relative to R. R determines which part of the preferential structure is visible.

Let $R(s) := \{t : sRt\}$.

Definition 7.3.2 We repeat here from the introduction, and assume $A_i = M(\alpha_i)$, $B = M(\beta)$, and μ expresses the minimality of the preferential structure.

$$t \models \alpha_i > \beta :\Leftrightarrow \mu(A_i) \cap R(t) \subseteq B.$$

We will also abuse notation and just write,

$$t \models A_i > B \text{ in this case.}$$

We then define:

$$t \models \mathcal{C} \text{ iff at the smallest is.t.} \mu(A_i) \cap R(t) \neq \emptyset, \mu(A_i) \cap R(t) \subseteq \boldsymbol{B} \text{ holds.}$$

This motivates Definition 5.2.1 (p. 125).

Note that automatically for $X \subseteq \boldsymbol{A}$, $\mu(X) \subseteq A_j$ when j is the smallest i s.t. $X \cap A_i \neq \emptyset$. The idea is now to make the A_i the layers, and "trigger" the first layer A_j s.t. $\mu(A_j) \cap R(x) \neq \emptyset$, and check whether $\mu(A_j) \cap R(x) \subseteq B_j$. A suitable ranked structure will automatically find this A_j. More definitions and results for such $\mathcal{A}$ and $\mathcal{C}$ will be found in Sect. 7.3.5 (p. 214).

7.3.4 Connections with Other Concepts

7.3.4.1 Hierarchical Conditionals and Programmes

Our situation is now very similar to a sequence of computer program instructions:

if A_1 then do B_1;
else if A_2 then do B_2;
else if A_3 then do B_3;

where we can see the B_i as subroutines.

We can deepen this analogy in two directions:

(1) connect it to update
(2) put an imperative touch to it.

In both cases, we differentiate between different degrees of fulfillment of $\mathcal{C}$: the lower the level is which is fulfilled, the better.

(1) We can consider all threads of reachability which lead to a model m, where $m \models \mathcal{C}$. Then we take as best threads those which lead to the best fulfillment of $\mathcal{C}$. So degree of fulfillment gives the order by which we should do the update. (This is then not update in the sense that we choose the most normal developments, but rather we actively decide for the most desirable ones.) We will not pursue this line any further here, but leave it for future research.
(2) We introduce an imperative operator, say!.! means that one should fulfill $\mathcal{C}$ as best as possible by suitable choices. We will elaborate this now.

First, we can easily compare the degree of satisfaction of $\mathcal{C}$ of two models:

Definition 7.3.3 Let $m, m' \models \mathcal{C}$, and define $m < m' :\Rightarrow \mu(\mu(\mathcal{A}_m) \cup \mu(\mathcal{A}_{m'})) \cap \mu(\mathcal{A}_{m'}) = \emptyset$. ($\mu$ is, as usual, relative to some fixed $\leq_t$.)

For two sets of models, X, X', the situation does not seem so easy. So suppose that $X, X' \models \mathcal{C}$. First, we have to decide how to compare this, we do by the maximum: $X < X'$ iff the worst satisfaction of all $x \in X$ is better than the worst satisfaction in X'. More precisely, we look at all $\gamma(\mathcal{C})$ for all $x \in X$, take the maximum (which exists, as $\mathcal{A}$ is finite), and then compare the maxima for X and for X'.

Suppose now that there are points where we can make decisions ("free will"), let m be such a point. We introduce a new relation D, and let mDm' iff we can decide to go from m to m'. The relation D expresses this possibility – it is our definition of "free will".

Definition 7.3.4 Consider now some formula ϕ, and define

$$m \models !\phi :\Rightarrow D(m) \cap M(\phi) < D(m) \cap M(\neg\phi).$$

(as defined in Definition 7.3.3 (p. 213)).

7.3.4.2 Connection with Theory Revision

In particular, the situation of contrary to duty obligations (see Sect. 7.3.1 (p. 205)) shows an intuitive similarity to revision. You have the duty not to have a fence. If this is impossible (read: inconsistent), then it should be white. So the duty is revised.

But there is also a formal analogy: As is well known, AGM revision (with fixed left hand side K) corresponds to a ranked order of models, where models of K have lowest rank (or distance 0 from K-models). The structures we consider ($\mathcal{A}$-rankings) are partially ranked, i.e. there is only a partial ranked preference, inside the layers, nothing is said about the ordering. This partial ranking is natural, as we have only a limited number of cases to consider.

But we use the revision order (based on K, so it is really a $\leq_K$ relation) differently: We do not revise K, but use only the order to choose the first layer which has nonempty intersection with the set of possible cases. Still, the spirit (and formal apparatus) of revision is there, just used somewhat differently. The K-relation expresses here deontic quality, and if the best situation is impossible, we choose the second best, etc.

Theory revision with variable K is expressed by a distance between models (see [LMS01]), where $K * \phi$ is defined by the set of ϕ models which have minimal distance from the set of K models.

We can now generalize our idea of layered structure to a partial distance as follows: For instance, $d(K, A)$ is defined, $d(K, B)$ too, and we know that all A models with minimal distance to K have smaller distance than the B models with minimal distance to K. But we do NOT know a precise distance for other A models, we can sometimes compare, but not always. We may also know that all A models are closer to K than any B model is, but for a and a', both A models, we might not know if one or the other is closer to K, or is they have the same distance.

The representation results for $\mathcal{A}$-ranked structures were shown already in Sect. 5.2 (p. 125), so we can turn immediately to the following.

7.3.5 Formal Results and Representation for Hierarchical Conditionals

We look here at the following problem:

Given

(1.1) a finite, ordered partition $\mathcal{A}$ of $\mathbf{A}$, $\mathcal{A} = \langle \{A_i : i \in I\}, < \rangle$,

(1.2) a normality relation $\prec$, which is an $\mathcal{A}$-ranking, defining a choice function μ on subsets of $\mathbf{A}$, (so, obviously, $A < A'$ iff $\mu(A \cup A') \cap A' = \emptyset$),

(1.3) a subset $\mathbf{B} \subseteq \mathbf{A}$, and we set $\mathcal{C} := \langle \mathcal{A}, \mathbf{B} \rangle$ (thus, the B_i are just $A_i \cap \mathbf{B}$, this way of writing saves a little notation),

(2.1) a set of models M,

(2.2) an accessibility relation R on M, with some finite upper bound on R-chains,

(2.3) an unknown extension of R to pairs (m, a), $m \in M$, $a \in \mathbf{A}$,

(3.1) a notion of validity $m \models \mathcal{C}$, for $m \in M$, defined by $m \models \mathcal{C}$ iff $\{a \in \mathbf{A} :$ $\exists A \in \mathcal{A}(a \in \mu(A), a \in R(m)$, and
$\neg \exists a'(\exists A' \in \mathcal{A}(a' \in \mu(A'), a' \in R(m), a' \prec a\} \subseteq \mathbf{B}$,

(3.2) a subset M' of M,
give a criterion which decides whether it is possible to construct the extension of R to pairs (m, a) s.t. $\forall m \in M(m \in M' \Leftrightarrow m \models \mathcal{C})$.

We first show some elementary facts on the situation, and give the criterion in Proposition 7.3.1 (p. 216), together with the proof that it does what is wanted.

Fact 7.3.1 Reachability for a transitive relation is characterized by

$$y \in R(x) \to R(y) \subseteq R(x).$$

Proof Define directly xRz iff $z \in R(x)$. This does it. □

Let now S be a set with an accessibility relation R', generated by transitive closure from the intransitive subrelation R. All modal notation will be relative to this R.

Let $\mathbf{A} = M(\alpha)$, $A_i = M(\alpha_i)$, the latter is justified by Fact 5.2.3 (p. 130).

Definition 7.3.5

(1) $\mu(\mathcal{A}) := \bigcup\{\mu(A_i) : i \in I\}$,
(warning: this is NOT $\mu(\mathbf{A})$)
(2) $\mathcal{A}_m := R(m) \cap \mathbf{A}$,
(3) $\mu(\mathcal{A}_m) := \bigcup\{\mu(A_i) \cap R(m) : i \in I\}$,
(3a) $\nu(\mathcal{A}_m) := \mu(\mu(\mathcal{A}_m))$,
(thus $\nu(\mathcal{A}_m) = \{a \in \mathbf{A} : \exists A \in \mathcal{A}(a \in \mu(A), a \in R(m)$, and $\neg\exists a'(\exists A' \in \mathcal{A}(a' \in \mu(A'), a' \in R(m), a' \prec a\}$.
(4) $m \models \mathcal{C} :\leftrightarrow \nu(\mathcal{A}_m)) \subseteq \mathbf{B}$.
See Diagram 7.3.3 (p. 215).

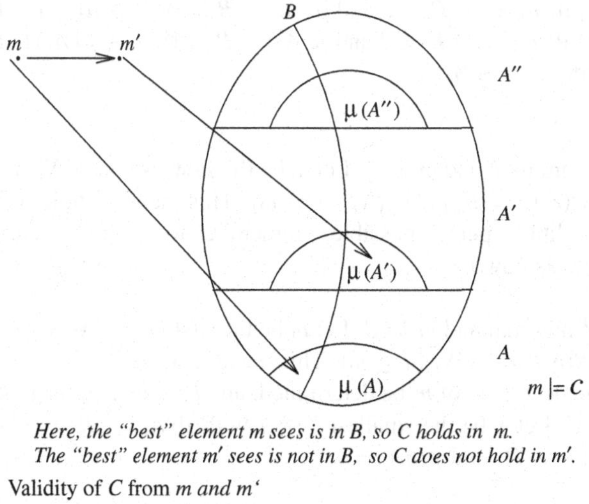
B
m
m'
A''
μ (A'')
A'
μ (A')
A
μ (A)
m |= C
Here, the "best" element m sees is in B, so C holds in m.
The "best" element m' sees is not in B, so C does not hold in m'.

Diagram 7.3.3 Validity of C from *m and m'*

We have the following fact for $m \models \mathcal{C}$:

Fact 7.3.2 Let $m, m' \in M$, $A, A' \in \mathcal{A}$.

(1) $m \models \Box\neg\alpha,\ mRm' \Rightarrow m' \models \Box\neg\alpha,$
(2) $mRm',\ \nu(\mathcal{A}_m) \cap A \neq \emptyset,\ \nu(\mathcal{A}_{m'}) \cap A' \neq \emptyset, \Rightarrow A \leq A',$
(3) $mRm',\ \nu(\mathcal{A}_m) \cap A \neq \emptyset,\ \nu(\mathcal{A}_{m'}) \cap A' \neq \emptyset,\ m \models \mathcal{C},\ m' \not\models \mathcal{C}, \Rightarrow A < A'.$

Proof Trivial. □

Fact 7.3.3 We can conclude from above properties that there are no arbitrarily long R-chains of models m, changing from $m \models \mathcal{C}$ to $m \not\models \mathcal{C}$ and back.

Proof Trivial: By Fact 7.3.2 (p. 216), (3), any change from $\models \mathcal{C}$ to $\not\models \mathcal{C}$ results in a strict increase in rank. □

We solve now the representation task described at the beginning of Sect. 7.3.5 (p. 214), all we need are the properties shown in Fact 7.3.2 (p. 216).

(Note that constructing R between the different m, m' is trivial: We could just choose the empty relation.)

Proposition 7.3.1 *If the properties of Fact* 7.3.2 *(p.* 216*) hold, we can extend R to solve the representation problem described at the beginning of this Sect.* 7.3.5 *(p.* 214*).*

Proof By induction on R. This is possible, as the depth of R on M was assumed to be finite. □

Construction 7.3.1 We choose now elements as possible, which ones are chosen exactly does not matter.

$X_i := \{b_i, c_i\}$ iff $\mu(A_i) \cap \mathbf{B} \neq \emptyset$ and $\mu(A_i) - \mathbf{B} \neq \emptyset$, $b_i \in \mu(A_i) \cap \mathbf{B}$, $c_i \in \mu(A_i) - \mathbf{B}$,
$X_i := \{c_i\}$ iff $\mu(A_i) \cap \mathbf{B} = \emptyset$ and $\mu(A_i) - \mathbf{B} \neq \emptyset$, $c_i \in \mu(A_i) - \mathbf{B}$,
$X_i := \{b_i\}$ iff $\mu(A_i) \cap \mathbf{B} \neq \emptyset$ and $\mu(A_i) - \mathbf{B} = \emptyset$, $b_i \in \mu(A_i) \cap \mathbf{B}$,
$X_i := \emptyset$ iff $\mu(A_i) = \emptyset$.

Case 1:

Let m be R-minimal and $m \models \mathcal{C}$. Let i_0 be the first i s.t. $b_i \in X_i$, make $\gamma(m) := i_0$, and make $R(m) := \{b_{i_0}\} \cup \bigcup\{X_i : i > i_0\}$. This makes $\mathcal{C}$ hold. (This leaves us as many possibilities open as possible – remember we have to decrease the set of reachable elements now.)

Case 2:

Let m be R-minimal and $m \not\models \mathcal{C}$. Let i_0 be the first i s.t. $c_i \in X_i$, make $\gamma(m) := i_0$, and make $R(m) := \bigcup\{X_i : i \geq i_0\}$. This makes $\mathcal{C}$ false.

Let all R-predecessors of m be determined, and $i := max\{\gamma(m') : m'Rm\}$.

Case 3: $m \models \mathcal{C}$. Let j be the smallest $i' \geq i$ with $\mu(A_{i'}) \cap \mathbf{B} \neq \emptyset$. Let $R(m) := \{b_j\} \cup \bigcup\{X_k : k > j\}$, and $\gamma(m) := j$.

Case 4: $m \not\models \mathcal{C}$.

Case 4.1: For all $m'Rm$ with $i = \gamma(m')$ $m' \not\models \mathcal{C}$.

Take one such m' and set $R(m) := R(m')$, $\gamma(m) := i$.

Case 4.2: There is $m'Rm$ with $i = \beta(m')$ $m' \models \mathcal{C}$.

Let j be the smallest $i' > i$ with $\mu(A_{i'}) - \boldsymbol{B} \neq \emptyset$. Let $R(m) := \bigcup\{X_k : k \geq j\}$. (Remark: To go from $\models$ to $\not\models$, we have to go higher in the hierarchy.)

Obviously, validity is done as it should be. It remains to show that the sets of reachable elements decrease with R.

Fact 7.3.4 In above construction, if mRm', then $R(m') \subseteq R(m)$.

Proof By induction, considering R. (Fact 7.3.4 (p. 217) and Proposition 7.3.1 (p. 216)). □

We consider an example for illustration.

Example 7.3.1 Let $a_1 R a_2 R c R c_1$, $b_1 R b_2 R b_3 R c R d_1 R d_2$.

Let $\mathcal{C} = (A_1 > B_1, \ldots, A_n > B_n)$ with the C_i consistency with $\mu(A_i)$.

Let $\mu(A_2) \cap B_2 = \emptyset$, $\mu(A_3) \subseteq B_3$, and for the other i hold neither of these two.

Let $a_1, a_2, b_2, c_1, d_2 \models \mathcal{C}$, the others $\not\models \mathcal{C}$.

Let $\mu(A_1) = \{a_{1,1}, a_{1,2}\}$, with $a_{1,1} \in B_1$, $a_{1,2} \notin B_1$,
$\mu(A_2) = \{a_{2,1}\}$, $\mu(A_3) = \{a_{3,1}\}$ (there is no reason to differentiate),
and the others like $\mu(A_1)$. Let $\mu A := \bigcup\{\mu(A_i) : i \leq n\}$.

We have to start at a_1 and b_1, and make $R(x)$ progressively smaller.

Let $R(a_1) := \mu A - \{a_{1,2}\}$, so $a_1 \models \mathcal{C}$. Let $R(a_2) = R(a_1)$, so again $a_2 \models \mathcal{C}$.

Let $R(b_1) := \mu A - \{a_{1,1}\}$, so $b_1 \not\models \mathcal{C}$. We now have to take $a_{1,2}$ away, but $a_{2,1}$ too to be able to change. So let $R(b_2) := R(b_1) - \{a_{1,2}, a_{2,1}\}$, so we begin at $\mu(A_3)$, which is a (positive) singleton. Then let $R(b_3) := R(b_2) - \{a_{3,1}\}$.

We can choose $R(c) := R(b_3)$, as $R(b_3) \subseteq R(a_2)$.

Let $R(c_1) := R(c) - \{a_{4,2}\}$ to make $\mathcal{C}$ hold again. Let $R(d_1) := R(c)$, and $R(d_2) := R(c_1)$.

Chapter 8
Theory Update and Theory Revision

8.1 Update

8.1.1 Introduction

We will treat here problems due to lack of information, i.e. we can "see" some dimensions, but not all.

8.1.2 Hidden Dimensions

8.1.2.1 Introduction

We look here at situations where only one dimension is visible in the results, and the other ones stay hidden. This is, e.g., the case when we can observe only the outcome of developments, but not the developments themselves.

It was the authors' intention to treat here the general infinite case, and then show that the problems treated in [BLS99] and [LMS01] (the not necessarily symmetric case there) are special cases thereof. Unfortunately, we failed in the attempt to solve the general infinite case; it seems that one needs new and quite different methods to solve it, so we will just describe modestly what we see that can be done, what the problems seem to be, and conclude with a very short remark on the situation described in [BLS99].

8.1.2.2 Situation, Definitions, and Basic Results

In several situations, we can observe directly only one dimension of a problem. In a classical ranked structure, we "see" everything about an optimal model. It is there and fully described. But look at a ranking of developments, where we can observe only the outcome. The earlier dimensions remain hidden, and when we see the outcome of the "best" developments, we do not directly see the threads which led there. A similar case is theory revision based on not necessarily symmetric distances, where we cannot "look back" from the result to see the closest elements of the former theory (see [LMS01] for details).

D.M. Gabbay, K. Schlechta, *Logical Tools for Handling Change in Agent-Based Systems*, Cognitive Technologies, DOI 10.1007/978-3-642-04407-6_8,

The non definable case and the case of hidden dimensions are different aspects of a common problem: In the case of non definability, any not too small subset might generate what we see, while in the case of hidden dimensions, any thread ending in an optimal outcome might be optimal.

The Situation, More Formally

The universe is a finite or even infinite product ΠU_i, $i \in I$. We will see that the finite case is already sufficiently nasty, so we will consider only the finite situation. If $X \subseteq \Pi U_i$, then possible results will be projections on some fixed coordinate, say j, of the best $\sigma \in X$, where best is determined by some ranking $\prec$ on ΠU_i, $\pi_j(\mu(X))$.

As input, we will usually not have arbitrary $X \subseteq \Pi U_i$, but again some product $X := \Pi U_i'$, with $U_i' \subseteq U_i$. Here is the main problem: We cannot use as input arbitrary sets, but only products. We will see that this complication will hide almost all information in sufficiently nasty situations.

We will make now some reasonable assumptions:

First, without loss of generality, we will always take the last dimension as outcome. Obviously, this does not change the general picture.

Second, the difficult situations are those where (some of) the U_i are infinite. We will take the infinite propositional case with theories as input as motivation, and assume that for each U_i, we can choose any finite U_i' as input, and the possible U_i' are closed under intersections and finite unions.

Of course, $\Pi U_i' \cup \Pi U_i''$ need not be a product ΠV_i – here lies the main problem, the domain is not closed under finite unions.

We will see that the case presents serious difficulties, which the authors do not know how to solve. The basic problem is that we do not have enough information to construct an order.

Notation 8.1.1 (.)! will be the projection on the fixed, (last) coordinate, so for $X \subseteq \Pi U_i, i \leq n$, $X! := \{\sigma(n) : \sigma \in X\}$, and analogously, $\sigma! := \sigma(n)$ for $\sigma \in \Pi U_i$.

To avoid excessive parentheses, we will also write $\mu X!$ for $(\mu(X))!$, where μ is the minimal choice operator, etc.

For any set of sequences X, $[X]$ will be the smallest product which contains X. Likewise, for σ, σ', $[\sigma, \sigma']$ will be the product (which will always be defined as we have singletons and finite unions in the components), $\Pi\{\sigma(i), \sigma'(i)\}$ $(i \in I)$. Analogously, $(\sigma, \sigma') := [\sigma, \sigma'] - \{\sigma, \sigma'\}$. (The interval notation is intended.)
If $\sigma \in X$, a σ-cover of X will be a set $\{X_k : k \in K\}$ s.t. $\sigma \notin X_k$ for all k, and $\bigcup\{X_k : k \in K\} \cup \{\sigma\} = X$.
A (finite) sequence $\sigma = \sigma_1, \ldots, \sigma_n$ will also be written $(\sigma_1, \ldots, \sigma_n)$.
The Hamming distance is very important in our situation. We can define the Hamming distance between two (finite) sequences as the number of arguments where

they disagree, or as the set of those arguments. In our context, it does not matter which definition we choose. "H-closest to σ" means thus "closest to σ, measured by one of the Hamming distances".

Direct and Indirect Observation

If $\sigma ! = \sigma' !$, we cannot compare them directly, as we always see the same projection. If $\sigma ! \neq \sigma' !$, we can compare them, if e.g. $X := \{\sigma, \sigma'\}$ is in the domain of μ (which usually is not the case, but only if they differ in the last coordinate), $\mu(X)!$ can be $\{\sigma !\}$, $\{\sigma' !\}$, $\{\sigma !, \sigma' !\}$, so σ is better, σ' is better, or they are equal.
Thus, to compare σ and σ' if $\sigma ! = \sigma' !$, we have to take a detour, via some τ with $\tau ! \neq \sigma !$. This is illustrated by the following example:

Example 8.1.1 Let a set A be given, $B := \{b, b'\}$, and $\mu(A \times B)! = \{b\}$. Of course, without any further information, we have no idea if for all $a \in A$ $\langle a, b \rangle$ is optimal, if only for some, etc.

Consider now the following situation: Let $a, a' \in A$ be given, $A' := \{a, a'\}$, and $C := \{b, c, c'\}$ – where c, c' need not be in B. Let $A' \times C$, $\{a\} \times C$, $\{a'\} \times C$ be defined and observable, and

(1) $\mu(A' \times C)! = \{b, c\}$,
(2) $\mu(\{a\} \times C)! = \{b, c\}$,
(3) $\mu(\{a'\} \times C)! = \{b, c'\}$.

Then by (3) $(a', b) \approx (a', c') \prec (a', c)$, by (2) $(a, b) \approx (a, c) \prec (a, c')$, by (1) (a', c') cannot be optimal, so neither is (a', b); thus (a, b) must be better than (a', b), as one of (a, b) and (a', b) must be optimal.

Thus, in an indirect way, using elements c and c' not in B, we may be able to find out which pairs in $A \times B$ are really optimal. Obviously, arbitrarily long chains might be involved, and such chains may also lead to contradictions.

So it seems difficult to find a general way to find the best pairs – and much will depend on the domain, how much we can compare, etc. It might also well be that we cannot find out – so we will branch into all possibilities, or choose arbitrarily one – loosing ignorance, of course.

Definition 8.1.1 Let $\{X_k : k \in K\}$ be a σ-cover of X.
Then we can define $\prec$ and $\preceq$ in the following cases:

(1) $\forall k(\mu X_k ! \not\subseteq \mu X!) \rightarrow \sigma \prec \sigma'$ for all $\sigma' \in X$, $\sigma' \neq \sigma$,
(2) $\mu X! \not\subseteq \bigcup \mu X_k ! \rightarrow \sigma \prec \sigma'$ for all $\sigma' \in X$ s.t. $\sigma' ! \notin \mu(X)!$, and $\sigma \preceq \sigma'$ for all $\sigma' \in X$ s.t. $\sigma' ! \in \mu X!$.

Explanation of the latter: If $\sigma' ! \in X!$, there is some σ'' s.t. σ'' is one of the best, and $\sigma'' ! = \sigma'$ – but we are not sure which one. So we really need $\preceq$ here.

Of course, we also know that for each $x \in \mu X!$ there is some σ s.t. $\sigma ! = x$ and $\sigma \preceq \sigma'$ for all $\sigma' \in X$, etc., but we do not know which σ.

We describe now the two main problems. We want to construct a representing relation. In particular,

(a) if $\tau \in X$, $\tau! \notin \mu X!$, we will need some $\sigma \in X$, $\sigma \prec \tau$ and
(b) if $\tau! \in \mu X!$, and $x' \neq \tau!$, $x' \in \mu X!$, we will need some $\sigma \in X$, $\sigma \preceq \tau$.

Consider now some such candidate, and the smallest set containing them, $X := [\tau, \sigma]$.

Problem (1): $\sigma \prec \tau$ might well be the case, but there is $\tau' \in (\tau, \sigma)$, $\tau'! = \tau!$, and $\tau' \prec \sigma$ – so we will not see this, as $\mu X! = \{\tau'!\}$.
Problem (2): $\sigma \prec \tau$ might well be the case, but there is $\sigma' \in (\tau, \sigma)$, $\sigma'! = \sigma!$, $\sigma' \prec \tau$ – so we will not see this, as already $\mu[\sigma', \tau]! = \sigma!$.

Problem (2) can be overcome for our purposes, by choosing a H-closest σ s.t. $\sigma \prec \tau$, and if (1) is not a problem, we can see now that $\sigma \prec \tau$: For all suitable ρ we have $\mu[\rho, \tau]! = \tau!$, and only $\mu[\tau, \sigma]! = \sigma!$.

Problem (1) can be overcome if we know that some σ minimizes all τ, e.g. if $\sigma! \in \mu X!$, $\tau! \notin \mu X!$. We choose then for some τ some such σ, and consider $[\tau, \sigma]$. There might now already be some $\sigma' \in (\tau, \sigma)$, $\sigma'! = \sigma!$, which minimizes τ and all elements of $[\sigma', \tau]$, so we have to take the H-closest one to avoid Problem (2), work with this one, and we will see that $\sigma' \prec \tau$.

Both cases illustrate the importance of choosing H-closest elements, which is, of course, possible by finiteness of the index set. Problem (1) can, however, be unsolvable if we take the limit approach, as we may be forced to do (see discussion in Sect. 8.1.2 (p. 219)), as we will not necessarily have anymore such σ which minimize all τ – see Example 8.1.4 (p. 224), Case (4). But in the other, i.e. non problematic, cases, this is the approach to take:
Choose for τ some σ s.t. $\sigma \preceq \rho$ for all ρ – if this exists – then consider $[\tau, \sigma]$, choose in $[\tau, \sigma]$ the H-closest (measured from τ) ρ s.t. ρ is smaller or equal to all $\rho' \in [\rho, \tau]$. This has to exist, as σ is a candidate, and we work with a finite product, so the H-closest has to exist.
We will then use a cover property.

Property 6.3.1

Let $\tau \in X$, $\tau! \notin \mu X!$, $\Sigma_x := \{\sigma \in X : \sigma! = x\}$ for some fixed $x \in \mu X!$. Let $\Sigma_x \subseteq \bigcup \Sigma_i$ for $i \in I$ and $X_i \supseteq \Sigma_i \cup \{\sigma\}$, then there is $i_0 \in I$ s.t. $\tau! \notin \mu X_i!$.
(This will also hold in the limit case.)
This allows to find witnesses, if possible:
Let $\tau \in X$, $\tau! \notin \mu X!$, then τ is minimized by one σ with $\sigma! \in \mu X!$. Take such σ with H-closest to τ; this exists by cover property. Then we see that $\tau! \notin \mu[\sigma, \tau]!$ and that σ does it, so we define $\sigma \prec \tau$.
We will also have something similar for two elements $\sigma, \sigma' \in X$, $\sigma! \neq \sigma'!$, $\sigma!, \sigma! \in \mu X!$: We will seek σ, σ' which have minimal H-distance among all τ, τ' with the same endpoints s.t. $\sigma \preceq \rho$, $\sigma' \preceq \rho$ for all $\rho \in [\sigma, \sigma']$. This must exist in the minimal variant, but, warning, we are not sure that they are really the minimal ones (by the order) – see Example 8.1.4 (p. 224), (1).

Small Sets and Easy Properties

As a first guess, one might think that small sets always suffice to define the relation. This is not true. First, if $\sigma! = \sigma'!$, then $[\sigma, \sigma']$ will give us no information. The following (also negative) example is more instructive:

Example 8.1.2 Consider $X := \{0, 1, 2\} \times \{0, 1\}$, with the ordering: $(0, 0) \prec (2.1) \prec (1, 0) \prec (0, 1) \sim (1, 1) \sim (2, 0)$. We want to find out that $(0, 0) \prec (1, 1)$ using the results $\mu Y!$.
Consider $X' := \{0, 1\} \times \{0, 1\}$, the smallest set containing both. Any $(0, 0)$-cover of X' has to contain some X'' with $(1, 0) \in X''$, but then $\mu X''! = \{0\}$, so we cannot see that $(0, 0) \prec (1, 1)$.
But consider the $(0, 0)$-cover $\{X', X''\}$ of X with $X' := \{1, 2\} \times \{0, 1\}$, $X'' := \{(0, 1)\}$, then $\mu X'! = \mu X''! = \{1\}$, but $\mu X! = \{0\}$, so we see that $(0, 0) \prec (1, 1)$.

(But we can obtain the same result through a chain of small sets: $(0, 0) \prec (2, 1)$: look at the $(0, 0)$-cover $\{\{2\} \times \{0, 1\}, \{(0, 1)\}\}$ of $\{0, 2\} \times \{0, 1\}$, $(2, 1) \prec (2, 0)$ is obvious, $(2, 0) \preceq (1, 1)$: look at the $(2, 0)$-cover $\{\{1\} \times \{0, 1\}, \{(2, 1)\}\}$ of $\{1, 2\} \times \{0, 1\}$, we see that $\mu(\{1\} \times \{0, 1\})! = \mu\{(2, 1)\}! = \{1\}$, and $\mu(\{1, 2\} \times \{0, 1\})! = \{0, 1\}$.)

Remark 8.1.1 The following are immediate:

(1) $\bigcap\{\Pi\{X_i^j : i \in I\} : j \in J\} = \Pi\{\bigcap\{X_i^j : j \in J\} : i \in I\}$,
(2) $\mu X! \subseteq X!$,
(3) $X = \bigcup X_k \to \mu X! \subseteq \bigcup(\mu X_k!)$,
(4) $X \subseteq X' \to X! - \mu X! \subseteq X'! - \mu X'!$ is in general wrong, though $X \subseteq X' \to X - \mu X \subseteq X' - \mu X'$ is correct: There might be a new minimal thread in X', which happens to have the same endpoint as discarded threads in X.

8.1.2.3 A First Serious Problem

We describe now the first serious problem and indicate how to work around it. So, we think this can still be avoided in a reasonable way. It is the next one, where we meet essentially the same difficulty and where we see no solution or detour.

We will complete the ranked relation (or, better, what we see of it) as usual, i.e. close under transitivity it has to be free of cycles, etc. Usually, the relation will not be fully determined, so, forgetting about ignorance, we complete it using our abstract nonsense lemma Fact 4.2.14 (p. 116). But there is a problem, as the following sketch of an example shows. We have to take care that we do not have infinite descending chains of threads σ with the same endpoint, as the endpoint will then not appear anymore, by definition of minimality. More precisely, the following example describes.

Example 8.1.3 Let $\sigma! = \tau!$ and $\sigma'! = \tau'!$, and $\sigma' \prec \sigma$, $\tau \prec \tau'$, then choosing $\sigma \preceq \tau$ will result in $\sigma' \prec \tau'$. Likewise, $\sigma'' \preceq \tau''$ with $\sigma''! = \tau''!$ may lead to $\rho' \prec \sigma'$ (with $\sigma'! = \rho'!$), etc., and to an infinite descending chain. This problem seems difficult to solve, and the authors do not know if there is always a solution, probably not, as we need just ω many pairs to create an infinite descending chain, but the universe can be arbitrarily big.

Two Ways to Work Around this Problem

(1) We may neglect such infinite descending chains, and do as if they were not there. Thus, we will take a limit approach (see Sect. 5.5 (p. 145)) locally, within the concerned $\sigma!$. This might be justified intuitively by the fact that we are really interested only in one dimension and do not really care about how we got there, so we neglect the other dimensions somewhat.
(2) We may consider immediately the global limit approach. In this case, it will probably be the best strategy to consider formula-defined model sets, in order to be able to go easily to the logical version – see Sect. 5.5 (p. 145).

For (1): We continue to write $\mu X!$ for the observed result, though this will not be a minimum anymore. We then have

$$\forall \sigma \in X\ \forall x \in \mu X!\ \exists \tau(\tau! = x \wedge \tau \preceq \sigma),$$

but we will not have anymore

$$\forall x \in \mu X!\ \forall \sigma \in X\ \exists \tau \in X(\tau! = x \wedge \forall \sigma \in X.\tau \preceq \sigma).$$

Note that in the limit case

(1) $X \neq \emptyset \rightarrow \mu X! \neq \emptyset$,
(2) if $x, x' \in \mu X!$ and x is a limit (i.e. there is an infinite descending chain of σs with $\sigma = x$), then so is x',
(3) by finiteness of the Hamming distance, e.g. there will be for all $\sigma \in X$ and all $x \in \mu X!$ cofinally often some H-closest (to σ) $\tau \in X$ s.t. $\tau! = x$ and $\tau \preceq \sigma$. (Still, this will not help us much, unfortunately.)

8.1.2.4 A Second Serious Problem

This problem can be seen in the (class of) unpleasant examples. It has to do with the fact that interesting elements may be hidden by other elements.

Example 8.1.4 We discuss various variants of this example and first present the common parts.
The domain U will be $2 \times \omega \times 2$. $X \subseteq U$ etc. will be legal products. We define suitable orders and will see that we can hardly get any information about the order.

$$\sigma_i := (0, i, 0), \ \tau_i := (1, i, 1).$$

The key property is as follows:

If $\sigma_i, \tau_j \in X \subseteq U$ for some $i, j \in \omega$, then $\sigma_k \in X \leftrightarrow \tau_k \in X$. (Proof : Let, e.g. $\sigma_k = (0, k, 0) \in X$, then, as $(1, j, 1) \in X$, $(1, k, 1) \in X$.)

In all variants, we make the top layer consisting of all $(1, i, 0)$ and $(0, i, 1)$, $i \in \omega$. They will play a minor role. All other elements will be strictly below this top layer. We turn to the variants.

(1) Let $\sigma_i \sim \tau_i$ for all i, and $\sigma_i \sim \tau_i \prec \sigma_{i+1} \sim \tau_{i+1}$ is close under transitivity. Thus, there is a minimal pair, σ_0, τ_0, and we have an "ascending ladder" of the other σ_i, τ_i. But we cannot see which pair is at the bottom.
Proof: Let $X \subseteq U$. If X contains only elements of the type $(a, b, 0)$ or $(a, b, 1)$, the result is trivial: $\mu X! = \{0\}$, etc. Suppose not. If X contains no σ_i and no τ_i, the result is trivial again and gives no information about our question. The same is true, if X contains only σ_is or only $\tau_i s$, but not both. If X contains some σ_i, and some τ_j, let k be the smallest such index i or j, then, by above remark, σ_k and τ_k will be in X, so $\mu X! = \{0, 1\}$ in any case, and this will not tell us anything. (More formally, consider in parallel the modified order, where the pair σ_1, τ_1 is the smallest one, instead of σ_0, τ_0, and order the others as before, then this different order will give the same result in all cases.)

(2) Order the elements as in (1), only make a descending ladder, instead of an ascending one. Thus, $\mu U! = \emptyset$, but $\mu X!$ in variant (1) and in variant (2) will not be different for any finite X.
Proof: As above, the interesting case is when there is some σ_i and some τ_j in X. We now take the biggest such index k, then σ_k and τ_k will both be in X, and thus $\mu X! = \{0, 1\}$. Consequently, only infinite X allow us to distinguish variant (1) from variant (2). But they give no information on how the relation is precisely defined, only that there are infinite descending chains.
Thus, finite X do not allow us to distinguish between absence and presence of infinite descending chains.

(3) Order the σ_i in an ascending chain, let $\sigma_i \prec \tau_i$, and no other comparisons between the σ_i and the τ_j. Thus, this is not a ranked relation. Again, the only interesting case is where some $\sigma_i \in X$ and some $\tau_j \in X$. But by above reasoning, we have always $\mu X! = \{0\}$.
Modifying the order such that $\tau_i \prec \sigma_{i+1}$ (making one joint ascending chain) is a ranked order with the same results $\mu X!$.
Thus, we cannot distinguish ranked from non ranked relations.

(4) Let now $\sigma_i \prec \tau_i$, and $\tau_{i+1} \prec \sigma_i$, is close under transitivity. As we have an infinite descending chain, we take the limit approach.
Let X be finite and some $\sigma_i = (0, i, 0) \in X$, then $\mu X! = \{0\}$. Thus, we can never see that some specific τ_j minimizes some σ_k.

Proof: All elements in the top layer are minimized by σ_i. If there is no $\tau_j \in X$, we are done. Let $\tau_k = (1, k, 1)$ be the smallest $\tau_j \in X$, then by the above $\sigma_k = (0, k, 0) \in X$, too, so $1 \notin \mu X!$.
But we can see the converse: Consider $X := [\sigma_i, \tau_i]$, then $\mu X! = \{0\}$, as we just saw. The only other $\sigma \in X$ with $\sigma! = 0$ is $(1, i, 0)$, but we see that $\tau_i \prec \sigma$, as $\mu[\tau_i, \sigma] = \{1\}$, and $[\tau_i, \sigma]$ contains only two elements, so σ_i must minimize τ_i.
Consequently, we have the information about $\sigma_i \prec \tau_i$, but not any information about any $\tau_i \prec \sigma_j$. Yet, taking the limit approach applied to U, we see that there must be below each σ_j some τ_i, otherwise we would see only $\{0\}$ as result. But we do not know how to choose the τ_i below the σ_j.

8.1.2.5 Resume

The problem is that we might not have enough information to construct the order. Taking any completion is not sure to work, as it might indirectly contradict existing limits or non limits, which give only very scant information (there is or not an infinite descending chain), but do not tell us anything about the order.
For such situations, cardinalities of the sets involved might play a role.
Unfortunately, the authors have no idea how to attack these problems. It seems quite certain that the existing techniques in the domain will not help us.
So this might be a reason (or pretext?) to look at simplifications.
It might be intuitively justifiable to impose a continuity condition:

(Cont) If τ is between σ and σ' in the Hamming distance, then so is its ranking, i.e. $\rho(\sigma) \leq \rho(\sigma') \rightarrow \rho(\sigma) \leq \rho(\tau) \leq \rho(\sigma')$, if ρ is the ranking function.

This new condition seems to solve above problems, and can perhaps be justified intuitively. But we have to be aware of the temptation to hide our ignorance and inability to solve problems behind flimsy intuitions. Yet, has someone ever come up with an intuitive justification of definability preservation? (The disastrous consequences of its absence were discussed in [Sch04].)
On the other hand, this continuity or interpolation property might be useful in other circumstances too, where we can suppose that small changes have small effects, see the second author's [Sch95-2] or Chap. 4 in [Sch97-1] for a broader and deeper discussion.
This additional property should probably be further investigated.
A word of warning: This condition is NOT compatible with distance-based reasoning: The distance from 0 to 2 is the same as the distance from 1 to 3, but the set $[(0, 2), (1, 3)]$ contains $(1, 2)$ with a strictly smaller distance. A first idea is then not to work on states, but on differences between states, considering the sum. But this does not work either. The sequences $(0, 1)$, $(1, 0)$ contain between them the smaller one $(0, 0)$ and the bigger one $(1, 1)$. Thus, the condition can be countertintuitive, and the authors do not know if there are sufficiently many natural scenarios where it holds.
It seems, however, that we do not need the full condition, but the following would suffice: If σ is such that there is τ with $\sigma! \neq \tau!$, and $\tau \prec \sigma$, then we find τ' s.t. $\tau! = \tau'!$ and τ' is smaller than all sequences in $[\sigma, \tau]$ (and perhaps a similar

condition for $\preceq$). Then, we would see smaller elements in Example 8.1.4 (p. 224). The above is a limit condition and is similar to the condition that there are cofinally many definable initial segments in the limit approach; see the discussion [there] – or in [Sch04].

8.1.2.6 A Comment on Former Work

We make here some very short remarks on our joint article with S. Berger and D. Lehmann, [BLS99].
Perhaps the central definition of the article is Definition 3.10 in [BLS99].
First, we see that the relation R is generated only by sets where one includes the other. Thus, if the domain does not contain such sets, the relation will be void, and trivial examples show that completeness may collapse in such cases.

More subtly, Case 3. Condition (a) is surprising, as $\forall i\,(\,[B_1^i \ldots B_{n-1}^i.C] \not\subseteq [A_1 \ldots .A_{n-1}.C]\,)$ is expected, and not the (in result) much weaker condition given. The proof shows that the very definition of a patch allows to put the interesting sequence in all elements of the patch (attention: elements of the patch are not mutually disjoint, they are only disjoint from the starting set), so that we can work with the intersection. The proof uses, however, that singletons are in the domain, and that the elements of arbitrary patches are in the domain. In particular, this will generally not be true in the infinite case, as patches work with set differences.
Thus, this proof is another good illustration that we may pay in domain conditions what we win in representation conditions.

8.2 Theory Revision

We begin with a very succinct introduction into AGM theory revision, and the subsequent results by the second author as published in [Sch04]. It is not supposed to be a self-contained introduction, but to help the readers recollect the situation.

Section 8.2.2 (p. 236) describes very briefly parts of the work by Booth and co-authors, and then solves a representation problem in the infinite case left open by Booth et al.

8.2.1 Introduction to Theory Revision

Recall from the introduction that theory revision was invented in order to "fuse" together two separately consistent, but together inconsistent theories or formulas to a consistent result. The by far best known approach is that by Alchourron, Gardenfors, and Makinson, and known as the AGM approach, see [AGM85]. They formulated "rationality postulates" for various variants of theory revision, which we give now in a very succinct form. Lehmann, Magidor, Schlechta, see [LMS01], gave a distance semantics for theory revision; this is further elaborated in [Sch04] and presented here in very brief outline too.

Definition 8.2.1 We present in parallel the logical and the semantic (or purely algebraic) side. For the latter, we work in some fixed universe U, and the intuition is $U = M_{\mathcal{L}}$, $X = M(K)$, etc. So, e.g. $A \in K$ becomes $X \subseteq B$.
(For reasons of readability, we omit most caveats about definability.)
$K_{\perp}$ will denote the inconsistent theory.
We consider two functions, – and $*$, taking a deductively closed theory and a formula as arguments, and returning a (deductively closed) theory on the logics side. The algebraic counterparts work on definable model sets. It is obvious that $(K-1)$, $(K*1)$, $(K-6)$, $(K*6)$ have vacuously true counterparts on the semantical side. Note that K (X) will never change, and everything is relative to fixed K (X). $K*\phi$ is the result of revising K with ϕ. $K-\phi$ is the result of subtracting enough from K to be able to add $\neg\phi$ in a reasonable way, called contraction.
Moreover, let $\leq_K$ be a relation on the formulas relative to a deductively closed theory K on the formulas of $\mathcal{L}$, and $\leq_X$ a relation on $\mathcal{P}(U)$ or a suitable subset of $\mathcal{P}(U)$ relative to fixed X. When the context is clear, we simply write $\leq$. $\leq_K$ $(\leq_X)$ is called a relation of epistemic entrenchment for K (X).
The following Table 8.1 presents the "rationality postulates" for contraction (–), rationality postulates revision ($*$), and epistemic entrenchment. In AGM tradition, K will be a deductively closed theory, and ϕ, ψ are formulas. Accordingly, X will be the set of models of a theory and A, B the model sets of formulas.

In the further development, formulas ϕ, etc. may sometimes also be full theories. As the transcription to this case is evident, we will not go into details (Table 8.1).

Remark 8.2.1

(1) Note that $(X \mid 7)$ and $(X \mid 8)$ express a central condition for ranked structures, see Sect. 3.10: If we note $X \mid .$ by $f_X(.)$, we then have: $f_X(A) \cap B \neq \emptyset \Rightarrow f_X(A \cap B) = f_X(A) \cap B$.
(2) It is trivial to see that AGM revision cannot be defined by an individual distance (see Definition 2.3.5 below): Suppose $X \mid Y := \{y \in Y : \exists x_y \in X\ (\forall y' \in Y\ d(x_y, y) \leq d(x_y, y'))\}$. Consider a, b, c. $\{a, b\} \mid \{b, c\} = \{b\}$ by $(X \mid 3)$ and $(X \mid 4)$, so $d(a, b) < d(a, c)$. But on the other hand, $\{a, c\} \mid \{b, c\} = \{c\}$, so $d(a, b) > d(a, c)$, *contradiction*.

Proposition 8.2.1 *Contraction, revision, and epistemic entrenchment are interdefinable by the following equations, i.e. if the defining side has the respective properties, so will the defined side.*

$K * \phi := \overline{(K - \neg\phi) \cup \phi}$	$X \mid A := (X \ominus \mathbf{C}A) \cap A$
$K - \phi := K \cap (K * \neg\phi)$	$X \ominus A := X \cup (X \mid \mathbf{C}A)$
$K - \phi := \{\psi \in K : (\phi <_K \phi \vee \psi \text{ or } \vdash \phi)\}$	$X \ominus A := \begin{cases} X & iff\ A = U, \\ \bigcap\{B : X \subseteq B \subseteq U, A <_X A \cup B\} & otherwise \end{cases}$
$\phi \leq_K \psi :\leftrightarrow \begin{cases} \vdash \phi \wedge \psi \\ or \\ \phi \notin K - (\phi \wedge \psi) \end{cases}$	$A \leq_X B :\leftrightarrow \begin{cases} A, B = U \\ or \\ X \ominus (A \cap B) \not\subseteq A \end{cases}$

Table 8.1 AGM theory revision

Contraction, $K - \phi$			
$(K-1)$	$K - \phi$ is deductively closed		
$(K-2)$	$K - \phi \subseteq K$	$(X \ominus 2)$	$X \subseteq X \ominus A$
$(K-3)$	$\phi \notin K \Rightarrow K - \phi = K$	$(X \ominus 3)$	$X \not\subseteq A \Rightarrow X \ominus A = X$
$(K-4)$	$\not\vdash \phi \Rightarrow \phi \notin K - \phi$	$(X \ominus 4)$	$A \neq U \Rightarrow X \ominus A \not\subseteq A$
$(K-5)$	$K \subseteq \overline{(K - \phi) \cup \{\phi\}}$	$(X \ominus 5)$	$(X \ominus A) \cap A \subseteq X$
$(K-6)$	$\vdash \phi \leftrightarrow \psi \Rightarrow K - \phi = K - \psi$		
$(K-7)$	$(K - \phi) \cap (K - \psi) \subseteq$ $K - (\phi \wedge \psi)$	$(X \ominus 7)$	$X \ominus (A \cap B) \subseteq$ $(X \ominus A) \cup (X \ominus B)$
$(K-8)$	$\phi \notin K - (\phi \wedge \psi) \Rightarrow$ $K - (\phi \wedge \psi) \subseteq K - \phi$	$(X \ominus 8)$	$X \ominus (A \cap B) \not\subseteq A \Rightarrow$ $X \ominus A \subseteq X \ominus (A \cap B)$
Revision, $K * \phi$			
$(K*1)$	$K * \phi$ is deductively closed	–	
$(K*2)$	$\phi \in K * \phi$	$(X \mid 2)$	$X \mid A \subseteq A$
$(K*3)$	$K * \phi \subseteq \overline{K \cup \{\phi\}}$	$(X \mid 3)$	$X \cap A \subseteq X \mid A$
$(K*4)$	$\neg\phi \notin K \Rightarrow$ $\overline{K \cup \{\phi\}} \subseteq K * \phi$	$(X \mid 4)$	$X \cap A \neq \emptyset \Rightarrow$ $X \mid A \subseteq X \cap A$
$(K*5)$	$K * \phi = K_\perp \Rightarrow \vdash \neg\phi$	$(X \mid 5)$	$X \mid A = \emptyset \Rightarrow A = \emptyset$
$(K*6)$	$\vdash \phi \leftrightarrow \psi \Rightarrow K * \phi = K * \psi$	–	
$(K*7)$	$K * (\phi \wedge \psi) \subseteq$ $\overline{(K * \phi) \cup \{\psi\}}$	$(X \mid 7)$	$(X \mid A) \cap B \subseteq$ $X \mid (A \cap B)$
$(K*8)$	$\neg\psi \notin K * \phi \Rightarrow$ $\overline{(K * \phi) \cup \{\psi\}} \subseteq K * (\phi \wedge \psi)$	$(X \mid 8)$	$(X \mid A) \cap B \neq \emptyset \Rightarrow$ $X \mid (A \cap B) \subseteq (X \mid A) \cap B$
Epistemic entrenchment			
$(EE1)$	$\leq_K$ is transitive	$(EE1)$	$\leq_X$ is transitive
$(EE2)$	$\phi \vdash \psi \Rightarrow \phi \leq_K \psi$	$(EE2)$	$A \subseteq B \Rightarrow A \leq_X B$
$(EE3)$	$\forall \phi, \psi$ $(\phi \leq_K \phi \wedge \psi$ or $\psi \leq_K \phi \wedge \psi)$	$(EE3)$	$\forall A, B$ $(A \leq_X A \cap B$ or $B \leq_X A \cap B)$
$(EE4)$	$K \neq K_\perp \Rightarrow$ $(\phi \notin K$ iff $\forall \psi\, \phi \leq_K \psi)$	$(EE4)$	$X \neq \emptyset \Rightarrow$ $(X \not\subseteq A$ iff $\forall B\, A \leq_X B)$
$(EE5)$	$\forall \psi \psi \leq_K \phi \Rightarrow \vdash \phi$	$(EE5)$	$\forall B\, B \leq_X A \Rightarrow A = U$

8.2.1.1 A Remark on Intuition

The idea of epistemic entrenchment is that ϕ is more entrenched than ψ (relative to K) iff $M(\neg\psi)$ is closer to $M(K)$ than $M(\neg\phi)$ is to $M(K)$. In shorthand, the more we can twiggle K without reaching $\neg\phi$, the more ϕ is entrenched. Truth is maximally entrenched – no twiggling whatever will reach falsity. The more ϕ is entrenched, the more we are certain about it. Seen this way, the properties of epistemic entrenchment relations are very natural (and trivial): As only the closest points of $M(\neg\phi)$ count (seen from $M(K)$), ϕ or ψ will be as entrenched as $\phi \wedge \psi$, and there is a logically strongest ϕ' which is as entrenched as ϕ – this is just the sphere around $M(K)$ with radius $d(M(K), M(\neg\phi))$.

Definition 8.2.2 $d : U \times U \to Z$ is called a pseudo-distance on U iff (d1) holds:

(d1) Z is totally ordered by a relation $<$.

If, in addition, Z has a $<$-smallest element 0, and (d2) holds, we say that d respects identity:

$$\text{(d2) } d(a, b) = 0 \text{ iff } a = b.$$

If, in addition, (d3) holds, then d is called symmetric:

$$\text{(d3) } d(a, b) = d(b, a).$$

(For any $a, b \in U$.)
Note that we can force the triangle inequality to hold trivially (if we can choose the values in the real numbers): It suffices to choose the values in the set $\{0\} \cup [0.5, 1]$, i.e. in the interval from 0.5 to 1, or as 0.

Definition 8.2.3 We define the collective and the individual variant of choosing the closest elements in the second operand by two operators, $|, \uparrow$: $\mathcal{P}(U) \times \mathcal{P}(U) \to \mathcal{P}(U)$:
Let d be a distance or pseudo-distance.

$$X \mid Y := \{y \in Y : \exists x_y \in X \, \forall x' \in X, \; \forall y' \in Y(d(x_y, y) \leq d(x', y'))\}$$

(the collective variant, used in theory revision)
and

$$X \uparrow Y := \{y \in Y : \exists x_y \in X \, \forall y' \in Y(d(x_y, y) \leq d(x_y, y'))\}$$

(the individual variant, used for counterfactual conditionals and theory update).
Thus, $A \mid_d B$ is the subset of B consisting of all $b \in B$ that are closest to A. Note that if A or B is infinite, $A \mid_d B$ may be empty, even if A and B are not empty. A condition assuring nonemptiness will be imposed when necessary.

Definition 8.2.4 An operation $|$: $\mathcal{P}(U) \times \mathcal{P}(U) \to \mathcal{P}(U)$ is representable iff there is a pseudo-distance $d : U \times U \to Z$ such that

$$A \mid B = A \mid_d B := \{b \in B : \exists a_b \in A \forall a' \in A \forall b' \in B(d(a_b, b) \leq d(a', b'))\}.$$

The following is the central definition; it describes the way a revision $*_d$ is attached to a pseudo-distance d on the set of models (Table 8.2).

Definition 8.2.5 $T *_d T' := Th(M(T) \mid_d M(T'))$.
$*$ is called representable iff there is a pseudo-distance d on the set of models s.t. $T * T' = Th\left(M(T) \mid_d M(T')\right)$.

Fact 8.2.1 A distance-based revision satisfies the AGM postulates provided

(1) it respects identity, i.e. $d(a, a) < d(a, b)$ for all $a \neq b$,
(2) it satisfies a limit condition: minima exist, and

Table 8.2 Distance representation conditions

		$(*Equiv)$ $\models T \leftrightarrow S, \models T' \leftrightarrow S', \Rightarrow T * T' = S * S'$
		$(*CCL)$ $T * T'$ is a consistent, deductively closed theory
	$(\mid Succ)$ $A \mid B \subseteq B$	$(*Succ)$ $T' \subseteq T * T'$
	$(\mid Con)$ $A \cap B \neq \emptyset \Rightarrow A \mid B = A \cap B$	$(*Con)$ $Con(T \cup T') \Rightarrow T * T' = \overline{T \cup T'}$
Intuitively, using symmetry $d(X_0, X_1) \leq d(X_1, X_2)$, $d(X_1, X_2) \leq d(X_2, X_3)$, $d(X_2, X_3) \leq d(X_3, X_4)$, ... $d(X_{k-1}, X_k) \leq d(X_0, X_k)$ $\Rightarrow$ $d(X_0, X_1) \leq d(X_0, X_k)$, i.e. transitivity or absence of loops involving $<$	$(\mid Loop)$ $(X_1 \mid (X_0 \cup X_2)) \cap X_0 \neq \emptyset$, $(X_2 \mid (X_1 \cup X_3)) \cap X_1 \neq \emptyset$, $(X_3 \mid (X_2 \cup X_4)) \cap X_2 \neq \emptyset$, ... $(X_k \mid (X_{k-1} \cup X_0)) \cap X_{k-1} \neq \emptyset$ $\Rightarrow$ $(X_0 \mid (X_k \cup X_1)) \cap X_1 \neq \emptyset$	$(*Loop)$ $Con(T_0, T_1 * (T_0 \vee T_2))$, $Con(T_1, T_2 * (T_1 \vee T_3))$, $Con(T_2, T_3 * (T_2 \vee T_4))$, ... $Con(T_{k-1}, T_k * (T_{k-1} \vee T_0))$ $\Rightarrow$ $Con(T_1, T_0 * (T_k \vee T_1))$

(3) it is definability preserving.

(It is trivial to see that the first two are necessary, and Example 8.2.1 (p. 233), (2) below shows the necessity of (3). In particular, (2) and (3) will hold for finite languages.)

Proof We use $|$ to abbreviate $|_d$. As a matter of fact, we show slightly more, as we admit also full theories on the right of $*$. $(K*1)$, $(K*2)$, $(K*6)$ hold by definition, $(K*3)$ and $(K*4)$ as d respects identity, and $(K*5)$ by existence of minima. It remains to show $(K*7)$ and $(K*8)$; we do them together and show: If $T*T'$ is consistent with T'', then $T*(T'\cup T'')=\overline{(T*T')\cup T''}$.
Note that $M(S\cup S')=M(S)\cap M(S')$, and that $M(S*S')=M(S)\mid M(S')$. (The latter is only true if $|$ is definability preserving.) By prerequisite, $M(T*T')\cap M(T'')\neq\emptyset$, so $(M(T)\mid M(T'))\cap M(T'')\neq\emptyset$. Let $A:=M(T)$, $B:=M(T')$, $C:=M(T'')$. "$\subseteq$": Let $b\in A\mid(B\cap C)$. By prerequisite, there is $b'\in(A\mid B)\cap C$. Thus $d(A,b')\geq d(A,B\cap C)=d(A,b)$. As $b\in B$, $b\in A\mid B$, but $b\in C$ too. "$\supseteq$": Let $b'\in(A\mid B)\cap C$. Thus $d(A,b')=d(A,B)\leq d(A,B\cap C)$, so by $b'\in B\cap C$ $b'\in A\mid(B\cap C)$. We conclude $M(T)\mid(M(T')\cap M(T''))=(M(T)\mid M(T'))\cap M(T'')$, thus that $T*(T'\cup T'')=\overline{(T*T')\cup T''}$. □

Definition 8.2.6 For $X,Y\neq\emptyset$, set $U_Y(X):=\{z:d(X,z)\leq d(X,Y)\}$.

Fact 8.2.2 Let $X,Y,Z\neq\emptyset$. Then

(1) $U_Y(X)\cap Z\neq\emptyset$ iff $(X\mid(Y\cup Z))\cap Z\neq\emptyset$ and
(2) $U_Y(X)\cap Z\neq\emptyset$ iff $\mathbf{C}Z\leq_X\mathbf{C}Y$ – where $\leq_X$ is epistemic entrenchement relative to X.

Proof

(1) Trivial.
(2) $\mathbf{C}Z\leq_X\mathbf{C}Y$ iff $X\ominus(\mathbf{C}Z\cap\mathbf{C}Y)\not\subseteq\mathbf{C}Z$. $X\ominus(\mathbf{C}Z\cap\mathbf{C}Y)=X\cup(X\mid\mathbf{C}(\mathbf{C}Z\cap\mathbf{C}Y))$ $=X\cup(X\mid(Z\cup Y))$. So $X\ominus(\mathbf{C}Z\cap\mathbf{C}Y)\not\subseteq\mathbf{C}Z\Leftrightarrow(X\cup(X\mid(Z\cup Y)))\cap Z\neq\emptyset$ $\Leftrightarrow X\cap Z\neq\emptyset$ or $(X\mid(Z\cup Y))\cap Z\neq\emptyset\Leftrightarrow d(X,Z)\leq d(X,Y)$. □

Definition 8.2.7 Let $U\neq\emptyset$, $\mathcal{Y}\subseteq\mathcal{P}(U)$ satisfy $(\cap)$, $(\cup)$, $\emptyset\notin\mathcal{Y}$.
Let $A,B,X_i\in\mathcal{Y}$, $|:\mathcal{Y}\times\mathcal{Y}\to\mathcal{P}(U)$.
Let $*$ be a revision function defined for arbitrary consistent theories on both sides. (This is thus a slight extension of the AGM framework, as AGM work with formulas only on the right of $*$.)

Proposition 8.2.2 The following connections between the logical and algebraic side might be the most interesting ones. We will consider in all cases also the variant with full theories.
Given $*$ which respects logical equivalence, let $M(T)\mid M(T'):=M(T*T')$; conversely, given $|$, let $T*T':=Th(M(T)\mid M(T'))$. We then have (Table 8.3)

Proof We consider the equivalence of $T*(T'\cup T'')\subseteq\overline{(T*T')\cup T''}$ and $(M(T)\mid M(T'))\cap M(T'')\subseteq M(T)\mid(M(T')\cap M(T''))$.

Table 8.3 Revision and definability preservation

(1.1)	$(K*7)$	$\Rightarrow$	$(X \mid 7)$
(1.2)		$\Leftarrow (\mu dp)$	
(1.3)		$\Leftarrow$ B is the model set for some ϕ	
(1.4)		$\not\Leftarrow$ in general	
(2.1)	$(*Loop)$	$\Rightarrow$	$(\mid Loop)$
(2.2)		$\Leftarrow (\mu dp)$	
(2.3)		$\Leftarrow$ all X_i are the model sets for some ϕ_i	
(2.4)		$\not\Leftarrow$ in general	

(1.1): $(M(T) \mid M(T')) \cap M(T'') = M(T*T') \cap M(T'') = M((T*T') \cup T'') \subseteq_{(K*7)} M(T*(T' \cup T'')) = M(T) \mid M(T' \cup T'') = M(T) \mid (M(T') \cap M(T''))$.
(1.2): $T*(T' \cup T'') = Th(M(T) \mid M(T' \cup T'')) = \overline{Th(M(T) \mid (M(T') \cap M(T'')))} \subseteq_{(X|7)} Th((M(T) \mid M(T')) \cap M(T'')) =_{(\mu dp)} \overline{Th(M(T) \mid M(T')) \cup T''} = \overline{Th(M(T*T') \cup T''} = \overline{(T*T') \cup T''}$.
(1.3): Let T'' be equivalent to ϕ''. We can then replace the use of (μdp) in the proof of (1.2) by Fact 2.2.3 (p. 35), (3).
(1.4): By Example 8.2.1 (p. 233), (2), $(K*7)$ may fail, though $(X \mid 7)$ holds.
(2.1) and (2.2): $Con(T_0, T_1 * (T_0 \vee T_2)) \Leftrightarrow M(T_0) \cap M(T_1 * (T_0 \vee T_2)) \neq \emptyset$.
$M(T_1 * (T_0 \vee T_2)) = M(Th(M(T_1) \mid M(T_0 \vee T_2))) = M(Th(M(T_1) \mid (M(T_0) \cup M(T_2)))) =_{(\mu dp)} M(T_1) \mid (M(T_0) \cup (T_2))$, so $Con(T_0, T_1 * (T_0 \vee T_2)) \Leftrightarrow M(T_0) \cap (M(T_1) \mid (M(T_0) \cup (T_2))) \neq \emptyset$.
Thus, all conditions translate one-to-one, and we use $(\mid Loop)$ and $(*Loop)$ to go back and forth.
(2.3): Let $A := M(Th(M(T_1) \mid (M(T_0) \cup M(T_2))))$, $A' := M(T_1) \mid (M(T_0) \cup (T_2))$, then we do not need $A = A'$, it suffices to have $M(T_0) \cap A \neq \emptyset \Leftrightarrow M(T_0) \cap A' \neq \emptyset$. $A = \widehat{A'}$, so we can use Fact 2.2.3 (p. 35), (4), if T_0 is equivalent to some ϕ_0.

This has to hold for all T_i, so all T_i have to be equivalent to some ϕ_i.
(2.4): By Proposition 8.2.3 (p. 235), all distance defined $\mid$ satisfy $(\mid Loop)$. By Example 8.2.1 (p. 233), (1), $(*Loop)$ may fail. □

The following Table 8.4 summarizes representation of theory revision functions by structures with a distance.

By "pseudo-distance" we mean here a pseudo-distance which respects identity and is symmetrical.

$(\mid \emptyset)$ means that if $X, Y \neq \emptyset$, then $X \mid_d Y \neq \emptyset$.

The following Example 8.2.1 (p. 233) shows that, in general, a revision operation defined on models via a pseudo-distance by $T * T' := Th\left(M(T) \mid_d M(T')\right)$ might not satisfy $(*Loop)$ or $(K*7)$, unless we require $\mid_d$ to preserve definability.

Example 8.2.1 Consider an infinite propositional language $\mathcal{L}$.

Table 8.4 Distance representation

$\mid -$ function		Distance structure		$*-$ function
$(\mid Succ) + (\mid Con)+$ $(\mid Loop)$	$\Leftrightarrow$, $(\cup) + (\cap)$ Proposition 8.2.3 (page 235)	Pseudo-distance	$\Leftrightarrow$ $(\mu dp) + (\mid \emptyset)$ Proposition 8.2.4 (page 235)	$(*Equiv) + (*CCL)+$ $(*Succ)+ (*Con) + (*Loop)$
Any finite characterization	$\not\Leftrightarrow$ Proposition 8.2.5 (page 236)		$\not\Rightarrow$ without (μdp) Example 8.2.1 (page 233)	

Let X be an infinite set of models, m, m_1, m_2 be models for $\mathcal{L}$. Arrange the models of $\mathcal{L}$ in the real plane s.t. all $x \in X$ have the same distance < 2 (in the real plane) from m, m_2 has distance 2 from m, and m_1 has distance 3 from m.
Let T, T_1, T_2 be complete (consistent) theories, T' a theory with infinitely many models, $M(T) = \{m\}$, $M(T_1) = \{m_1\}$, and $M(T_2) = \{m_2\}$. The two variants diverge now slightly:

(1) $M(T') = X \cup \{m_1\}$. T, T', T_2 will be pairwise inconsistent.

(2) $M(T') = X \cup \{m_1, m_2\}$, $M(T'') = \{m_1, m_2\}$.

Assume in both cases $Th(X) = T'$, so X will not be definable by a theory.
Now for the results:
Then $M(T) \mid M(T') = X$, but $T * T' = Th(X) = T'$.
(1) We easily verify $Con(T, T_2 * (T \vee T))$, $Con(T_2, T * (T_2 \vee T_1))$, $Con(T, T_1 * (T \vee T))$, $Con\left(T_1, T * \left(T_1 \vee T'\right)\right)$, $Con(T, T' * (T \vee T))$, and conclude by Loop (i.e. $(*Loop)$) $Con\left(T_2, T * \left(T' \vee T_2\right)\right)$, which is wrong.

(2) So $T * T'$ is consistent with T'' and $\overline{(T * T') \cup T''} = T''$. But $T' \cup T'' = T''$ and $T * (T' \cup T'') = T_2 \neq T''$, contradicting $(K * 7)$.

Proposition 8.2.3 *Let* $U \neq \emptyset$, $\mathcal{Y} \subseteq \mathcal{P}(U)$ *be closed under finite* $\cap$ *and finite* $\cup$, $\emptyset \notin \mathcal{Y}$.

(a) $\mid$ *is representable by a symmetric pseudo-distance* $d : U \times U \to Z$ *iff* $\mid$ *satisfies* ($\mid$ Succ) *and* ($\mid$ Loop) in *Definition* 8.2.7 (*p.* 232).

(b) $\mid$ *is representable by an identity respecting symmetric pseudo-distance* d: $U \times U \to Z$ *iff* $\mid$ *satisfies* ($\mid$ Succ) , ($\mid$ Con) , *and* ($\mid$ Loop) *in Definition* 8.2.7 (*p.* 232).

See [LMS01] *or* [Sch04].

Proposition 8.2.4 *Let* $\mathcal{L}$ *be a propositional language.*

(a) A revision operation $*$ *is representable by a symmetric consistency and definability preserving pseudo-distance iff* $*$ *satisfies* ($*$Equiv), ($*$CCL), ($*$Succ), ($*$Loop).

(b) A revision operation $*$ *is representable by a symmetric consistency and definability preserving, identity respecting pseudo-distance iff* $*$ *satisfies* ($*$Equiv), ($*$CCL), ($*$Succ), ($*$Con), ($*$Loop).

See [LMS01] *or* [Sch04].

Example 8.2.2 This example shows the expressive weakness of revision based on distance: Not all distance relations can be reconstructed from the revision operator. Thus, a revision operator does not allow to "observe" all distances relations; so transitivity of $\leq$ cannot necessarily be captured in a short condition, requiring arbitrarily long conditions, see Proposition 8.2.5 (p. 236).
Note that even when the pseudo-distance is a real distance, the resulting revision operator $\mid_d$ does not always permit to reconstruct the relations of the distances: Revision is a coarse instrument to investigate distances.
Distances with common start (or end, by symmetry) can always be compared by looking at the result of revision:

$a \mid_d \{b, b'\} = b$ iff $d(a, b) < d(a, b')$,
$a \mid_d \{b, b'\} = b'$ iff $d(a, b) > d(a, b')$,
$a \mid_d \{b, b'\} = \{b, b'\}$ iff $d(a, b) = d(a, b')$.

This is not the case with arbitrary distances $d(x, y)$ and $d(a, b)$, as this example will show.

We work in the real plane, with the standard distance, and the angles have 120 degrees. a' is closer to y than x is to y, a is closer to b than x is to y, but a' is farther away from b' than x is from y. Similarly for b, b'. But we cannot distinguish the situation $\{a, b, x, y\}$ and the situation $\{a', b', x, y\}$ through $\mid_d$ (see Diagram 8.2.1 (p. 236).):

Seen from a, the distances are in that order: y, b, x.
Seen from a', the distances are in that order: y, b', x.
Seen from b, the distances are in that order: y, a, x.
Seen from b', the distances are in that order: y, a', x.
Seen from y, the distances are in that order: $a/b, x$.
Seen from y, the distances are in that order: $a'/b', x$.
Seen from x, the distances are in that order: $y, a/b$.
Seen from x, the distances are in that order: $y, a'/b'$.

Thus, any $c \mid_d C$ will be the same in both situations (with a interchanged with a', b with b'). The same holds for any $X \mid_d C$ where X has two elements.
Thus, any $C \mid_d D$ will be the same in both situations, when we interchange a with a', and b with b'. So we cannot determine by $\mid_d$ whether $d(x, y) > d(a, b)$ or not.

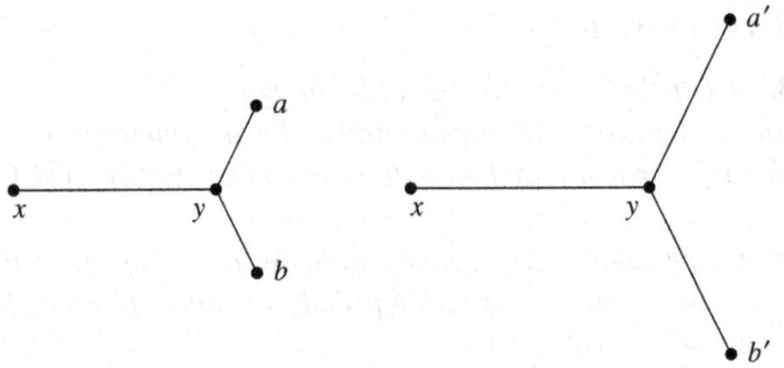

Diagram 8.2.1 The two diagrams are Indiscernible by revision

Proposition 8.2.5 *There is no finite characterization of distance-based* $\mid$*-operators. (Attention: this is, of course, false when we fix the left hand side: the AGM axioms give a finite characterization. So this also shows the strength of being able to change the left hand side.)*
See [Sch04].

8.2.2 Booth Revision

The material in this section is taken from [GS08g].

8.2.2.1 Introduction

This material is due to Booth and co-authors.

The Problem We Solve

Booth and his co-authors have shown in very interesting papers, see [BN06] and [BCMG06], that many new approaches to theory revision (with fixed K) can be represented by two relations, $<$ and $\lhd$, where $<$ is the usual ranked relation and $\lhd$ is a sub relation of $<$. They have, however, left open the characterization of the infinite case, which we treat here.
The, for us, main definition they give is (in slight modification, we use the strict subrelations) as follows:

Definition 8.2.8 Given K and $<$ and $\lhd$, we define
$K \ominus \phi := Th(\{w : w \lhd w'$ for some $w' \in min(M(\neg\phi), <)\})$,
i.e. $K \ominus \phi$ is given by all those worlds, which are below the closest ϕ-worlds, as seen from K.

We want to characterize $K \ominus \phi$, for fixed K. Booth et al. have done the finite case by working with complete consistent formulas, i.e. single models. We want to do the infinite case without using complete consistent theories, i.e. in the usual style of completeness results in the area.
Our approach is basically semantic, though we use sometimes the language of logic, on the one hand to show how to approximate with formulas a single model, and on the other hand when we use classical compactness. This is, however, just a matter of speaking, and we could translate it into model sets too, but we do not think that we would win much by doing so. Moreover, we will treat only the formula case, as this seems to be the most interesting (otherwise the problem of approximation by formulas would not exist), and restrict ourselves to the definability preserving case. The more general case is left open for a young researcher who wants to sharpen his tools by solving it. Another open problem is to treat the same question for variable K for distance-based revision.

The Framework

For the reader's convenience, and to put our work a bit more into perspective, we repeat now some of the definitions and results given by Booth and his co-authors. Consequently, all material in this section is due to Booth and his co-authors.
$\leq$ will be a total pre order, anchored on $M(K)$, the models of K, i.e. $M(K) = min(W, \leq)$, the set of $\leq$-minimal worlds.
We have a second binary relation $\preceq$ on W, which is a reflexive subrelation of $\leq$.

Definition 8.2.9

(1) $(\leq, \preceq)$ is a K-context iff $\leq$ is a total pre order on W, anchored on $M(K)$, and $\preceq$ is a reflexive sub relation of $\leq$.

(2) $K \ominus \phi := Th(\{w : w \preceq w' \text{ for some } w' \in min(M(\neg\phi), \leq)\})$ is called a basic removal operator.

Theorem 8.2.1 *Basic removal is characterized by*

$(B1)$ $K \ominus \phi = Cn(K \ominus \phi)$ – *Cn classical consequence,*
$(B2)$ $\phi \notin K \ominus \phi$,
$(B3)$ If $\models \phi \leftrightarrow \phi'$, *then* $K \ominus \phi = K \ominus \phi'$,
$(B4)$ $K \ominus \bot = K$,
$(B5)$ $K \ominus \phi \subseteq Cn(K \cup \{\neg\phi\})$,
$(B6)$ if $\sigma \in K \ominus (\sigma \wedge \phi)$, *then* $\sigma \in K \ominus (\sigma \wedge \phi \wedge \psi)$,
$(B7)$ if $\sigma \in K \ominus (\sigma \wedge \phi)$, *then* $K \ominus \phi \subseteq K \ominus (\sigma \wedge \phi)$,
$(B8)$ $(K \ominus \sigma) \cap (K \ominus \phi) \subseteq K \ominus (\sigma \wedge \phi)$,
$(B9)$ if $\phi \notin K \ominus (\sigma \wedge \phi)$, *then* $K \ominus (\sigma \wedge \phi) \subseteq K \ominus \phi$.

$(B1)$–$(B3)$ belong to the basic AGM contraction postulates, $(B4)$ and $(B5)$ are weakened versions of another basic AGM postulate:

$$(Vacuity) \text{ If } \phi \notin K, \text{ then } K \ominus \phi = K,$$

which does not necessarily hold for basic removal operators.
The same holds for the remaining two basic AGM contraction postulates:

$$(Inclusion) K \ominus \phi \subseteq K,$$
$$(Recovery) K \subseteq Cn((K \ominus \phi) \cup \{\phi\}).$$

The main definition towards the completeness result of Booth et al. is given below.

Definition 8.2.10 Given K and $\ominus$, the structure $C(K, \ominus)$ is defined by

$$(\leq) w \leq w' \text{ iff } \neg\alpha \notin K \ominus (\neg\alpha \wedge \neg\alpha') \text{ and}$$
$$(\preceq) \text{ w} \preceq w' \text{ iff } \neg\alpha \notin K \ominus \neg\alpha',$$

where α is a formula which holds exactly in w, analogously for w' and α'.

Booth et al. then give a long list of theorems showing equivalence between various postulates and conditions on the orderings $\leq$ and $\preceq$. This, of course, shows the power of their approach.
We give three examples:

Condition 8.2.1

(a) If (for each $i = 1, 2$) $w_i \leq w'$ for all w', then $w_1 \preceq w_2$.
(b) If $w_1 \leq w_2$ for all w_2, then $w_1 \preceq w_2$ for all w_2.
(c) If $w_1 \preceq w_2$, then $w_1 = w_2$ or $w_1 \leq w'$ for all w'.

Theorem 8.2.2 *Let* $\ominus$ *be a basic removal operator as defined above.*

(1) $\ominus$ *satisfies one half of* $(Vacuity)$: If $\phi \notin K$, *then* $K \subseteq K \ominus \phi$,

(2.1) *If* $(\leq, \preceq)$ *satisfies* (a), *then* $\ominus$ *satisfies* (Vacuity).
(2.2) *If* $\ominus$ *satisfies* (Vacuity), *then* $C(K, \ominus)$ *satisfies* (a).
(3.1) *If* $(\leq, \preceq)$ *satisfies* (b), *then* $\ominus$ *satisfies* (Inclusion).
(3.2) *If* $\ominus$ *satisfies* (Inclusion), *then* $C(K, \ominus)$ *satisfies* (b).
(4.1) *If* $(\leq, \preceq)$ *satisfies* (c), *then* $\ominus$ *satisfies* (Recovery).
(4.2) *If* $\ominus$ *satisfies* (Recovery), *then* $C(K, \ominus)$ *satisfies* (c).
(5) *The following are equivalent:*
(5.1) $\ominus$ *is a full AGM contraction operator,*
(5.2) $\ominus$ *satisfies* $(B1)$–$(B9)$, (Inclusion), *and* (Recovery)
(5.3) $\ominus$ *is generated by some* $(\leq, \preceq)$ *satisfying* (b) *and* (c).

8.2.2.2 Construction and Proof

We change perspective a little, and work directly with a ranked relation, so we forget about the (fixed) K of revision, and have an equivalent, ranked structure. We are then interested in an operator ν, which returns a model set $\nu(\phi) := \nu(M(\phi))$, where $\nu(\phi) \cap M(\phi)$ is given by a ranked relation $<$, and $\nu(\phi) - M(\phi) := \{x \notin M(\phi) : \exists y \in \nu(\phi) \cap M(\phi)(x \lhd y)\}$, and $\lhd$ is an arbitrary subrelation of $<$. The essential problem is to find such y, as we have only formulas to find it. (If we had full theories, we could just look at all $Th(\{y\})$ if $x \in \nu(Th(\{y\}))$.) There is still some more work to do, as we have to connect the two relations, and simply taking a ready representation result will not do, as we shall see.
We first introduce some notation, then a set of conditions, and formulate the representation result. Soundness will be trivial. For completeness, we construct first the ranked relation $<$, show that it does what it should do, and then the subrelation $\lhd$.

Notation 8.2.1 We set

$$\mu^+(X) := \nu(X) \cap X$$
$$\mu^-(X) := \nu(X) - X$$

where $X := M(\phi)$ for some ϕ.

Condition 8.2.2

(μ^-1) $Y \cap \mu^-(X) \neq \emptyset \rightarrow \mu^+(Y) \cap X = \emptyset$
(μ^-2) $Y \cap \mu^-(X) \neq \emptyset \rightarrow \mu^+(X \cup Y) = \mu^+(Y)$
(μ^-3) $Y \cap \mu^-(X) \neq \emptyset \rightarrow \mu^-(Y) \cap X = \emptyset$
(μ^-4) $\mu^+(A) \subseteq \mu^+(B) \rightarrow \mu^-(A) \subseteq \mu^-(B)$
(μ^-5) $\mu^+(X \cup Y) = \mu^+(X) \cup \mu^+(Y) \rightarrow \mu^-(X \cup Y) = \mu^-(X) \cup \mu^-(Y)$

Fact 8.2.3 (μ^-1) and $(\mu\emptyset)$, $(\mu \subseteq)$ for μ^+ imply

(1) $\mu^+(X) \cap Y \neq \emptyset \rightarrow \mu^+(X) \cap \mu^-(Y) = \emptyset$
(2) $X \cap \mu^-(X) = \emptyset$.

Proof

(1) Let $\mu^+(X) \cap \mu^-(Y) \neq \emptyset$, then $X \cap \mu^-(Y) \neq \emptyset$, so by (μ^-1) $\mu^+(X) \cap Y = \emptyset$.
(2) Set $X := Y$, and use $(\mu\emptyset)$, $(\mu \subseteq)$, (μ^-1), (1). □

Proposition 8.2.6 $\nu : \{M(\phi) : \phi \in F(\mathcal{L})\} \rightarrow \boldsymbol{D}_{\mathcal{L}}$ *is representable by* $<$ *and* $\lhd$*, where* $<$ *is a smooth-ranked relation,* $\lhd$ *a subrelation of* $<$*,* $\mu^+(X)$ *is the usual set of* $<$*-minimal elements of X, and* $\mu^-(X) = \{x \notin X : \exists y \in \mu^+(X).(x \lhd y)\}$*, iff the following conditions hold:* $(\mu \subseteq)$, $(\mu\emptyset)$, $(\mu =)$ for μ^+*, and* (μ^-1) – (μ^-5) for μ^+ *and* μ^-*.*

Proof

Soundness

The first three hold for smooth-ranked structures, and others are easily verified.

Completeness

We first show how to generate the ranked relation $<$:
There is a small problem.
The authors first thought that one may take any result for ranked structures off the shelf, plug in the other relation somehow (see the second half), and that's it. No, that is not it: Suppose there is x, and a sequence x_i converging to x in the usual topology. Thus, if $x \in M(\phi)$, then there will always be some x_i in $M(\phi)$ too. Take now a ranked structure $\mathcal{Z}$, where all the x_i are strictly smaller than x. Consider $\mu(\phi)$, which will usually not contain x (avoid some nasty things with definability); so in the usual construction ($\preceq_1$ below), x will not be forced to be below any element y, how high up $y > x$ might be. However, there is ψ separating x and y, e.g. $x \models \neg\psi$, $y \models \psi$, and if we take as the second relation just the ranking again, $x \in \mu^-(\psi)$, so this becomes visible.
Consequently, considering μ^- may give strictly more information, and we have to put in a little more work. We just patch a proof for simple ranked structures, adding information obtained through μ^-.
We follow closely the strategy of the proof of 3.10.11 in [Sch04]. We will, however, change notation at one point: the relation R in [Sch04] is called $\preceq$ here. The proof goes over several steps, which we will enumerate.
Note that by Fact 2.3.1 (p. 40), taken from [Sch04], see also [GS08c], $(\mu \parallel)$, $(\mu\cup)$, $(\mu\cup')$, $(\mu =')$ hold, as the prerequisites about the domain are valid.

(1) To generate the ranked relation $<$, we define two relations, $\preceq_1$ and $\preceq_2$, where $\preceq_1$ is the usual one for ranked structures, as defined in the proof of 3.10.11 of [Sch04], $a \preceq_1 b$ iff $a \in \mu^+(X)$, $b \in X$, or $a = b$, and $a \preceq_2 b$ iff $a \in \mu^-(X)$, $b \in X$.
Moreover, we set $a \preceq b$ iff $a \preceq_1 b$ or $a \preceq_2 b$.
(2) Obviously, $\preceq$ is reflexive, we show that $\preceq$ is transitive by looking at the four different cases.

(2.1) In [Sch04], it was shown that $a \preceq_1 b \preceq_1 c \rightarrow a \preceq_1 c$. For completeness' sake, we repeat the argument: Suppose $a \preceq_1 b$, $b \preceq_1 c$, let $a \in \mu^+(A)$, $b \in A$, $b \in \mu^+(B)$, $c \in B$. We show $a \in \mu^+(A \cup B)$. By $(\mu \parallel)$, $a \in \mu^+(A \cup B)$ or $b \in \mu^+(A \cup B)$. Suppose $b \in \mu^+(A \cup B)$, then $\mu^+(A \cup B) \cap A \neq \emptyset$, so by $(\mu =)$ $\mu^+(A \cup B) \cap A = \mu^+(A)$, so $a \in \mu^+(A \cup B)$.

(2.2) Suppose $a \preceq_1 b \preceq_2 c$, we show $a \preceq_1 c$: Let $c \in Y$, $b \in \mu^-(Y) \cap X$, $a \in \mu^+(X)$. Consider $X \cup Y$. As $X \cap \mu^-(Y) \neq \emptyset$, by (μ^-2) $\mu^+(X \cup Y) = \mu^+(X)$, so $a \in \mu^+(X \cup Y)$ and $c \in X \cup Y$, so $a \preceq_1 c$.

(2.3) Suppose $a \preceq_2 b \preceq_2 c$, we show $a \preceq_2 c$: Let $c \in Y$, $b \in \mu^-(Y) \cap X$, $a \in \mu^-(X)$. Consider $X \cup Y$. As $X \cap \mu^-(Y) \neq \emptyset$, by (μ^-2) $\mu^+(X \cup Y) = \mu^+(X)$, so by (μ^-5) $\mu^-(X \cup Y) = \mu^-(X)$, so $a \in \mu^-(X \cup Y)$ and $c \in X \cup Y$, so $a \preceq_2 c$.

(2.4) Suppose $a \preceq_2 b \preceq_1 c$, we show $a \preceq_2 c$: Let $c \in Y$, $b \in \mu^+(Y) \cap X$, $a \in \mu^-(X)$. Consider $X \cup Y$. As $\mu^+(Y) \cap X \neq \emptyset$, $\mu^+(X) \subseteq \mu^+(X \cup Y)$. (Here is the argument: By $(\mu \parallel)$, $\mu^+(X \cup Y) = \mu^+(X) \parallel \mu^+(Y)$, so, if $\mu^+(X) \not\subseteq \mu^+(X \cup Y)$, then $\mu^+(X) \cap \mu^+(X \cup Y) = \emptyset$, so $\mu^+(X) \cap (X \cup Y - \mu^+(X \cup Y)) \neq \emptyset$ by $(\mu\emptyset)$, so by $(\mu\cup')$ $\mu^+(X \cup Y) = \mu^+(Y)$. But if $\mu^+(Y) \cap X = \mu^+(X \cup Y) \cap X \neq \emptyset$, $\mu^+(X) = \mu^+(X \cup Y) \cap X$ by $(\mu =)$, so $\mu^+(X) \cap \mu^+(X \cup Y) \neq \emptyset$, *contradiction*.) So $\mu^-(X) \subseteq \mu^-(X \cup Y)$ by (μ^-4), so $c \in X \cup Y$, $a \in \mu^-(X \cup Y)$, and $a \preceq_2 c$.

(3) We also see

(3.1) $a \in \mu^+(A)$, $b \in A - \mu^+(A) \rightarrow b \not\preceq a$.
(3.2) $a \in \mu^-(A)$, $b \in A \rightarrow b \not\preceq a$.

Proof of (3.1):

(a) $\neg(b \preceq_1 a)$ was shown in [Sch04], we repeat again the argument: Suppose there is B s.t. $b \in \mu^+(B)$, $a \in B$. Then by $(\mu\cup)$ $\mu^+(A \cup B) \cap B = \emptyset$, and by $(\mu\cup')$ $\mu^+(A \cup B) = \mu^+(A)$, but $a \in \mu^+(A) \cap B$, *contradiction*.
(b) Suppose there is B s.t. $a \in B$, $b \in \mu^-(B)$. But $A \cap \mu^-(B) \neq \emptyset$ implies $\mu^+(A) \cap B = \emptyset$ by (μ^-1).

Proof of (3.2):

(a) Suppose $b \preceq_1 a$, so there is B s.t. $a \in B$, $b \in \mu^+(B)$, so $B \cap \mu^-(A) \neq \emptyset$, so $\mu^+(B) \cap A = \emptyset$ by (μ^-1).
(b) Suppose $b \preceq_2 a$, so there is B s.t. $a \in B$, $b \in \mu(B)$, so $B \cap \mu\,(A) \neq \emptyset$, so $\mu^-(B) \cap A = \emptyset$ by (μ^-3).

(4) Let, by Fact 4.2.14 (p. 116), S be a total, transitive, reflexive relation on U, which extends $\preceq$ s.t. $xSy, ySx \rightarrow x \preceq y$ (recall that $\preceq$ is transitive and reflexive). But note that we loose ignorance here. Define $a < b$ iff aSb, but not bSa. If $a \perp b$ (i.e. neither $a < b$ nor $b < a$), then, by totality of S, aSb and bSa. $<$ is ranked: If $c < a \perp b$, then by transitivity of S cSb, but if bSc, then again by transitivity of S aSc. Similarly for $c > a \perp b$.
(5) It remains to show that $<$ represents μ and is $\mathcal{Y}$-smooth.

Let $a \in A - \mu^+(A)$. By $(\mu\emptyset)$, $\exists b \in \mu^+(A)$, so $b \preceq_1 a$, but by case (3.1) above $a \not\preceq b$, so bSa, but not aSb, so $b < a$, so $a \in A - \mu_<(A)$. Let $a \in \mu^+(A)$, then for all $a' \in A$ $a \preceq a'$, so aSa', so there is no $a' \in A$ $a' < a$, so $a \in \mu_<(A)$. Finally, $\mu^+(A) \neq \emptyset$, all $x \in \mu^+(A)$ are minimal in A as we just saw, and for $a \in A - \mu^+(A)$ there is $b \in \mu^+(A)$, $b \preceq_1 a$, so the structure is smooth.
The subrelation $\lhd$:
Let $x \in \mu^-(X)$, we look for $y \in \mu^+(X)$ s.t. $x \lhd y$, where $\lhd$ is the smaller, additional relation. By the definition of the relation $\preceq_2$ above, we know that $\lhd \subseteq \preceq$ and by (3.2) above $\lhd \subseteq <$.
Take an arbitrary enumeration of the propositional variables of $\mathcal{L}$, $p_i : i < \kappa$. We will inductively decide for p_i or $\neg p_i$. σ, etc. will denote a finite subsequence of the choices made so far, i.e. $\sigma = \pm p_{i_0}, \ldots, \pm p_{i_n}$ for some $n < \omega$. Given such σ, $M(\sigma) := M(\pm p_{i_0}) \cap \ldots \cap M(\pm p_{i_n})$. $\sigma + \sigma'$ will be the union of two such sequences. This is again one such sequence.
Take an arbitrary model m for $\mathcal{L}$, i.e. a function $m : v(\mathcal{L}) \to \{t, f\}$. We will use this model as a "strategy", which will tell us how to decide, if we have some choice.
We determine y by an inductive process, essentially cutting away $\mu^+(X)$ around y. We choose p_i or $\neg p_i$ preserving the following conditions inductively: For all finite sequences σ as above we have:

(1) $M(\sigma) \cap \mu^+(X) \neq \emptyset$,
(2) $x \in \mu^-(X \cap M(\sigma))$.

For didactic reasons, we do the case p_0 separately.
Consider p_0. Either $M(p_0) \cap \mu^+(X) \neq \emptyset$, or $M(\neg p_0) \cap \mu^+(X) \neq \emptyset$, or both. If, e.g. $M(p_0) \cap \mu^+(X) \neq \emptyset$, but $M(\neg p_0) \cap \mu^+(X) = \emptyset$, then we have no choice, and we take p_0, in the opposite case, we take $\neg p_0$. For example, in the first case, $\mu^+(X \cap M(p_0)) = \mu^+(X)$, so $x \in \mu^-(X \cap M(p_0))$ by (μ^-4). If both intersections are non empty, then by (μ^-5) $x \in \mu^-(X \cap M(p_0))$ or $x \in \mu^-(X \cap M(\neg p_0))$, or both. Only in the last case, we use our strategy to decide whether to choose p_0 or $\neg p_0$: If $m(p_0) = t$, we choose p_0, if not, we choose $\neg p_0$.
Obviously, (1) and (2) above are satisfied.
Suppose we have chosen p_i or $\neg p_i$ for all $i < \alpha$, i.e. defined a partial function from $v(\mathcal{L})$ to $\{t, f\}$, and the induction hypotheses (1) and (2) hold. Consider p_α. If there is no finite subsequence σ of the choices done so far s.t. $M(\sigma) \cap M(p_\alpha) \cap \mu^+(X) = \emptyset$, then p_α is a candidate. Likewise for $\neg p_\alpha$.
One of p_α or $\neg p_\alpha$ is a candidate. Suppose not, then there are σ and σ' subsequences of the choices done so far, and $M(\sigma) \cap M(p_\alpha) \cap \mu^+(X) = \emptyset$ and $M(\sigma') \cap M(\neg p_\alpha) \cap \mu^+(X) = \emptyset$. But then $M(\sigma + \sigma') \cap \mu^+(X) = M(\sigma) \cap M(\sigma') \cap \mu^+(X) \subseteq M(\sigma) \cap M(p_\alpha) \cap \mu^+(X) \cup M(\sigma') \cap M(\neg p_\alpha) \cap \mu^+(X) = \emptyset$, contradicting (1) of the induction hypothesis.
So induction hypothesis (1) will hold again.
Recall that for each candidate and any σ by induction hypothesis (1) $M(\sigma) \cap M(p_\alpha) \cap \mu^+(X) = \mu^+(M(\sigma) \cap M(p_\alpha) \cap X)$ by $(\mu =')$, and also for $\sigma \subseteq \sigma'$ $\mu^+(M(\sigma') \cap M(p_\alpha) \cap X) \subseteq \mu^+(M(\sigma) \cap M(p_\alpha) \cap X)$ by $(\mu =')$ and $M(\sigma') \subseteq M(\sigma)$, and thus by (μ^-4) $\mu^-(M(\sigma') \cap M(p_\alpha) \cap X) \subseteq \mu^-(M(\sigma) \cap M(p_\alpha) \cap X)$.

If we have only one candidate left, say, e.g. p_α, then for each sufficiently big sequence σ $M(\sigma) \cap M(\neg p_\alpha) \cap \mu^+(X) = \emptyset$, thus for such σ $\mu^+(M(\sigma) \cap M(p_\alpha) \cap X) = M(\sigma) \cap M(p_\alpha) \cap \mu^+(X) = M(\sigma) \cap \mu^+(X) = \mu^+(M(\sigma) \cap X)$, and thus by (μ^-4) $\mu^-(M(\sigma) \cap M(p_\alpha) \cap X) = \mu^-(M(\sigma) \cap X)$, so $\neg p_\alpha$ plays no really important role. In particular, induction hypothesis (2) holds again.
Suppose now that we have two candidates, thus for p_α and $\neg p_\alpha$ and each σ $M(\sigma) \cap M(p_\alpha) \cap \mu^+(X) \neq \emptyset$ and $M(\sigma) \cap M(\neg p_\alpha) \cap \mu^+(X) \neq \emptyset$.
By the same kind of argument as above we see that either for p_α or for $\neg p_\alpha$, or for both, and for all σ $x \in \mu^-(M(\sigma) \cap M(p_\alpha) \cap X)$ or $x \in \mu^-(M(\sigma) \cap M(\neg p_\alpha) \cap X)$. If not, there are σ and σ' and $x \notin \mu^-(M(\sigma) \cap M(p_\alpha) \cap X) \supseteq \mu^-(M(\sigma + \sigma') \cap M(p_\alpha) \cap X)$ and $x \notin \mu^-(M(\sigma') \cap M(\neg p_\alpha) \cap X) \supseteq \mu^-(M(\sigma + \sigma') \cap M(\neg p_\alpha) \cap X)$, but $\mu^-(M(\sigma + \sigma') \cap X) = \mu^-(M(\sigma + \sigma') \cap M(p_\alpha) \cap X) \cup \mu^-(M(\sigma + \sigma') \cap M(\neg p_\alpha) \cap X)$, so $x \notin \mu^-(M(\sigma + \sigma') \cap X)$, contradicting the induction hypothesis (2).
If we can choose both, we let the strategy decide, as for p_0.
So induction hypotheses (1) and (2) will hold again.
This gives a complete description of some y (relative to the strategy!), and we set $x \lhd y$. We have to show: for all $Y \in \mathcal{Y}$ $x \in \mu^-(Y) \leftrightarrow x \in \mu_\lhd(Y) :\leftrightarrow \exists y \in \mu^+(Y), x \lhd y$. "$\rightarrow$": As we will do above construction for all Y, it suffices to show that $y \in \mu^+(X)$. "$\leftarrow$": Conversely, if the y constructed above is in $\mu^+(Y)$, then x has to be in $\mu^-(Y)$.
If $y \notin \mu^+(X)$, then $Th(y)$ is inconsistent with $Th(\mu^+(X))$, as μ^+ is definability preserving. So by classical compactness there is a suitable finite sequence σ with $M(\sigma) \cap \mu^+(X) = \emptyset$, but this was excluded by the induction hypothesis (1). So $y \in \mu^+(X)$.
Suppose $y \in \mu^+(Y)$, but $x \notin \mu^-(Y)$. So $y \in \mu^+(Y)$ and $y \in \mu^+(X)$, and $Y = M(\phi)$ for some ϕ, so there will be a suitable finite sequence σ s.t. for all σ' with $\sigma \subseteq \sigma'$ $M(\sigma') \cap X \subseteq M(\phi) = Y$, and by our construction $x \in \mu^-(M(\sigma') \cap X)$. As $y \in \mu^+(X) \cap \mu^+(Y) \cap (M(\sigma') \cap X)$, $\mu^+(M(\sigma') \cap X) \subseteq \mu^+(Y)$, so by (μ^-4) $\mu^-(M(\sigma') \cap X) \subseteq \mu^-(Y)$, so $x \in \mu^-(Y)$, *contradiction*.
We now do this construction for all strategies. Obviously, this does not modify our results.
This finishes the completeness proof. □
As we postulated definability preservation, there are no problems to translate the result into logic. (Note that ν was applied to formula-defined model sets, but the resulting sets were perhaps theory-defined model sets.)
Comment:
One might try a construction similar to the one for counterfactual conditionals, see [SM94], and try to patch together several ranked structures, one for each K on the left, to obtain a general distance by repeating elements.
So we would have different "copies" of A, say A_i, more precisely of its elements, and the natural definition seems to be: $A * \phi \vdash \psi$ iff for all i $A_i * \phi \vdash \psi$, so $A \mid B = \bigcup\{A_i \mid B : i \in I\}$.
But this does not work: Take $A := \{a, a', a''\}$, $B := \{b, b'\}$, with $A \mid B := \{b, b'\}$, and $a \mid B = a' \mid B = a'' \mid B = \{b\}$. Then for all copies of the singletons, the result

cannot be empty, but must be $\{b\}$. But $A \mid B$ can only be a "partial" union of the $x \mid B, x \in A$, so it must be $\{b\}$ for all copies of A, contradiction.
(Alternative definitions with copies fail too, but no systematic investigation was done.)

8.2.3 Revision and Independence

The material in this section is taken from [GS09b].

8.2.3.1 Introduction

We give some results on

(1) Theory revision:
Parikh and co-authors (see, [CP00]), and, independently, Rodrigues (see, [Rod97]) have investigated a notion of logical independence, based on the sharing of essential propositional variables. We do a semantical analogue here. What Parikh et al. call splitting on the logical level, we call factorization (on the semantical level).
A comparison of the work by Parikh and Rodrigues can be found in [Mak09]. Note that many of our results are valid for arbitrary products, not only for classical model sets.
We go very slightly beyond Parikh's work. As a matter of fact, our generalization is already contained in the axiom *P2g*, due to Georgatos (see, [CP00]):
$(P2g)$ If T is split between $\mathcal{L}_1$ and $\mathcal{L}_2$, and α, β are in $\mathcal{L}_1$ and $\mathcal{L}_2$, respectively, then $T * \alpha * b = T * \beta * \alpha = T * (\alpha \wedge \beta)$.
On the other hand, we stay below Parikh's work and do *not* investigate partial overlap (see $\mathcal{B}$-structures model in [CP00]).
We claim no originality of the basic ideas, just our proofs and perhaps an example might be new – but they are always elementary and very easy.
(2) Preferential reasoning:
We shortly discuss preferential structures which have properties of defaults in the fact that they permit to treat sub ideal information. Usually, we have only the ideal case, where all "normal" information holds, and the classical case. (Reiter) defaults, and also, e.g. inheritance systems permit to satisfy only some, but not necessarily all default rules, and are thus more flexible. We show how to construct preferential structures with the same properties.

The Situation in the Case of Theory Revision

We work here with arbitrary, non empty products. Intuitively, $\mathcal{Y}$ is the set of models for the propositional variable set U.

Definition 8.2.11 Let U be an index set, $\mathcal{Y} = \Pi\{Y_k : k \in U\}$, let all $Y_k \neq \emptyset$, and $\mathcal{X} \subseteq \mathcal{Y}$. Thus, $\sigma \in \mathcal{X}$ is a function from U to $\bigcup\{Y_k : k \in U\}$ s.t. $\sigma(k) \in Y_k$. We then note $X_k := \{y \in Y_k : \exists \sigma \in \mathcal{X}, \sigma(k) = y\}$.

If $U' \subseteq U$, then $\sigma \lceil U'$ will be the restriction of σ to U', and $\mathcal{X} \lceil U' := \{\sigma \lceil U' : \sigma \in \mathcal{X}\}$.
If $\mathcal{A} := \{A_i : i \in I\}$ is a partition of U, $U' \subseteq U$, then $\mathcal{A} \lceil U' := \{A_i \cap U' \neq \emptyset : i \in I\}$.
Let $\mathcal{A} := \{A_i : i \in I\}$, $\mathcal{B} := \{B_j : j \in J\}$ both be partitions of U, then $\mathcal{A}$ is called a refinement of $\mathcal{B}$ iff for all $i \in I$ there is $j \in J$ s.t. $A_i \subseteq B_j$.
A partition $\mathcal{A}$ of U will be called a factorization of $\mathcal{X}$ iff $\mathcal{X} = \{\sigma \in \mathcal{Y} : \forall i \in I(\sigma \lceil A_i \in \mathcal{X} \lceil A_i)\}$; we will also sometimes say for clarity that $\mathcal{A}$ is a partition of $\mathcal{X}$ over U.
We will adhere to above notations throughout these pages.
If $\mathcal{X}$ is as above, $U' \subseteq U$ and $\sigma \in \mathcal{X} \lceil U'$, then there is obviously some (usually not unique) $\tau \in \mathcal{X}$ s.t. $\tau \lceil U' = \sigma$. This trivial fact will be used repeatedly in the following pages. We will denote by σ^+ some such τ – context will tell which are the U' and U. (To be more definite, we may take the first such τ in some arbitrary enumeration of $\mathcal{X}$.)
Given a propositional language $\mathcal{L}$, $v(\mathcal{L})$ will be the set of its propositional variables, and $v(\phi)$ the set of variables occuring in ϕ. A model set C is called definable iff there is a theory T s.t. $C = M(T)$ – the set of models of T.

We treat here the following:

(1) We give a purely algebraic description of factorization.
This is the algebraic analogue of work by Parikh and co-authors, and we claim almost no originality, perhaps with the exception of an example and the remark on language independence.
(2) We generalize slightly the Parikh approach so that it can be described as commuting with decomposition into sublanguages and addition.
We show by a trivial argument that this corresponds to a generalized Hamming distance between models.
(3) We go beyond rational monotony and show how to construct a preferential structure from a set of (normal) defaults. Thus, we give an independent semantics to normal defaults, translating their usual treatment into a homogenous construction of the preferential structure.
In the general case, this gives nothing new, as any preferential structure can be constructed this way. (We consider the one-copy case only.) Most of the time, it will result in a special structure which automatically takes into account the specificity criterion to resolve conflicts. The essential idea is to take a modified Hamming distance on the set of satisfied defaults, modified as we do not count the defaults, but look at them as sets, together with the subset relation.
We also show that our approach can be seen as a revision of the ideal, perhaps non-existant case, or as an approach to this ideal case as the limit. Of course, when the ideal case is consistent, then this will be our result.
(4) Independence in the case of theory revision is treated by looking at "independent" parts "independently", and later summing up. In the case of defaults, we treat the defaults independently, just as in the Reiter approach, but also "inside" the model sets, we treat subsets just as the sets themselves, resulting in a partial kind of rankedness (by default).

(5) We conclude by giving a simple informal argument why the theory revision situation is more complicated than the default situation.

8.2.3.2 Factorization

Fact 8.2.4 If $\mathcal{A}$, $\mathcal{B}$ are two partitions of U, $\mathcal{A}$ a factorization of $\mathcal{X}$, and $\mathcal{A}$ a refinement of $\mathcal{B}$, then $\mathcal{B}$ is also a factorization of $\mathcal{X}$.

Proof Trivial by definition. □

Fact 8.2.5 Let $\mathcal{A}$ be a factorization of $\mathcal{X}$ over U, $U' \subseteq U$. Then $\mathcal{A}\lceil U'$ is a factorization of $\mathcal{X}\lceil U'$ over U'.

Proof If $A_i \cap U' \neq \emptyset$, let $\sigma_i' \in \mathcal{X}\lceil(A_i \cap U')$. Let then $\sigma_i := \sigma_i'^{+}\lceil A_i$. If $A_i \cap U' = \emptyset$, let $\sigma_i := \tau\lceil A_i$ for any $\tau \in \mathcal{X}$. Then $\sigma := \bigcup\{\sigma_i : i \in I\} \in \mathcal{X}$ by hypothesis, so $\sigma\lceil U' \in \mathcal{X}\lceil U'$, and $\sigma\lceil(A_i \cap U'\} = \sigma_i'$. □

Fact 8.2.6 If $\{A, A'\}$ is a factorization of $\mathcal{X}$ over U, $\mathcal{A}$ a factorization of $\mathcal{X}\lceil A$ over A, $\mathcal{A}'$ a factorization of $\mathcal{X}\lceil A'$ over A', then $\mathcal{A} \cup \mathcal{A}'$ is a factorization of $\mathcal{X}$ over U.

Proof Trivial. □

Fact 8.2.7 If $\mathcal{A}$, $\mathcal{B}$ are two factorizations of $\mathcal{X}$, then there is a common refining factorization.

Proof Let σ s.t. $\forall i \in I\ \forall j \in J\ (\sigma\lceil(A_i \cap B_j) \in \mathcal{X}\lceil(A_i \cap B_j))$, show $\sigma \in \mathcal{X}$. Fix $i \in I$. By Fact 8.2.5 (p. 246), $\mathcal{B}\lceil A_i$ is a factorization of $\mathcal{X}\lceil A_i$, so $\bigcup\{\sigma\lceil(A_i \cap B_j) : j \in J,\ A_i \cap B_j \neq \emptyset\} = \sigma\lceil A_i \in \mathcal{X}\lceil A_i$. As $\mathcal{A}$ is a factorization of $\mathcal{X}$, $\sigma \in \mathcal{X}$. □

This does not generalize to infinitely many factorizations.

Example 8.2.3 Take as index set $\omega + 1$, all $Y_k := \{0, 1\}$. Take $\mathcal{X} := \{\sigma : \sigma\lceil\omega$ arbitrary, and $\sigma(\omega) := 0$ iff $\sigma\lceil\omega$ is finally constant$\}$. Consider the partitions $\mathcal{A}_n := \{n, (\omega+1) - n\}$, they are all factorizations of $\mathcal{X}$, as it suffices to know the sequence from $n + 1$ on to know its value on ω. A common refinement $\mathcal{A}$ will have some $A \in \mathcal{A}$ s.t. $\omega \in A$. Suppose there is some $n \in \omega \cap A$, then $A \not\subseteq n + 1$, $A \not\subseteq (\omega + 1) - (n + 1)$. This is impossible, so $A = \{\omega\}$. If $\mathcal{A}$ were a factorization of $\mathcal{X}$, so would be $\{\omega, \{\omega\}\}$ by Fact 8.2.4 (p. 246), but $\mathcal{X}$ does not factor into $\mathcal{X}\lceil\omega$ and $\mathcal{X}\lceil\{\omega\}$.

Comment 8.2.1 Above set $\mathcal{X}$ is not definable as a model set of a corresponding language $\mathcal{L}$: If ϕ is not a tautology, there is a model m s.t. $m \models \neg\phi$. ϕ is finite, let its variables be among $p_1, \ldots, p_n$ and perhaps p_ω. If p_ω is not among its variables, it is trivially also false in some m' in $\mathcal{X}$. If it is, then modify m accordingly beyond n. Thus, exactly all tautologies are true in $\mathcal{X}$, but $\mathcal{X} \neq \mathcal{Y}$ = the set of all $\mathcal{L}$-models.

We have, however the following fact:

Fact 8.2.8 Let $\mathcal{X} = \bigcap\{\mathcal{X}_m : m \in M\}$ and $\mathcal{X}, \mathcal{X}_m \subseteq \mathcal{Y}$, for all $m \in M$.

Let $\mathcal{A}$ be a partition of U and a factorization of all $\mathcal{X}_m$.

Then $\mathcal{A}$ is also a factorization of $\mathcal{X}$.

Proof Let σ s.t. $\forall i \in I\ \sigma\lceil A_i \in \mathcal{X}\lceil A_i$.
But $\mathcal{X}\lceil A_i = (\bigcap\{\mathcal{X}_m : m \in M\})\lceil A_i \subseteq \bigcap\{\mathcal{X}_m\lceil A_i : m \in M\}$: Let $\tau \in \mathcal{X}\lceil A_i$, so by $\mathcal{X} = \bigcap\{\mathcal{X}_m : m \in M\}$ $\tau^+ \in \mathcal{X}_m$ for all $m \in M$, so $\tau \in \mathcal{X}_m\lceil A_i$ for all $m \in M$.
Thus, $\forall i \in I, \forall m \in M : \sigma\lceil A_i \in \mathcal{X}_m\lceil A_i$, so $\forall m \in M, \sigma \in \mathcal{X}_m$ by prerequisite, so $\sigma \in \mathcal{X}$. □

Fact 8.2.9 Let $A \cup A'$ be a partition of U, and for all $\sigma \in \mathcal{X}\lceil A$ and all $\tau : A' \to \bigcup\{X_k : k \in A'\}$ with $\tau(k) \in X_k, \sigma \cup \tau \in \mathcal{X}$. Then

(1) $A \cup A'$ is a factorization of $\mathcal{X}$ over U.
(2) Any partition $\mathcal{A}' = \{A'_k : k \in I'\}$ of A' is a factorization of $\mathcal{X}\lceil A'$ over A'.
(3) If $\mathcal{A}$ is a factorization of $\mathcal{X}\lceil A$ over A, and $\mathcal{A}'$ a partition of A', then $\mathcal{A} \cup \mathcal{A}'$ is a factorization of $\mathcal{X}$.

Proof (1) and (2) are trivial, (3) follows from (1), (2), and Fact 8.2.6 (p. 246). □

Corollary 8.2.1 *Let $U = v(\mathcal{L})$ for some language $\mathcal{L}$. Let $\mathcal{X}$ be definable, and $\{\mathcal{A}_m : m \in M\}$ be a set of factorizations of $\mathcal{X}$ over U. Then $\mathcal{A} := \cup\{\mathcal{A}_m : m \in M\}$ is also a factorization of $\mathcal{X}$.*

Proof Let $\mathcal{X} = M(T)$. Consider $\phi \in T$. $v(\phi)$ is finite, consider $\mathcal{X}\lceil v(\phi)$. There are only finitely many different ways. $v(\phi)$ is partitioned by the $\mathcal{A}_m$, let them all be among $\mathcal{A}_{m_0}, \ldots, \mathcal{A}_{m_p}$. $M(\phi)\lceil v(\phi)$ might not be factorized by all $\mathcal{A}_{m_0}\lceil v(\phi), \ldots, \mathcal{A}_{m_p}\lceil v(\phi)$, but $M(T)\lceil v(\phi)$ is by Fact 8.2.5 (p. 246). By Fact 8.2.7 (p. 246), $\mathcal{A}\lceil v(\phi)$ is a factorization of $M(T)\lceil v(\phi)$.
Consider now $\mathcal{X}_\phi := (M(T)\lceil v(\phi)) \times \Pi\{(0, 1) : k \in v(\mathcal{L}) - v(\phi)\}$.
By Fact 8.2.9 (p. 247), (1) $\{v(\phi), v(\mathcal{L}) - v(\phi)\}$ is a factorization of $\mathcal{X}_\phi$ over $v(\mathcal{L})$.
By Fact 8.2.9 (p. 247), (2) $\mathcal{A}\lceil(v(\mathcal{L}) - v(\phi))$ is a factorization of $\mathcal{X}_\phi\lceil(v(\mathcal{L}) - v(\phi))$ over $v(\mathcal{L}) - v(\phi)$.
By Fact 8.2.9 (p. 247), (3) $\mathcal{A}$ is a factorization of $\mathcal{X}_\phi$ over $v(\mathcal{L})$.
$M(T) = \bigcap\{(M(T)\lceil v(\phi)) \times \Pi\{(0, 1) : k \in v(\mathcal{L}) - v(\phi)\}: \phi \in T\}$, so by Fact 8.2.8 (p. 246), $\mathcal{A}$ is a factorization of $M(T)$. □

Comment 8.2.2 Obviously, it is unimportant here that we have only two truth values, the proof would just as well work with any, even an infinite, number of truth values. What we really need is the fact that a formula affects only finitely many propositional variables, and the rest are free.

Unfortunately, the manner of coding can determine if there is a factorization, as can be seen by the following example:

Example 8.2.4

1. p= "*blue*", q= "*round*", q'= "*blue iff round*".
   Then
   $p \wedge q = blue\ and\ round, \neg p \wedge \neg q = \neg blue\ and\ \neg round,$

$p \wedge q' = blue\ and\ round, \neg p \wedge q' = \neg blue\ and\ \neg round.$
Thus, both code the same (*meta-*) situation, the first cannot be factorized, but the second can.
Our example (first presented in [Sch07a]) is discussed in more detail in [Mak09], see, Sect. 5 there.

2. More generally, we can code, e.g. the non factorizing situation $\{p \wedge q \wedge r, \neg p \wedge \neg q \wedge \neg r\}$ also using $q' = p \leftrightarrow q$, $r' = p \leftrightarrow r$, and have then the factorizing situation $\{p \wedge q \wedge r, \neg p \wedge q' \wedge r'\}$.
3. The following situation cannot be made factorizing: $\{p \wedge q,\ p \wedge \neg q,\ \neg p \wedge \neg q\}$. Suppose there were some such solution. Then we need some p' and q', and all four possibilities $\{p' \wedge q',\ p' \wedge \neg q',\ \neg p' \wedge q',\ \neg p' \wedge \neg q'\}$. If we do not admit impossible situations (i.e. one of the four possibilities is a contradictory coding), then the two possibilities have to contain the same situation, e.g. $p \wedge q$. But they are mutually exclusive (as they are negations), so this is impossible.

Remark 8.2.2 As we worked with abstract sequences, which need not be models, we can apply our results and the ideas behind them (essentially due to Parikh/Rodrigues), e.g. to

- Update: If a set of sequences (where the points are now models, and not true/false) factorizes, then we can update the components (i.e. look for the locally "best" subsequences), and then compose them to the globally "best" sequences.
- Utility streams: If a set of utility streams factorizes, we can do the same, commutativity and associativity of addition will guarantee the desired result.
- Preferential reasoning: Again, we factorize, and choose the locally best which we compose to the globally best.

The idea is always the same: If a set factorizes, choose locally, and compose to the global choice – provided this is the desired result!

8.2.3.3 Factorization and Hamming Distance

Both Hamming distances cooperate well with factorization, as we will see now. This is not surprising, as Hamming distances work componentwise.

Definition 8.2.12 We say that a revision function $*$ factorizes iff for all T and ϕ and joint factorizations, which we write for simplicity (and immediately for models) $M(T) = M(T)\lceil\mathcal{L}_1 \times, \ldots, \times M(T)\lceil\mathcal{L}_n$, $M(\phi) = M(\phi)\lceil\mathcal{L}_1 \times, \ldots, \times M(\phi)\lceil\mathcal{L}_n$, $M(T * \phi) = M(T) \mid M(\phi) = ((M(T)\lceil\mathcal{L}_1) \mid (M(\phi)\lceil\mathcal{L}_1)) \times, \ldots, \times((M(T)\lceil\mathcal{L}_n) \mid (M(\phi)\lceil\mathcal{L}_n))$.
To simplify notation, we will speak about $\Sigma \mid T$, $\Sigma_i \mid T_i$, $\sigma_1 \times \ldots \sigma_n$, etc.

The advantage is that, when factorization is possible, we can work with smaller theories, formulas, and languages, and then do a trivial composition operation by considering the product.

Fact 8.2.10 If $*$ is defined by the counting or the set variant of the Hamming distance, then $*$ factorizes.

Proof We do the proof for the set variant, and the counting variant proof is similar. Let a factorization as in the definition be given, and suppose $\tau \in T$ has minimal distance from Σ, i.e. $\tau \in \Sigma \mid T$. We show that each τ_i has minimal distance from Σ_i. If not, there is τ_i' closer to Σ_i, but then τ', which is like τ, only τ_i is replaced by τ_i' is also in T, by factorization. By definition of the Hamming distance, τ' is closer to Σ than τ is, *contradiction*. Thus $\tau \in (\Sigma_1 \mid T_1) \times \ldots \times (\Sigma_n \mid T_n)$. Conversely, let all $\tau_i \in (\Sigma_i \mid T_i)$, we have to show that $\tau := \tau_1 \times \ldots \times \tau_n \in \Sigma \mid T$. By factorization, $\tau \in T$. If there were a closer $\tau' \in T$, then at least one of the components τ_i' would be closer than τ_i, *contradiction*. □

The authors do not know if all factorizing distance-defined revisions can be defined by one of the above Hamming distances.

8.2.4 Preferential Modelling of Defaults

Reiter defaults have the advantage to give results also for non ideal cases. If, by default, α and α' hold, but α is inconsistent with the current situation, then α' will still "fire". Preferential structures say nothing about non ideal cases. We construct special preferential structures which have the same behaviour as Reiter defaults. In addition, specificity will be used to solve conflicts.
The idea is simple.
For simplicity, we admit direct contradictions: $\phi \mid\sim \psi$ and $\phi \mid\sim \neg\psi$. This is done only to make the representation proof simple. One can do without, but pays with more complexity (see below). We also use structures with one copy of each model only.

Definition 8.2.13

(1) We call the default $\phi \mid\sim \psi$ more specific than the default $\phi' \mid\sim \psi'$ iff $\phi \vdash \phi'$ – $\vdash$ is classical consequence.

(2) We say that the default $\phi \mid\sim \psi$ separates m and m' iff $m, m' \models \phi$, but only one of m, m' satisfies ψ.

Consider two models, m, m'. Take the most specific defaults which separate them. For each such default $\phi \mid\sim \psi$, if $m \models \psi$, $m' \not\models \psi$, then set $m \prec m'$. We might introduce cycles of length 2 here.

Remark 8.2.3

(1) The construction has a flavour of rankedness, as, if possible, we make each "good" element smaller than each "less good" element. If, e.g. $\phi \mid\sim \psi$ is the default, $m, m', n, n' \models \phi$, and $m, n \models \psi$, but $m', n' \not\models \psi$, then $m \prec m'$, $n \prec m'$, $m \prec n'$, $n \prec n'$.

(2) We may create indirectly loops, as we may have $m \prec m' \prec m''$, and also see $m'' \prec m$.

(3) Let $m \in X$ be minimal in our construction. Then there is no default $\phi \mid\sim \psi$ s.t. $m \models \phi \wedge \neg\psi$, and there is $m' \in X$, $m' \models \phi \wedge \psi$. Thus, minimal elements

are "as good as possible", i.e. there is no better one in X. Of course, there might be a default $\phi \mid\sim \psi$ with $m \models \phi \wedge \neg\psi$, but "better" elements are outside X. Thus, minimal elements are an approximation of the ideal case. We can also consider minimal elements as a revision of the ideal case by X, in the sense that we cannot get closer to the ideal within X.
Conversely, given *any* preferential structure, we take any two models m, m', if $m \prec m'$ (which we see by $m' \notin \mu(\{m, m'\})$), we add the default $Th\{m, m'\} \mid\sim Th\{m\}$. By the basic law of 1-copy preferential structures, we create the structure again. (Note that $\{m, m'\}$ is the most specific set containing both.) Thus, our approach cannot result in new structural rules for preferential structures, like smoothness, rankedness.

8.2.5 Remarks on Independence

The idea of independence was realized for defaults by trying to satisfy them independently, so if one fails, the others still have a chance. This is a simple idea.

The case of theory revision is more complicated, as we have no predefined structure. In particular, the starting theory T can be just a "blob" which makes independence difficult to realize. Moreover, we may try to revise just once, so no multiple default satisfaction or so is needed. Perhaps, the Parikh idea and its refinement through (modified) Hamming distances is all one can achieve.

It is evident how to treat our form of independent revision with IBRS; it can just be written as a diagram.

Perhaps, the best way to write defaults as a diagram is the trivial one: $\alpha \mid\sim \beta$ will be written $\alpha \Rightarrow \beta$, and the treatment is in the evaluation of the diagram – as outlined above.

8.2.5.1 Epistemic States and Independence

It is probably adequate to say that [AGM85] consider the revision function $*$ an epistemic state (depending on K), as revealed by the notion of epistemic entrenchment, and its equivalence to a revision function. In [LMS01], the global distance can probably be seen as (fixed, global) epistemic state. Essentially the critique of such too rigid, fixed, epistemic states resulted in dynamic states of [DP94] and [DP97]. The approach in [Spo88] incorporated already a dynamic approach. An excellent short overview of such dynamic revision approaches can be found in [Ker99].

In preferential structures, we might see the relation choosing the normal situations as (again fixed) epistemic state. In counterfactual conditionals, the distance can again be seen as the underlying epistemic state.

The present authors see the higher-order arrows of reactive structures (see, e.g. [GS08b]) as expressing epistemic states or changes of epistemic states. This will be explored in future research by the present authors.

But, we can also see the approaches discussed in this article as an epistemic state, which can perhaps be resumed as: "divide and conquer".

Chapter 9
An Analysis of Defeasible Inheritance Systems

The material in this chapter is taken from [GS08e].

9.1 Introduction

9.1.1 Terminology

"Inheritance" will stand here for "nonmonotonic or defeasible inheritance". We will use indiscriminately "inheritance system", "inheritance diagram", "inheritance network", "inheritance net".

In this introduction, we first give the connection to reactive diagrams, then give the motivation, then describe in very brief terms some problems of inheritance diagrams, and mention the basic ideas of our analysis.

9.1.2 Inheritance and Reactive Diagrams

Inheritance systems or diagrams have an intuitive appeal. They seem close to human reasoning, natural, and are also implemented (see [Mor98]). Yet, they are a more procedural approach to nonmonotonic reasoning, and, to the authors' knowledge, a conceptual analysis, leading to a formal semantics, as well as a comparison to more logic-based formalisms like the systems P and R of preferential systems are lacking. We attempt to reduce the gap between the more procedural and the more analytical approaches in this particular case. This will also give indications how to modify the systems P and R to approach them more to actual human reasoning. Moreover, we establish a link to multi-valued logics and the logics of information sources (see, e.g. [ABK07] and forthcoming work of the same authors, and also [BGH95]).

An inheritance net is a directed graph with two types of connections between nodes, $x \rightarrow y$ and $x \not\rightarrow y$. Diagram 9.1.1 (p. 252) is such an example. The meaning of $x \rightarrow y$ is that x is also a y and the meaning of $x \not\rightarrow y$ is that x is not a y.

D.M. Gabbay, K. Schlechta, *Logical Tools for Handling Change in Agent-Based Systems*, Cognitive Technologies, DOI 10.1007/978-3-642-04407-6_9,

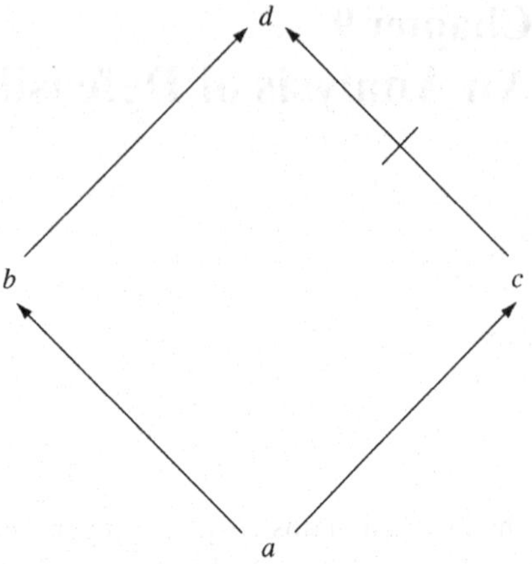

Diagram 9.1.1 The Nixon Diamond

We do not allow the combinations $x \not\rightarrow y \not\rightarrow z$ or $x \not\rightarrow y \rightarrow z$, but we do allow $x \rightarrow y \rightarrow z$ and $x \rightarrow y \not\rightarrow z$.

Given a complex diagram such as Diagram 9.1.2 (p. 252) and two points say z and y, the question we ask is to determine from the diagram whether the diagram says that

(1) z is y,
(2) z is not y,
(3) nothing to say.

Since in Diagram 9.2.2 (p. 259) there are paths to y from z either through x or through v, we need to have an algorithm to decide. Let $\mathcal{A}$ be such an algorithm.

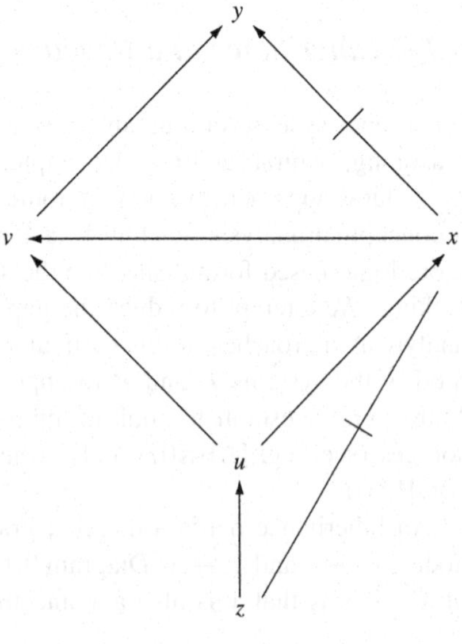

Diagram 9.1.2 The problem of downward chaining

We need $\mathcal{A}$ to decide

(1) Are there valid paths from z to y?
(2) Of the opposing paths (one which supports "z is y" and one which supports "z is not y"), which one wins (usually winning makes use of being more specific, but there are other possible options)?

So, for example, in Diagram 9.1.2 (p. 252), the connection $x \to v$ makes paths through x more specific than paths through v. The question is whether we have a valid path from z to x.

In the literature, as well as in this chapter, there are algorithms for deciding the valid paths and the relative specificity of paths. These are complex inductive algorithms, which may need the help of a computer for the case of the more complex diagrams.

It seems that for inheritance networks we cannot adopt a simple minded approach and just try to "walk" on the graph from z to y, and depending on what happens during this "walk" decide whether z is y or not. To explain what we mean, suppose we give the network a different meaning, that of fluid flow. $x \to y$ means there is an open pipe from x to y and $x \not\to y$ means there is a blocked pipe from x to y.

To the question "can fluid flow from z to y in Diagram 9.1.2" (p. 252), there is a simple answer:

Fluid can flow iff there is a path comprising of $\to$ only (without any $\not\to$ among them).

Similarly, we can ask in the inheritance network something like (*) below:

(*) z is (resp. is not) y according to diagram D, iff there is a path π from z to y in D such that some noninductive condition $\psi(\pi)$ holds for the path π.

Can we offer the reader such a ψ?

If we do want to help the user to "walk" the graph and get an answer, we can proceed as one of the following options:

Option 1. Add additional annotations to paths to obtain D^* from D, so that a predicate ψ can be defined on D^* using these annotations. Of course, these annotations will be computed using the inductive algorithm in $\mathcal{A}$, i.e. we modify $\mathcal{A}$ to $\mathcal{A}^*$ which also executes the annotations.

Option 2. Find a transformation τ on diagrams D to transform D to $D' = \tau(D)$, such that a predicate ψ can be found for D'. So we work on D' instead of on D.

We require a compatibility condition on options 1 and 2:

(C1) If we apply $\mathcal{A}^*$ to D^* we get D^* again.
(C2) $\tau(\tau(D)) = \tau(D)$.

We now present the tools we use for our annotations and transformation. These are the reactive double arrows.

Consider the following Diagram 9.1.3 (p. 254):

We want to walk from a to e. If we go to c, a double arrow from the arc $a \to c$ blocks the way from d to e. So the only way to go to e is through b. If we start at a' there is no such block. It is the travelling through the arc (a, c) that triggers the double arrow $(a, c) \twoheadrightarrow (d, e)$.

We want to use $\twoheadrightarrow$ in $\mathcal{A}^*$ and in τ. So in Diagram 9.1.2 (p. 259) the path $u \to x \not\to y$ is winning over the path $u \to v \to y$, because of the specificity arrow $x \to v$. However, if we start at z then the path $z \to u \to v \to y$ is valid because of $z \not\to x$. We can thus add the following double arrows to the diagram to get Diagram 9.1.4 (p. 255).

If we start from u and go to $u \to v$, then $v \to y$ is cancelled. Similarly, $u \to x \to v$ cancelled $v \to y$. So the only path is $u \to x \not\to y$.

If we start from z, then $u \to x$ is cancelled and so is the cancellation $(u, v) \twoheadrightarrow (v, y)$. Hence the path $z \to u \to v \to y$ is open.

We are not saying that τ(Diagram 9.1.3)= Diagram 9.1.4, but something effectively similar will be done by τ.

We emphasize that the construction depends on the point of departure. Consider Diagram 9.1.2 (p. 259). Starting at u, we will have to block the path uvy. Starting

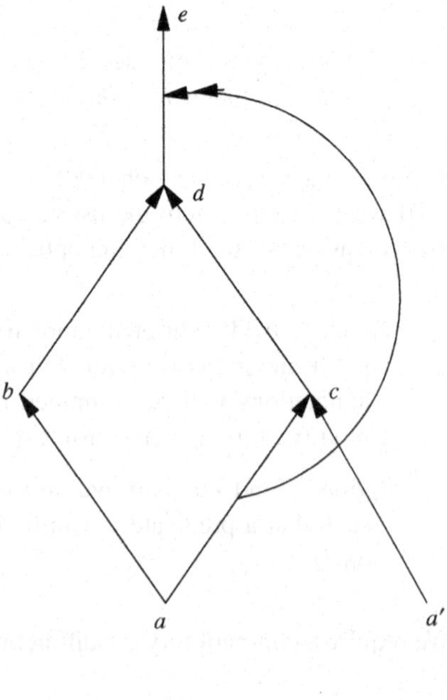

Diagram 9.1.3 Reactive graph

Diagram 9.1.4 Walking through the diagram

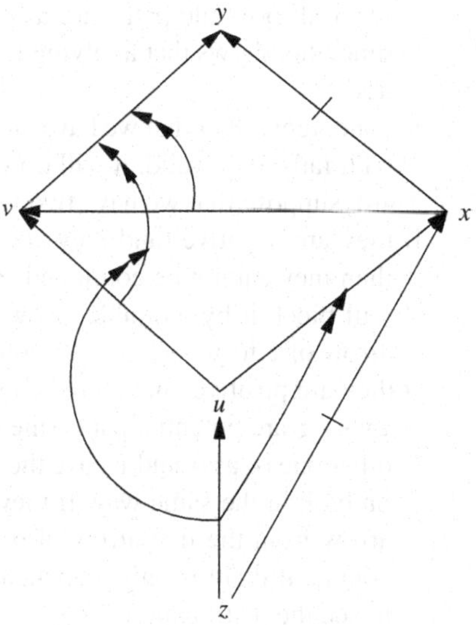

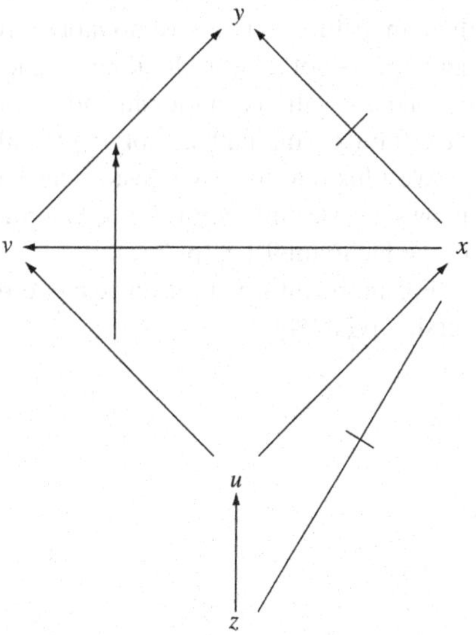

Diagram 9.1.5 The problem of downward chaining – reactive

at z, the path $zuvy$ has to be free. See Diagram 9.1.5 (p. 255). So we cannot just add a double arrow from $u \to v$ to $v \to y$, blocking $v \to y$, and leave it there when we start from z. We will have to erase it when we change the origin. At the same time, this shows an advantage over just erasing the arrow $v \to y$:

When we change the starting point, we can erase simply all double arrows, and do not have to remember the original diagram.

How do we construct the double arrows given some origin x?

First, if all possible paths are also valid, there is nothing to do. (At the same time, this shows that applying the procedure twice will not result in anything new.)

Second, remember that we have an upward chaining formalism. So if a potential path fails to be valid, it will do so at the end.

Third, suppose that we have two valid paths $\sigma : x \to y$ and $\tau : x \to y$.

If they are negative (and they are either both negative or positive, of course), then they cannot be continued. So if there is an arrow $y \to z$ or $y \not\to z$, we will block it by a double arrow from the first arrow of σ and from the first arrow of τ to $y \to z$ ($y \not\to z$, respectively).

If they are positive, and there is an arrow $y \to z$ or $y \not\to z$, both $\sigma y \to z$ and $\tau y \to z$ are potential paths (the case $y \not\to z$ is analogue). One is valid iff the other one is, as σ and τ have the same endpoint; so preclusion, if present, acts on both in the same way. If they are not valid, we block $y \to z$ by a double arrow from the first arrow of σ and from the first arrow of τ to $y \to z$. Of course, if there is only one such σ, we do the same, there is just to consider to see the justification.

We summarize: Our algorithm switches impossible continuations off by making them invisible; there is just no more arrow to concatenate. As validity in inheritance networks is not forward looking – validity of $\sigma : x \to y$ does not depend on what is beyond y – validity in the old and in the new network starting at x are the same. As we left only valid paths, applying the algorithm twice will not give anything new.

We illustrate this by considering Diagram 9.1.6 (p. 256). First, we add double arrows for starting point c, see Diagram 9.1.7 (p. 257), and then for starting point x, see Diagram 9.1.8 (p. 257).

For more information on reactive diagrams, see also [Gab08c], and an earlier version [Gab04].

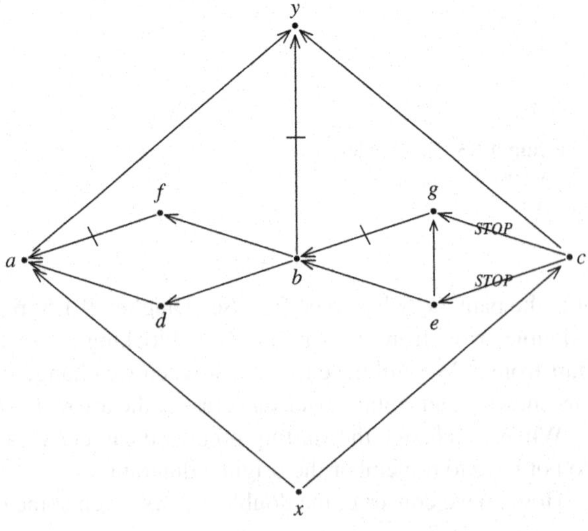

Diagram 9.1.6 STOP signs

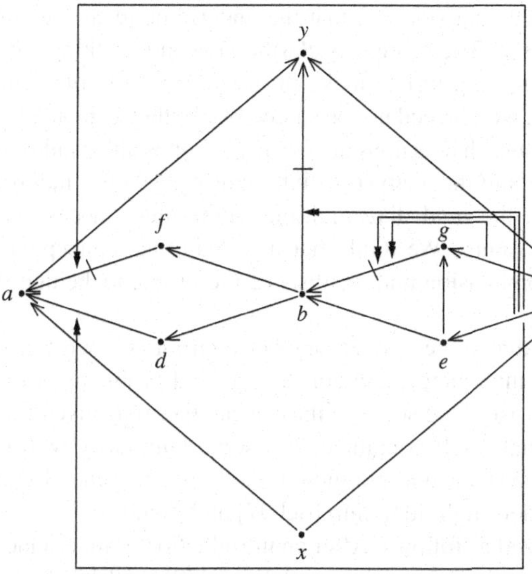

Diagram 9.1.7 Starting point c

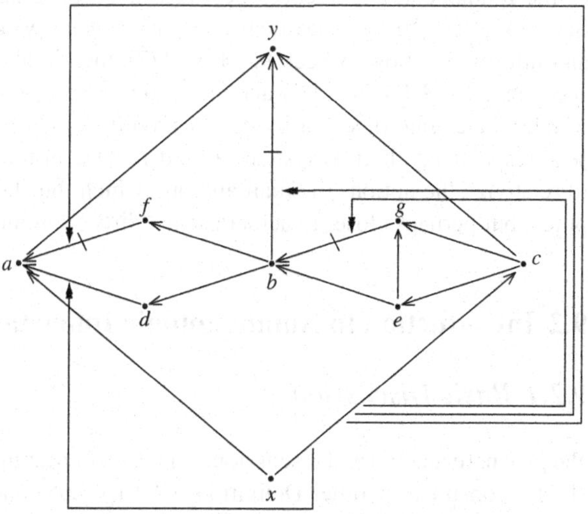

Diagram 9.1.8 Starting point x

9.1.3 Conceptual Analysis

Inheritance diagrams are deceptively simple. Their conceptually complicated nature is seen by, e.g. the fundamental difference between direct links and valid paths, and the multitude of existing formalisms, upward vs. downward chaining, intersection of extensions vs. direct scepticism, on-path vs. off-path preclusion (or pre-emption), split validity vs. total validity preclusion, etc. to name a few; see the discussion in Sect. 9.2.3 (p. 269). Such a proliferation of formalisms usually hints at deeper problems on the conceptual side, i.e. that the underlying ideas are ambiguous, and not sufficiently analysed. Therefore, any clarification and resulting reduction of possible formalisms seems a priori to make progress. Such clarification will involve conceptual decisions, which need not be shared by all, they can only be suggestions. Of course, a proof that such decisions are correct is impossible, and so is its contrary.

We will introduce into the analysis of inheritance systems a number of concepts not usually found in the field, like multiple truth values, access to information, comparison of truth values. We think that this additional conceptual burden pays off by a better comprehension and analysis of the problems behind the surface of inheritance.

We will also see that some distinctions between inheritance formalisms go far beyond questions of inheritance, and concern general problems of treating contradictory information – isolating some of these is another objective of this article.

The text is essentially self-contained, still some familiarity with the basic concepts of inheritance systems and nonmonotonic logics in general is helpful. For a presentation, the reader might look into [Sch97] and [Sch04].

The text is organized as follows. After an introduction to inheritance theory, connections with reactive diagrams in Sect. 9.3 (p. 270), and big and small subsets and the systems P and R in Sect. 9.2 (p. 258), we turn to an informal description of the fundamental differences between inheritance and the systems P and R in Sect. 9.4.2 (p. 276), give an analysis of inheritance systems in terms of information and information flow in Sect. 9.4.3 (p. 277), then in terms of reasoning with prototypes in Sect. 9.4.4 (p. 281), and conclude in Sect. 9.5 (p. 284) with a translation of inheritance into (necessarily deeply modified) coherent systems of big and small sets, respectively, logical systems P and R. One of the main modifications will be to relativize the notions of small and big, which thus become less "mathematically pure", but perhaps closer to actual use in "dirty" common sense reasoning.

9.2 Introduction to Nonmonotonic Inheritance

9.2.1 Basic Discussion

We give here an informal discussion. The reader unfamiliar with inheritance systems should consult in parallel Definition 9.2.3 (p. 265) and Definition 9.2.4 (p. 266). As there are many variants in the definitions, it seems reasonable to discuss them before a formal introduction, which, otherwise, would seem to pretend to be definite without being so.

9.2.1.1 (Defeasible or Nonmonotonic) Inheritance Networks or Diagrams

Nonmonotonic inheritance systems describe situations like "normally, birds fly", written $birds \rightarrow fly$. Exceptions are permitted, "normally penguins don't fly", $penguins \not\rightarrow fly$.

Definition 9.2.1 A nonmonotonic inheritance net is a finite DAG, directed, acyclic graph, with two types of arrows or links, $\rightarrow$ and $\not\rightarrow$, and labelled nodes. We will use Γ, etc. for such graphs, and σ, etc. for paths – the latter to be defined below.

Roughly (and to be made precise and modified below, we try to give here just a first intuition), $X \rightarrow Y$ means that "normal" elements of X are in Y, and $X \not\rightarrow Y$ means that "normal" elements of X are not in Y. In a semi-quantitative set interpretation, we will read "most" for "normal", thus "most elements of X are in Y", "most elements of X are not in Y", etc. These are by no means the only interpretations, as we will see – we will use these expressions for the moment just to help the reader's intuition. We should add immediately a word of warning: "most" is here not necessarily, but only by default, transitive, in the following sense. In the Tweety diagram, see Diagram 9.2.1 (p. 259) below, most penguins are birds, most birds fly, but it is not the case that most penguins fly. This is the problem of transfer of relative size which will be discussed extensively, especially in Sect. 9.5 (p. 284).

According to the set interpretation, we will also use informally expressions like $X \cap Y$, $X - Y$, $\boldsymbol{C}X$ – where $\boldsymbol{C}$ stands for set complement. But we will also use nodes informally as formulas, like $X \wedge Y$, $X \wedge \neg Y$, $\neg X$. All this will only be used here as an appeal to intuition.

Nodes at the beginning of an arrow can also stand for individuals, so $Tweety \not\rightarrow fly$ means something like: "Normally, Tweety will not fly". As always in nonmonotonic systems, exceptions are permitted, so the soft rules "birds fly", "penguins don't fly", and (the hard rule) "penguins are birds" can coexist in one diagram, penguins are then abnormal birds (with respect to flying). The direct link $penguins \not\rightarrow fly$ will thus be accepted, or considered valid, but not the composite path $penguins \rightarrow birds \rightarrow fly$, by specificity – see below. This is illustrated by Diagram 9.2.1 (p. 252), where a stands for Tweety, c for penguins, b for birds, and d for flying animals or objects.

(Remark: The arrows $a \rightarrow c$, $a \rightarrow b$, and $c \rightarrow b$ can also be composite paths – see below for the details.)

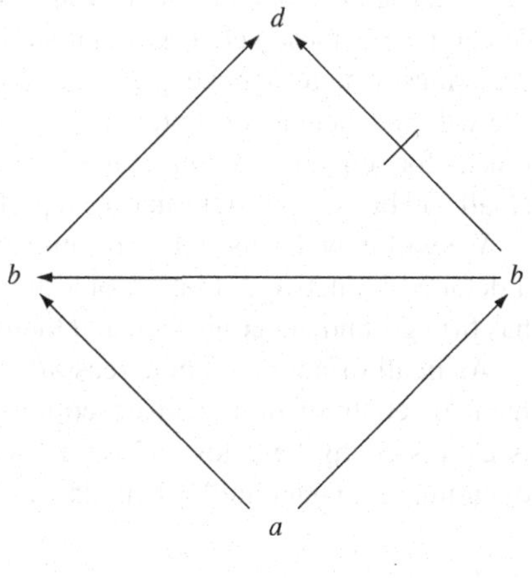

Diagram 9.2.1 Tweety diagram

(Of course, there is an analogous case for the opposite polarity, i.e. when the arrow from b to d is negative, and the one from c to d is positive.)

The main problem is to define in an intuitively acceptable way a notion of valid path, i.e. concatenations of arrows satisfying certain properties. We will write $\Gamma \models \sigma$, if σ is a valid path in the network Γ, and if x is the origin, and y the endpoint of σ, and σ is positive, we will write $\Gamma \models xy$, i.e. we will accept the conclusion that xs are ys, and analogously $\Gamma \models x\overline{y}$ for negative paths. Note that we will not accept any other conclusions, only those established by a valid path; so many questions about conclusions have a trivial negative answer: There is obviously no path from x to y. For example, there is no path from b to c in Diagram 9.2.1 (p. 256). Likewise, there are no disjunctions, conjunctions, etc. in our conclusions, and negation is present only in a strong form: "It is not the case that xs are normally ys" is not a possible conclusion, only "xs are normally not ys" is one. Also, possible contradictions are contained, there is no EFQ.

To simplify matters, we assume that for no two nodes $x, y \in \Gamma$ $x \rightarrow y$ and $x \not\rightarrow y$ are both in Γ, intuitively, that Γ is free from (hard) contradictions. This restriction is inessential for our purposes. We admit, however, soft contradictions and preclusion, which allows us to solve some soft contradictions – as we already did in the penguins example. We will also assume that all arrows stand for rules with possibly exceptions; again, this restriction is not important for our purposes. Moreover, in the abstract treatment, we will assume that all nodes stand for (nonempty) sets, though this will not be true for all examples discussed.

This might be the place for a remark on absence of cycles. Suppose we also have a positive arrow from b to c in Diagram 9.1.1 (p. 256). Then, the concept of preclusion collapses, as there are now equivalent arguments to accept $a \rightarrow b \rightarrow d$ and $a \rightarrow c \not\rightarrow d$. Thus, if we do not want to introduce new complications, we cannot rely on preclusion to decide conflicts. It seems that this would change the whole outlook on such diagrams. The interested reader will find more on the subject in [Ant97], [Ant99], [Ant05].

Inheritance networks were introduced about 20 years ago (see, e.g. [Tou84], [Tou86], [THT87]), and exist in a multitude of more or less differing formalisms; see, e.g. [Sch97] for a brief discussion. There still does not seem to exist a satisfying semantics for these networks. The authors' own attempt [Sch90] is an a posteriori semantics, which cannot explain or criticize or decide between different formalisms. We will give here a conceptual analysis, which provides also at least some building blocks for semantics, and a translation into (a modified version of) the language of small and big subsets, familiar from preferential structures; see, Sect. 3.2.2.6 (p. 58).

We will now discuss the two fundamental situations of contradictions, then give a detailed inductive definition of valid paths for a certain formalism so the reader has firm ground under his feet, and then present briefly some alternative formalisms.

As in all of nonmonotonic reasoning, the interesting questions arise in the treatment of contradictions and exceptions. The difference in quality of information is expressed by "preclusion" (or pre-emption). The basic diagram is the Tweety diagram, see, Diagram 9.2.1 (p. 256).

Unresolved contradictions give either rise to a branching into different extensions, which may roughly be seen as maximal consistent subsets, or to mutual cancellation in directly sceptical approaches. The basic diagram for the latter is the Nixon Diamond, see, Diagram 9.1.1 [(p. 257)], where $a = Nixon$, $b = Quaker$, $c = Republican$, $d = pacifist$.

In the directly sceptical approach, we will not accept any path from a to d as valid, as there is an unresolvable contradiction between the two candidates. The extensions approach can be turned into an indirectly sceptical one, by forming first all extensions, and then taking the intersection of either the sets of valid paths, or of valid conclusions, see [MS91] for a detailed discussion. See also Sect. 9.2.3 (p. 269) for more discussion on directly vs. indirectly sceptical approaches.

[We describe this now in more detail:]

9.2.1.2 Preclusion

In the above example, our intuition tells us that it is not admissible to conclude from the fact that penguins are birds, and that most birds fly that most penguins fly. The horizontal arrow $c \to b$ together with $c \not\to d$ bars this conclusion; it expresses specificity. Consequently, we have to define the conditions under which two potential paths neutralize each other, and when one is victorious. The idea is as follows: (1) We want to be sceptical, in the sense that we do not believe every potential path. We will not arbitrarily chose one either. (2) Our scepticism will be restricted, in the sense that we will often make well-defined choices for one path in the case of conflict: (*a*) If a compound potential path is in conflict with a direct link, the direct link wins. (*b*) Two conflicting paths of the same type neutralize each other, as in the Nixon Diamond, where neither potential path will be valid. (*c*) More specific information will win over less specific one.

(It is essential in the Tweety diagram that the arrow $c \not\to d$ is a direct link; so it is in a way stronger than compound paths.) The arrows $a \to b$, $a \to c$, $c \to b$ can also be composite paths: The path from c to b (read $c \subseteq \ldots \subseteq b$, where $\subseteq$ stands here for soft inclusion), however, tells us that the information coming from c is more specific (and thus considered more reliable), so the negative path from a to d via c will win over the positive one via b. The precise inductive definition will be given below. This concept is evidently independent of the length of the paths, $a \cdots \to c$ may be much longer than $a \cdots \to b$, so this is not shortest path reasoning (which has some nasty drawbacks, discussed in, e.g. [HTT87]).

A final remark: Obviously, in some cases, it need not be specificity, which decides conflicts. Consider the case where Tweety is a bird, but a dead animal. Obviously, Tweety will not fly, here because the predicate "dead" is very strong and overrules many normal properties. When we generalize this, we might have a hierarchy of causes, where one overrules the other, or the result may be undecided. For instance, a falling object might be attracted in a magnetic field, but a gusty wind might prevent this, sometimes, with unpredictable results. This is then additional information (strength of cause), and this problem is not addressed directly in traditional inheritance networks. We would have to introduce a subclass "dead bird" – and subclasses often have properties of "pseudo-causes", being a penguin probably is not a "cause" for not flying, nor bird for flying; still, things change from class to subclass for a reason. Before we give a formalism based on these ideas, we refine them, adopt one possibility (but indicate some modifications), and discuss alternatives later.

9.2.2 Directly Sceptical Split Validity Upward Chaining Off-Path Inheritance

Our approach will be directly sceptical, i.e. unsolvable contradictions result in the absence of valid paths, it is upward chaining, and split validity for preclusions (discussed below, in particular, in Sect. 9.2.3 (p. 269)). We will indicate modifications to make it extension based, as well as for total validity preclusion. This approach is strongly inspired by classical work in the field by Horty, Thomason, Touretzky, and others, and we claim no priority whatever. If it is new at all, it is a very minor modification of existing formalisms.

Our conceptual ideas to be presented in detail in Sect. 9.4.3 (p. 277) make split validity, off-path preclusion, and upward chaining a natural choice. For the reader's convenience, we give here a very short resume of these ideas: We consider only arrows as information, e.g. $a \to b$ will be considered information b valid at or for a. Valid composed positive paths will not be considered information in our sense. They will be seen as a way to obtain information, so a valid path $\sigma : x \ldots \to a$ makes information b accessible to x, and, second, as a means of comparing information strength, so a valid path $\sigma : a \ldots . \to a'$ will make information at a stronger than information at a'. Valid negative paths have no function; we will only consider the positive initial part as discussed above, and the negative end arrow as information, but never the whole path.

Choosing direct scepticism is a decision beyond the scope of this article, and we just make it. It is a general question how to treat contradictory and absent information, and if they are equivalent or not, see the remark in Sect. 9.4.4 (p. 281). (The fundamental difference between intersection of extensions and direct scepticism for defeasible inheritance was shown in [Sch93].) See also Sect. 9.2.3 (p. 269) for more discussion.

We now turn to the announced variants as well as a finer distinction within the directly sceptical approach. Again, see also Sect. 9.2.3 (p. 269) for more discussion. Our approach generates another problem, essentially that of the treatment of a mixture of contradictory and concordant information of multiple strengths or truth values. We bundle the decision of this problem with that for direct scepticism into a "plug-in" decision, which will be used in three approaches: the conceptual ideas, the inheritance algorithm, and the choice of the reference class for subset size (and implicitly also for the treatment as a prototype theory). It is thus well encapsulated and independent from the context.

These decisions (but, perhaps to a lesser degree, (1)) concern a wider subject than only inheritance networks. Thus, it is not surprising that there are different formalisms for solving such networks, deciding one way or the other. But this multitude is not the fault of inheritance theory, it is only a symptom of a deeper question. We first give an overview for a clearer overall picture, and discuss them in detail below, as they involve sometimes quite subtle questions.

(1) Upward chaining against downward or double chaining.
(2.1) Off-path against on-path preclusion.
(2.2) Split validity preclusion against total validity preclusion.
(3) Direct scepticism against intersection of extensions.
(4) Treatment of mixed contradiction and preclusion situations, no preclusion by paths of the same polarity.

(1) When we interpret arrows as causation (in the sense that $X \to Y$ expresses that condition X usually causes condition Y to result), this can also be seen as a difference in reasoning from cause to effect vs. backward reasoning, looking for causes for an effect. (A word of warning: There is a well-known article [SL89] from which a superficial reader might conclude that upward chaining is tractable and downward chaining is not. A more careful reading reveals that, on the negative side, the authors only show that double chaining is not tractable.) We will adopt upward chaining in all our approaches. See Sect. 9.4.4 (p. 281) for more remarks.
(2.1) and (2.2) Both are consequences of our view – to be discussed below in Sect. 4.3 – to see valid paths also as an absolute comparison of truth values, independent of reachability of information. Thus, in Diagram 9.2.1 (p. 256), the comparison between the truth values "penguin" and "bird" is absolute, and does not depend on the point of view "Tweety", as it can in total validity preclusion – if we continue to view preclusion as a comparison of information strength (or truth value). This question of absoluteness transcends obviously inheritance networks. Our decision is, of course, again uniform for all our approaches.
(3) This point, too, is much more general than the problems of inheritance. It is, among other things, a question of whether only the two possible cases (positive and negative) may hold, or whether there might be still other possibilities. See Sect. 9.4.4 (p. 281).
(4) This concerns the treatment of truth values in more complicated situations, where we have a mixture of agreeing and contradictory information. Again, this problem reaches far beyond inheritance networks.

We will group (3) and (4) together in one general, "plug-in" decision, to be found in all approaches we discuss.

Definition 9.2.2 This is an informal definition of a plug-in decision:

We describe now more precisely a situation which we will meet in all contexts discussed, and whose decision goes beyond our problem – thus, we have to adopt one or several alternatives, and translate them into the approaches we will discuss. There will be one global decision, which is (and can be) adapted to the different contexts.

Suppose we have information about ϕ and ψ, where ϕ and ψ are presumed to be independent – in some adequate sense. Suppose then that we have information sources $A_i : i \in I$ and $B_j : j \in J$, where the A_i speak about ϕ (they say ϕ or $\neg\phi$) and the B_j speak about ψ in the same way. Suppose further that we have a partial, not necessarily transitive (!), ordering $<$ on the information sources A_i and B_j together. $X < Y$ will say that X is better (intuition: more specific) than Y. (The potential lack of transitivity is crucial, as valid paths do not always concatenate to valid paths – just consider the Tweety diagram.)

We also assume that there are contradictions, i.e. some A_i say ϕ, some $\neg\phi$, likewise for the B_j – otherwise, there are no problems in our context. We can now take several approaches, all taking contradictions and the order $<$ into account.

- (P1) We use the global relation $<$, and throw away all information coming from sources of minor quality, i.e. if there is X such that $X < Y$, then no information coming from Y will be taken into account. Consequently, if Y is the only source of information about ϕ, then we will have no information about ϕ. This seems an overly radical approach, as one source might be better for ϕ, but not necessarily for ψ too.

  If we adopt this radical approach, we can continue as below, and can even split in analogue ways into (P1.1) and (P1.2), as we do below for (P2.1) and (P2.2).
- (P2) We consider the information about ϕ separately from the information about ψ. Thus, we consider for ϕ only the A_i, for ψ only the B_j. Take now, e.g. ϕ and the A_i. Again, there are (at least) two alternatives.

  – (P2.1) We eliminate again all sources among the A_i for which there is a better $A_{i'}$, irrespective of whether they agree on ϕ or not.

    - (a) If the sources left are contradictory, we conclude nothing about ϕ, and accept for ϕ none of the sources. (This is a directly sceptical approach of treating unsolvable contradictions, following our general strategy.)
    - (b) If the sources left agree for ϕ, i.e. all say ϕ or all say $\neg\phi$, then we conclude ϕ (or $\neg\phi$), and accept for ϕ all the remaining sources.

  – (P2.2) We eliminate again all sources among the A_i for which there is a better $A_{i'}$, but only if A_i and $A_{i'}$ have contradictory information. Thus, more sources may survive than in approach (P2.1).

We now continue as for (P2.1):

- (a) If the sources left are contradictory, we conclude nothing about ϕ, and accept for ϕ none of the sources.
- (b) If the sources left agree for ϕ, i.e. all say ϕ or all say $\neg\phi$, then we conclude ϕ (or $\neg\phi$), and accept for ϕ all the remaining sources.

The difference between (P2.1) and (P2.2) is illustrated by the following simple example. Let $A < A' < A''$, but $A \not< A''$ (recall that $<$ is not necessarily transitive), and $A \models \phi$, $A' \models \neg\phi$, $A'' \models \neg\phi$. Then (P2.1) decides for ϕ (A is the only survivor), (P2.2) does not decide, as A and A'' are contradictory, and both survive in (P2.2).

There are arguments for and against either solution: (P2.1) gives a uniform picture, more independent from ϕ, (P2.2) gives more weight to independent sources, it "adds" information sources, and thus gives potentially more weight to information from several sources. (P2.2) seems more in the tradition of inheritance networks, so we will consider it in the further development.

The reader should note that our approach is quite far from a fixed-point approach in two ways: First, fixed-point approaches seem more appropriate for extension-based approaches, as both try to collect a maximal set of uncontradictory information. Second, we eliminate information when there is better, contradicting information, even if the final result agrees with the first. This, too, contradicts in spirit the fixed-point approach.

After these preparations, we turn to a formal definition of validity of paths.

9.2.2.1 The Definition of $\models$ (i.e. of Validity of Paths)

All definitions are relative to a fixed diagram Γ. The notion of degree will be defined relative to all nodes of Γ, as we will work with split validity preclusion, so the paths to consider may have different origins. For simplicity, we consider Γ to be just a set of points and arrows, thus, e.g. $x \to y \in \Gamma$ and $x \in \Gamma$ are defined, when x is a point in Γ, and $x \to y$ an arrow in Γ. Recall that we have two types of arrows, positive and negative ones.

We first define generalized and potential paths, then the notion of degree, and finally validity of paths, written $\Gamma \models \sigma$, if σ is a path, as well as $\Gamma \models xy$, if $\Gamma \models \sigma$ and $\sigma : x \ldots. \to y$.

Definition 9.2.3 (1) Generalized paths:

A generalized path is an uninterrupted chain of positive or negative arrows pointing in the same direction; more precisely:

$x \to p \in \Gamma \to x \to p$ is a generalized path,

$x \not\to p \in \Gamma \to x \not\to p$ is a generalized path.

If $x \cdots \to p$ is a generalized path, and $p \to q \in \Gamma$, then $x \cdots \to p \to q$ is a generalized path,

if $x \cdots \to p$ is a generalized path, and $p \not\to q \in \Gamma$, then $x \cdots \to p \not\to q$ is a generalized path.

(2) Concatenation:

If σ and τ are two generalized paths, and the end point of σ is the same as the starting point of τ, then $\sigma \circ \tau$ is the concatenation of σ and τ.

(3) Potential paths (pp.):

A generalized path contains atmost one negative arrow, and this at the end, is a potential path. If the last link is positive, it is a positive potential path, if not, a negative one.

(4) Degree:
As already indicated, we shall define paths inductively. As we do not admit cycles in our systems, the arrows define a well-founded relation on the vertices. Instead of using this relation for the induction, we shall first define the auxiliary notion of degree, and do induction on the degree. Given a node x (the origin), we need a (partial) mapping f from the vertices to natural numbers such that $p \to q$ or $p \not\to q \in \Gamma$ implies $f(p) < f(q)$, and define (relative to x) :
Let σ be a generalized path from x to y, then $deg_{\Gamma,x}(\sigma) := deg_{\Gamma,x}(y) :=$ the maximal length of any generalized path parallel to σ, i.e. beginning in x and ending in y.

Definition 9.2.4 Inductive definition of $\Gamma \models \sigma$:
Let σ be a potential path.

- Case I:
  σ is a direct link in Γ. Then $\Gamma \models \sigma$
  (Recall that we have no hard contradictions in Γ.)
- Case II:
  σ is a compound potential path, $deg_{\Gamma,a}(\sigma) = n$, and $\Gamma \models \tau$ is defined for all τ with degree less than n – whatever their origin and endpoint.
- Case II.1: Let σ be a positive pp. $x \cdots \to u \to y$, let $\sigma' := x \cdots \to u$, so $\sigma = \sigma' \circ u \to y$.
  Then, informally, $\Gamma \models \sigma$ iff

  (1) σ is a candidate by upward chaining,
  (2) σ is not precluded by more specific contradicting information,
  (3) all potential contradictions are themselves precluded by information contradicting them.

  Note that (2) and (3) are the translation of (P2.2) in Definition 9.2.2 (p. 263).
  Formally, $\Gamma \models \sigma$ iff

  (1) $\Gamma \models \sigma'$ and $u \to y \in \Gamma$.
  (The initial segment must be a path, as we have an upward chaining approach. This is decided by the induction hypothesis.)
  (2) There are no v, τ, τ' such that $v \not\to y \in \Gamma$ and $\Gamma \models \tau := x \cdots \to v$ and $\Gamma \models \tau' := v \cdots \to u$. ($\tau$ may be the empty path, i.e. $x = v$.)
  (σ itself is not precluded by split validity preclusion and a contradictory link. Note that $\tau \circ v \not\to y$ need not be valid; it suffices that it is a better candidate (by τ').)
  (3) All potentially conflicting paths are precluded by information contradicting them:
  For all v and τ such that $v \not\to y \in \Gamma$ and $\Gamma \models \tau := x \cdots \to v$ (i.e. for all potentially conflicting paths $\tau \circ v \not\to y$) there is z such that $z \to y \in \Gamma$ and either
  $z = x$
  (the potentially conflicting pp. is itself precluded by a direct link, which is thus valid)

or
there are $\Gamma \models \rho := x \cdots \to z$ and $\Gamma \models \rho' := z \cdots \to v$ for suitable ρ and ρ'.

- Case II.2: The negative case, i.e. σ a negative pp. $x \cdots \to u \not\to y$, $\sigma' := x \cdots \to u$, $\sigma = \sigma' \circ u \not\to y$ is entirely symmetrical.

Remark 9.2.1 The following remarks all concern preclusion.

(1) Thus, in the case of preclusion, there is a valid path from x to z, and z is more specific than v, so $\tau \circ v \not\to y$ is precluded. Again, $\rho \circ z \to y$ need not be a valid path, but it is a better candidate than $\tau \circ v \not\to y$ is, and as $\tau \circ v \not\to y$ is in simple contradiction, this suffices.
(2) Our definition is stricter than many popular ones in the following sense: We require – according to our general picture to treat only direct links as information – that the preclusion "hits" the precluded path at the end, i.e. $v \not\to y \in \Gamma$ and ρ' hits $\tau \circ v \not\to y$ at v. In other definitions it is possible that the preclusion hits at some v', which is somewhere on the path τ, and not necessarily at its end. For instance, in the Tweety diagram, see, Diagram 9.2.1 (p. 256), if there were a node b' between b and d, we will need the path $c \to b \to b'$ to be valid, (obvious) validity of the arrow $c \to b$ will not suffice.
(3) If we allow ρ to be the empty path, then the case $z = x$ is a subcase of the present one.
(4) Our conceptual analysis has led to a very important simplification of the definition of validity. If we adopt on-path preclusion, we have to remember all paths which led to the information source to be considered: In the Tweety diagram, we have to remember that there is an arrow $a \to b$; it is not sufficient to note that we somehow came from a to b by a valid path, as the path $a \to c \to b \to d$ is precluded, but not the path $a \to b \to d$. If we adopt total validity preclusion, see also Sect. 9.2.3 (p. 269) for more discussion, we have to remember the valid path $a \to c \to b$ to see that it precludes $a \to c \to d$. If we allow preclusion to "hit" below the last node, we also have to remember the entire path which is precluded. Thus, in all those cases, whole paths (which can be very long) have to be remembered, but NOT in our definition.
We only need to remember (consider the Tweety diagram)

   (a) we want to know if $a \to b \to d$ is valid, so we have to remember a, b, d. Note that the (valid) path from a to b can be composed and very long.
   (b) we look at possible preclusions, so we have to remember $a \to c \not\to d$; again the (valid) path from a to c can be very long.
   (c) we have to remember that the path from c to b is valid (this was decided by induction before).

So in all cases (the last one is even simpler), we need only to remember the starting node, a (or c), the last node of the valid paths, b (or c), and the information $b \to d$ or $c \not\to d$ – i.e. the size of what has to be recalled is ≤ 3. (Of course, there may be many possible preclusions, but in all cases we have to look at a very limited situation, and not arbitrarily long paths.)

We take a fast look forward to Sect. 4.3, where we describe diagrams as information and its transfer, and nodes also as truth values. In these terms – and the reader is asked to excuse the digression – we may note above point (a) as $a \Rightarrow_b d$ – expressing that, seen from a, d holds with truth value b, (b) as $a \Rightarrow_c \neg d$, (c) as $c \Rightarrow_c b$ – and this is all we need to know.

We indicate here some modifications of the definition without discussion, which is to be found below.

(1) For on-path preclusion only: Modify condition (2) in Case II.1 to: (2′) There is no v on the path σ (i.e. $\sigma : x \cdots \to v \cdots \to u$) such that $v \not\to y \in \Gamma$.
(2) For total validity preclusion: Modify condition (2) in Case II.1 to: (2′) There are no v, τ, τ' such that $v \not\to y \in \Gamma$ and $\tau := x \cdots \to v$ and $\tau' := v \cdots \to u$ such that $\Gamma \models \tau \circ \tau'$.
(3) For extension-based approaches: Modify condition (3) in Case II.1 as follows: (3′) If there are conflicting paths, which are not precluded themselves by contradictory information, then we branch recursively (i.e. for all such situations) into two extensions: one, where the positive non-precluded paths are valid second; where the negative non precluded paths are valid.

Definition 9.2.5 Finally, define $\Gamma \models xy$ iff there is $\sigma : x \to y$ s.t. $\Gamma \models \sigma$, likewise for $x\overline{y}$ and $\sigma : x \cdots \not\to y$.

Diagram 9.2.2 (p. 257) shows the most complicated situation for the positive case.

We have to show now that the above approach corresponds to the preceding discussion.

Fact 9.2.1 The above definition and the informal one outlined in Definition 9.2.2 (p. 263) correspond, when we consider valid positive paths as access to information and comparison of information strength, as indicated at the beginning of Sect. 9.2.2 (p. 262) and elaborated in Sect. 9.4.3 (p. 277).

Proof As Definition 9.2.2 (p. 263) is informal, this cannot be a formal proof, but it is obvious how to transform it into one.

We argue for the result, the argument for valid paths is similar. Consider then case (P2.2) in Definition 9.2.2 (p. 263), and start from some x.

Case 1: Direct links, $x \to z$ or $x \not\to z$.
By comparison of strength via preclusion, as a direct link starts at x, the information z or $\neg z$ is stronger than all other accessible information. Thus, the link and the information will be valid in both approaches. Note that we assumed Γ free from hard contradictions.

Case 2: Composite paths.
In both approaches, the initial segment has to be valid, as information will otherwise not be accessible. Also, in both approaches, information will have

the form of direct links from the accessible source. Thus, condition (1) in Case II.1 corresponds to condition (1) in Definition 9.2.2 (p. 263).
In both approaches, information contradicted by a stronger source (preclusion) is discarded, as well as information which is contradicted by other, not precluded sources; so (P2.2) in Definition 9.2.2 (p. 263) and II.1 (2) + (3) correspond. Note that variant (P2.1) of Definition 9.2.2 (p. 263) would give a different result – which we could, of course, also imitate in a modified inheritance approach.

Case 3: Other information.
Inheritance nets give no other information, as valid information is deduced only through valid paths by Definition 9.2.5 (p. 268), and we did not add any other information either in the approach in Definition 9.2.2 (p. 263). But as is obvious in Case 2, valid paths coincide in both cases.

Thus, both approaches are equivalent. □

9.2.3 Review of Other Approaches and Problems

We now discuss shortly in more detail some of the differences between various major definitions of inheritance formalisms. Diagram 6.8, p. 179, in [Sch97-1] (which is probably due to folklore of the field) shows requiring downward chaining would be wrong. We repeat it here, see, Diagram 9.1.2 (p. 259).

Preclusions valid above (here at u) can be invalid at lower points (here at z), as part of the relevant information is no longer accessible (or becomes accessible). We have $u \rightarrow x \not\rightarrow y$ valid, by downward chaining. Any valid path $z \rightarrow u \ldots . y$ has to have a valid final segment $u \ldots . y$, which can only be $u \rightarrow x \not\rightarrow y$, but intuition says that $z \rightarrow u \rightarrow v \rightarrow y$ should be valid. Downward chaining prevents such changes, and thus seems inadequate, so we decide for upward chaining. (Already preclusion itself underlines upward chaining: In the Tweety diagram, we have to know that the path from bottom up to penguins is valid. So at least some initial subpaths have to be known – we need upward chaining.) (The rejection of downward chaining seems at first sight to be contrary to the intuitions carried by the word "inheritance".) See also the remarks in Sect. 9.4.4 (p. 281).

9.2.3.1 Extension-Based vs. Directly Sceptical Definitions

As this distinction has already received detailed discussion in the literature, we shall be very brief here. An extension of a net is essentially a maximally consistent and in some appropriate sense reasonable subset of all its potential paths. This can, of course, be presented either as a liberal conception (focussing on individual extensions) or as a sceptical one (focusing on their intersection – or, the intersection of their conclusion sets). The seminal presentation is that of [Tou86], as refined by [San86]. The directly sceptical approach seeks to obtain a notion of sceptically accepted path and conclusion, but without detouring through extensions. Its classic presentation is that of [HTT87]. Even while still searching for fully adequate definitions of either kind, we may use the former approach as a useful "control" on the latter. For if we can find an intuitively possible and reasonable extension supporting a conclusion $x\overline{y}$, whilst a proposed definition for a directly sceptical notion of legitimate inference yields xy as a conclusion, then the counterexemplary extension seems to call into question the adequacy of the directly sceptical construction, more readily than inversely. It has been shown in [Sch93] that the intersection of extensions is fundamentally different from the directly sceptical approach. See also the remark in Sect. 9.4.4 (p. 281).

From now on, all definitions considered shall be (at least) upward chaining.

9.2.3.2 On-Path vs. Off-Path Preclusion

This is a rather technical distinction discussed in [THT87]. Briefly, a path $\sigma: x \to \ldots \to y \to \ldots \to z$ and a direct link $y \not\to u$ is an off-path preclusion of $\tau: x \to \ldots \to z \to \ldots \to u$, but an on-path preclusion only iff all nodes of τ between x and z lie on the path σ.

For instance, in the Tweety diagram, the arrow $c \not\to d$ is an on-path preclusion of the path $a \to c \to b \to d$, but the paths $a \to c$ and $c \to b$, together with $c \not\to d$, is an (split validity) off-path preclusion of the path $a \to b \to d$.

9.2.3.3 Split Validity vs. Total Validity Preclusion

Consider again a preclusion $\sigma : u \to \ldots \to x \to \ldots \to v$ and $x \not\to y$ of $\tau : u \to \ldots \to v \to \ldots \to y$. Most definitions demand for the preclusion to be effective – i.e. to prevent τ from being accepted – that the total path σ is valid. Some ([GV89], [KK89], [KKW89a], [KKW89b]) content themselves with the combinatorially simpler separate (split) validity of the lower and upper parts of σ: $\sigma' : u \to \ldots \to x$ and $\sigma'' : x \to \ldots \to v$. In Diagram 9.2.3 (p. 271), taken from [Sch97-1], the path $x \to w \to v$ is valid, so is $u \to x$, but not the whole preclusion path $u \to x \to w \to v$.

Thus, split validity preclusion will give here the definite result $u\overline{y}$. With total validity preclusion, the diagram has essentially the form of a Nixon Diamond.

9.3 Defeasible Inheritance and Reactive Diagrams

Before we discuss the relationship in detail, we first summarize our algorithm.

9.3.1 Summary of Our Algorithm

We look for valid paths from x to y.

(1) Direct arrows are valid paths.

Diagram 9.2.2 The complicated case

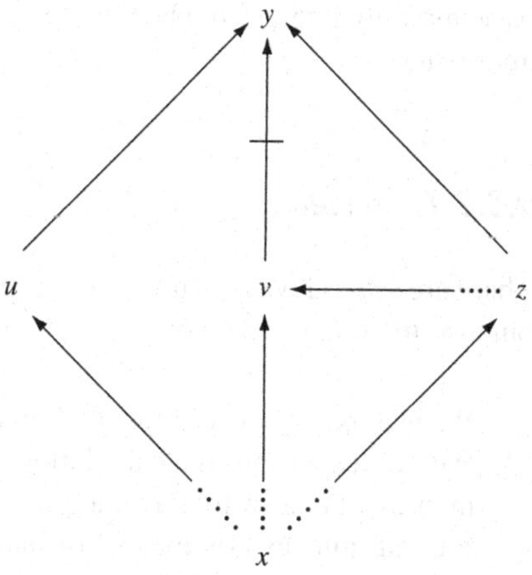

Diagram 9.2.3 Split vs. total validity preclusion

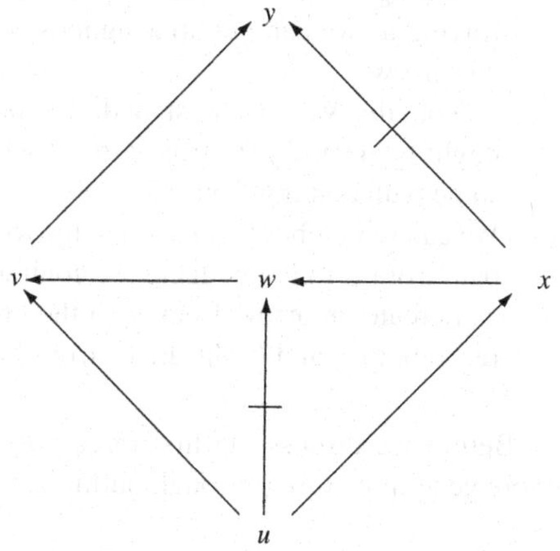

(2) Consider the set C of all direct predecessors of y, i.e. all c such that there is a direct link from c to y.

(2.1) Eliminate all those to which there is no valid positive path from x (found by induction); let the new set be $C' \subseteq C$.
If the existence of a valid path has not yet been decided, we have to wait.

(2.2) Eliminate from C' all c such that there is $c' \in C'$ and a valid positive path from c' to c (found by induction) – unless the arrows from c and from c' to y are of the same type. Let the new set be $C'' \subseteq C'$ (this handles preclusion).
If the existence of such valid paths has not yet been decided, we have to wait.

(2.3) If the arrows from all elements of C'' to y have same polarity, we have a valid path from x to y, if not, we have an unresolved contradiction, and there is no such path.

Note that we were a bit sloppy here. It can be debated whether preclusion by some c' such that c and c' have the same type of arrow to y should be accepted. As we are basically only interested whether there is a valid path, but not in its details, this does not matter.

9.3.2 Overview

There are several ways to use reactive graphs to help us solve inheritance diagrams – but also to go beyond them.

(1) We use them to record results found by adding suitable arrows to the graph.
(2) We go deeper into the calculating mechanism, and use new arrows not only as memory, but also for calculation.
(3) We can put up "signposts" to mark dead ends in the following sense: If we memorize valid paths from x to y, then, anytime we are at a branching point u coming from x, trying to go to y, and there is valid path through an arrow leaving u, we can put up a signpost saying "no valid path from x to y through this arrow".
Note that we have to state destination y (of course), but also outset, x: There might be a valid path from u to y, which may be precluded or contradicted by some path coming from x.
(4) We can remember preclusion in the following sense: If we found a valid positive path from a to b, and there are contradicting arrows from a and b to c, then we can create an arrow from a to the arrow from b to c. So, if, from x, we can reach both a and b, the arrow from b to c will be incapacitated.

Before we discuss the first three possibilities in detail, we shortly discuss the more general picture (in rough outline).

(1) Replacing labels by arrows and vice versa.
As we can switch arrows on and off, an arrow carries a binary value – even without any label. So the idea is obvious:
If an arrow has one label with n possible values, we can replace it with n parallel arrows (i.e. same source and destination), where we switch exactly one of them on – this is the label.
Conversely, we can replace n parallel arrows without labels, where exactly one is active, by one arrow with n labels.
We can also code labels of a node x by an arrow $\alpha : x \to x$, which has the same labels.
(2) Coding logical formulas and truth in a model.
We take two arrows for each propositional variable, one stands for true, the other for false. Negation blocks the positive arrow, enables the negative one. Conjunction is solved by concatenation, disjunction by parallel paths. If a variable occurs more than once, we make copies, which are "switched" by the "master arrow".

We come back to the first three ways to treat inheritance by reactive graphs, and also mention a way to go beyond usual inheritance.

9.3.3 Compilation and Memorization

When we take a look at the algorithm deciding which potential paths are valid, we see that, with one exception, we only need the results already obtained, i.e. whether there is a valid positive/negative path from a to b, and not the actual paths themselves. (This is, of course, due to our split validity definition of preclusion.) The exception is that preclusion works "in the upper part" with direct links. But this is a local problem: We only have to look at the direct predecessors of a node.

Consequently, we can do most of the calculation just once, in an induction of increasing "path distance", memorize valid positive (the negative ones cannot be used) paths with special arrows which we activated once their validity is established, and work now as follows with the new arrows:

Suppose we want to know if there is a valid path from x to y. We look backwards at all predecessors b of y (using a simple backward pointer), and look whether there is a valid positive path from x to b, using the new arrows. We then look at all arrows going from such b to y. If they all agree (i.e. all are positive or all are negative), we need not look further, and have the result. If they disagree, we have to look at possible comparisons by specificity. For this, we see whether there are new arrows between the bs. All such b to which there is a new arrow from another b are out of consideration. If the remaining agree, we have a valid path (and activate a new arrow from x to y if the path is positive), if not, there is no such path. (Depending on the technical details of the induction, it can be useful to note this also by activating a special arrow.)

9.3.4 Executing the Algorithm

Consider any two points x, y. There can be no path, a positive potential path, a negative potential path, both potential paths, a valid positive path, or a valid negative path (from x to y). Once a valid path is found, we can forget potential paths; so we can code the possibilities by $\{*, p+, p-, p+-, v+, v-\}$ in above order. We saw that we can either work with labels, or with a multitude of parallel arrows; we choose the first possibility. We create for each pair of nodes a new arrow, (x, y), which we intialize with label $*$.

First, we look for potential paths. If there is a direct link from x to y, we change the value $*$ of (x, y) directly to $v+$ or $v-$. If (x, y') has value $p+$, or $p+-$, or $v+$, and there is a direct link from y' to y, we change the value of (x, y) from $*$ to $p+$ or $p-$, depending on the link from y' to y, from $p+$ or $p-$ to $p+-$ if adequate (we found both possibilities), and leave the value unchanged otherwise. This determines all potential paths.

We make for each pair x, y at y a list of its predecessors, i.e. of all c s.t. there is a direct link from c to y. We do this for all x, so we can work in parallel. A list is, of course, a concatenation of suitable arrows. Suppose we want to know if there is valid path from x to y.

First, there might be a direct link, and we are done. Next, we look at the list of predecessors of y for x. If one c in the list has value $*$, $p-$, v$-$ for (x, c), it is eliminated from the list. If one c has $p+$ or $p+-$, we have to wait. We do this until all (x, c) have $*$, $v+$, or $v-$, so those remaining in the list will have $v+$. We look at all pairs in the list. While at least one (c, c') has $p+$ or $p+-$, we have to wait. Finally, all pairs will have $*$, $p-$, $v+$ or $v-$. Eliminate all c' s.t. there is (c, c') with value $v+$ – unless the arrows from c and c' to y are of the same type. Finally, we look at the list of the remaining predecessors, if they all have the same link to y. We set (x, y) to $v+$ or $v-$, otherwise to $*$. All such operations can be done by suitable operations with arrows, but it is very lengthy and tedious to write down the details.

9.3.5 Signposts

Putting up signposts requires memorizing all valid paths, as leaving one valid path does not necessarily mean that there is no alternative valid path. The easiest thing to do is probably to put up a warning post everywhere, and collect the wrong ones going backwards through the valid paths.

We illustrate this with Diagram 9.3.5 (p. 256):

There are the following potential paths from x to y : xcy, $xceb$-y, $xcebday$, xay.

The paths xc, xa, and $xceb$ are valid. The latter, xce is valid. $xceb$ is in competition with $xcg-b$ and $xceg-b$, but both are precluded by the arrow (valid path) eg. ys predecessors on those paths are a, b, c. a and b are not comparable, as there is no valid path from b to a, as bda is contradicted by $bf-a$. None is more specific than the other one. b and c are comparable, as the path ceb is valid, since $cg-b$ and $ceg-b$ are precluded by the valid path (arrow) eg. So c is more specific than b. Thus, b is out of consideration, and we are left with a and c. They agree, so there is positive valid path from x to y; more precisely, one through a, one through c. We have put STOP signs on the arrows ce and cg, as we cannot continue via them to y.

9.3.6 Beyond Inheritance

We can also go beyond usual inheritance networks.

Consider the following scenario:

- Museum airplanes usually will not fly, but usual airplanes will fly.
- Penguins don't fly, but birds do.
- Nonflying birds usually have fat wings.

- Nonflying aircraft usually are rusty.
- But it is not true that usually nonflying things are rusty and have fat wings.

We can model this with higher-order arrows as follows:

- Penguins $\rightarrow$ birds, museum airplanes $\rightarrow$ airplanes, birds $\rightarrow$ fly, airplanes $\rightarrow$ fly.
- Penguins $\not\rightarrow$ fly, museum airplanes $\not\rightarrow$ fly.
- Flying objects $\not\rightarrow$ rusty, flying objects $\not\rightarrow$ have fat wings.
- We allow concatenation of two negative arrows:

  For example, coming from penguins, we want to concatenate penguins $\not\rightarrow$ fly and fly $\not\rightarrow$ fat wings. Coming from museum aircraft, we want to concatenate museum airfcraft $\not\rightarrow$ fly and fly $\not\rightarrow$ rusty.

  We can enable this as follows: We introduce a new arrow α : (penguin $\not\rightarrow$ fly) $\rightarrow$ (fly $\not\rightarrow$ fat wings), which when traversing penguin $\not\rightarrow$ fly enables the algorithm to concatenate with the arrow it points to, using the rule "$- * - = +$", giving the result that penguins usually have fat wings.

See [Gab08d] for deeper discussion.

9.4 Interpretations

9.4.1 Introduction

We will discuss in this section three interpretations of inheritance nets.

First, we will indicate fundamental differences between inheritance and the systems P and R. They will be elaborated in Sect. 9.5 (p. 284), where an interpretation in terms of small sets will be tried nonetheless, and its limitations are explored.

Second, we will interpret inheritance nets as systems of information and information flow.

Third, we will interpret inheritance nets as systems of prototypes.

Inheritance nets present many intuitively attractive properties; thus, it is not surprising that we can interpret them in several ways. Similarly, preferential structures can be used as a semantics of deontic and of nonmonotonic logic; they express a common idea: choosing a subset of models by a binary relation. Thus, such an ambiguity need not be a sign for a basic flaw.

9.4.2 Informal Comparison of Inheritance with the Systems *P* and *R*

9.4.2.1 The Main Issues

In the authors' opinion, the following two properties of inheritance diagrams show the deepest difference to preferential and similar semantics, and the first even to classical logic. They have to be taken seriously, as they are at the core of inheritance systems, are independent of the particular formalism, and show that there is a fundamental difference between the former and the latter. Consequently, any attempt at translation will have to stretch one or both sides perhaps beyond the breaking point.

(1) Relevance and
(2) subideal situations or relative normality.

Both (and more) can be illustrated by the following simple Diagram 9.4.1 (p. 277) (which also shows conflict resolution by specificity).

(1) Relevance: As there is no monotonous path whatever between e and d, the question whether es are ds, or not, or vice versa, does not even arise. For the same reason, there is no question whether bs are cs, or not. (As a matter of fact, we will see below in Fact 9.5.1 (p. 286) that bs are *non*-cs in system P – see Definition 2.3 (p. 35).) In upward chaining formalisms, as there is no valid positive path from a to d, there is no question either whether as are fs or not. Of course, in classical logic, all information is relevant to the rest, so we can say, e.g. that es are ds, or es are *non*-ds, or some are ds, some are not, but there is a connection. As preferential models are based on classical logic, the same argument applies to them.
(2) In our diagram, as are bs, but not ideal bs, as they are not ds, the more specific information from c wins. But they are es, as ideal bs are. So they are not perfectly ideal bs, but as ideal bs as possible. Thus, we have graded ideality, which does not exist in preferential and similar structures. In those structures, if an element is an ideal element, it has all properties as such; if one such property is lacking, it is not ideal, and we can't say anything anymore. Here, however, we sacrifice as little normality as possible; it is thus a minimal change formalism.

In comparison, questions of information transfer and strength of information seem lesser differences. Already systems P and R (see Definition 2.3 (p. 35)) differ on information transfer. In both cases, transfer is based on the same notion of smallness, which describes ideal situations. But, as said in Remark 3.2.1 (p. 58), this is conceptually very different from the use of smallness, describing normal situations. Thus, it can be considered also on this level an independent question, and we can imagine systems based on absolutely ideal situations for normality, but with a totally different transfer mechanism.

For these reasons, extending preferential and related semantics to cover inheritance nets seems to stretch them to the breaking point. Thus, we should also look

Diagram 9.4.1 Information transfer

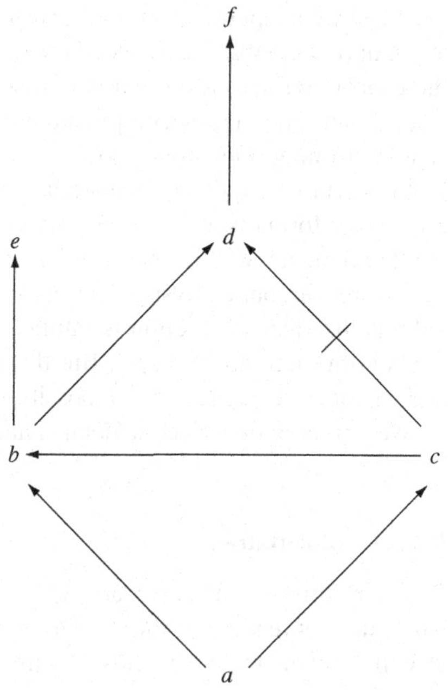

for other interpretations. (The term "interpretation" is used here in a nontechnical sense.) In particular, it seems worthwhile to connect inheritance systems to other problems, and see whether there are similarities there. This is what we do now. We come back to the systems P and R in Sect. 9.5 (p. 284).

Note that Reiter defaults behave much more like inheritance nets than like preferential logics.

9.4.3 Inheritance as Information Transfer

An informal argument showing parallel ideas common to inheritance with an upward chaining formalism and information transfer is as follows: First, arrows represent certainly some kind of information, of the kind "most as are bs" or so. (See, Diagram 9.4.1 (p. 277).) Second, to be able to use information, e.g. "ds are fs" at a, we have to be able to connect from a to d by a valid path. This information has to be made accessible to a, or, in other terms, a working information channel from a to d has to be established. Third, specificity (when present) decides conflicts (we take the split validity approach). This can be done procedurally, or, perhaps simpler and certainly in a more transparent way, by assigning a comparison of information strength to valid paths. Now, information strength may also be called truth value (to use a term familiar in logic) and the natural entity at hand is the node itself – this is just a cheap formal trick without any conceptual meaning.

When we adopt this view, nodes, arrows, and valid paths have multiple functions, and it may seem that we overload the (deceptively) simple picture. But, it is perhaps the charm and the utility and naturalness of inheritance systems that they are not "clean", and hide many complications under a simple surface, as human common sense reasoning often does, too.

In a certain way, this is a poor man's interpretation, as it does not base inheritance on another formalism, but gives only an intuitive reading. Yet, it gives a connection to other branches of reasoning, and is as such already justified – in the authors' opinion. Moreover, our analysis makes a clear distinction between arrows and composite valid paths. This distinction is implicit in inheritance formalisms, and we make it explicit through our concepts. But this interpretation is by no means the only one, and can only be suggested as a possibility.

We will now first give the details, and then discuss our interpretation.

9.4.3.1 Information

Direct positive or negative arrows represent information, valid for their source. Thus, in a set reading, if there is an arrow $A \to B$ in the diagram, most elements of A will be in B. In short: "most As are Bs" – and $A \not\to B$ will mean that most As are not Bs.

9.4.3.2 Information Sources and Flow

Nodes are information sources. If $A \to B$ is in the diagram, A is the source of the information "most As are Bs". A valid, composed or atomic positive path σ from U to A makes the information of source A accessible to U. One can also say that As information becomes relevant to U. Otherwise, information is considered independent – only (valid) positive paths create the dependencies.

(If we want to conform to inheritance, we must not add trivialities like "xs are xs", as this would require $x \to x$ in the corresponding net, which, of course, will not be there in an acyclic net.)

9.4.3.3 Information Strength

A valid, composed or atomic positive path σ from A' to A allows us to compare the strength of information source A' with that of A: A' is stronger than A. (In the set reading, this comparison is the result of specificity: more specific information is considered more reliable.) If there is no such valid path, we cannot resolve contradictions between information from A and A'. This interpretation results in split validity preclusion: the comparison between information sources A' and A is absolute, and does NOT depend on the U from which both may be accessible – as can be the case with total validity preclusion. Of course, if desired, we can also adopt the much more complicated idea of relative comparison.

Nodes are also truth values. They are the strength of the information whose source they are. This might seem an abuse of nodes, but we already have them, so why not use them?

9.4.3.4 Discussion

Considering direct arrows as information meets probably with little objection. The conclusion of a valid path (e.g. if $\sigma : a \ldots \to b$ is valid, then its conclusion is "as are bs") is certainly also information, but it has a status different from the information of a direct link, so we should distinguish it clearly. At least in upward chaining formalisms, using the path itself as some channel through which information flows, and not the conclusion, seems more natural. The conclusion says little about the inner structure of the path, which is very important in inheritance networks, e.g. for preclusion. When calculating validity of paths, we look at (sub- and other) paths, but not just their results, and should also express this clearly.

Once we accept this picture of valid positive paths as information channels, it is natural to see their upper ends as information sources. Our interpretation supports upward chaining, and vice versa, upward chaining supports our interpretation.

One of the central ideas of inheritance is preclusion, which, in the case of split validity preclusion, works by an absolute comparison between nodes. Thus, if we accept split validity preclusion, it is natural to see valid positive paths as comparisons between information of different strengths. Conversely, if we accept absolute comparison of information, we should also accept split validity preclusion – these interpretations support each other.

Whatever type of preclusion we accept, preclusion clearly compares information strength, and allows us to decide for the stronger one. We can see this procedurally, or by giving different values to different information, depending on their sources, which we can call truth values to connect our picture to other areas of logic. It is then natural – as we have it already – to use the source node itself as truth value, with comparison via valid positive paths.

9.4.3.5 Illustration

Thus, in a given node U, information from A is accessible iff there is a valid positive path from U to A, and if information from A' is also accessible, and there is a valid positive path from A' to A, then, in case of conflict, information from A' wins over that from A, as A' has a better truth value. In the Tweety diagram, see Diagram 9.2.1 (p. 256), Tweety has access to penguins and birds, the horizontal link from penguin to bird compares the strengths, and the fly/not fly arrows are the information we are interested in.

Note that negative links and (valid) paths have much less function in our picture than positive links and valid paths. In a way, this asymmetry is not surprising, as there are no negative nodes (which would correspond to something like the set complement or negation). To summarize: A negative direct link can only be information. A positive direct link is information at its source, but it can also be a comparison of truth values, or it can give access from its source to information at its end. A valid, positive, composed path can only be comparison of truth values, or give access to information; it is NOT information itself in the sense of direct links. This distinction is very important, and corresponds to the different treatment of direct arrows and valid paths in inheritance, as it appears, e.g. in the definition of preclusion. A valid, negative, composed path has no function, only its parts have.

We obtain automatically that direct information is stronger than any other information: If A has information ϕ, and there is a valid path from A to B, making Bs information accessible to A, then this same path also compares strength, and As information is stronger than Bs information.

Inheritance diagrams in this interpretation represent not only reasoning with many truth values, but also reasoning *about* those truth values: Their comparison is done by the same underlying mechanism.

We should perhaps note in this context a connection to an area currently en vogue: The problem of trust, especially in the context of web information. We can see our truth values as the degrees of trust we put into information coming from this node, and, we not only use, but also reason about them.

9.4.3.6 Further Comments

Our reading also covers enriched diagrams, where arbitrary information can be "appended" to a node. An alternative way to see a source of information is to see it as a reason to believe the information it gives. U needs a reason to believe something, i.e. a valid path from U to the source of the information, and also a reason to disbelieve, i.e. if U' is below U, and U believes and U' does NOT believe some information of A, then either U' has stronger information to the contrary, or there is not a valid path to A anymore (and neither to any other possible source of this information). ("Reason", a concept very important in this context, was introduced by Bochman into the discussion.)

The restriction that negative links can only be information applies to traditional inheritance networks, and the authors make no claim whatever that it should also hold for modified such systems, or in still other contexts. One of the reasons why we do not have "negative nodes", and thus negated arrows also in the middle of paths might be the following (with $\boldsymbol{C}$ complementation): If, for some X, we also have a node for $\boldsymbol{C}X$, then we should have $X \not\rightarrow \boldsymbol{C}X$ and $\boldsymbol{C}X \not\rightarrow X$, thus a cycle, and arrows from Y to X should be accompanied by their opposite to $\boldsymbol{C}X$, etc.

We translate the analysis and decision of Definition 9.2.2 (p. 263) now into the picture of information sources, accessibility, and comparison via valid paths. This is straightforward:

(1) We have that information from A_i, $i \in I$, about B is accessible from U, i.e. there are valid positive paths from U to all A_i. Some A_i may say $\neg B$, some B.
(2) If information from A_i is comparable with information from A_j (i.e. there is a valid positive path from A_i to A_j or the other way around), and A_i contradicts A_j with respect to B, then the weaker information is discarded.

(3) There remains a (nonempty, by lack of cycles) set of A_i such that for no such A_i there is A_j with better contradictory information about B. If the information from this remaining set is contradictory, we accept none (and none of the paths either), if not, we accept the common conclusion and all these paths.

We now continue Remark 9.2.1 (p. 267), (4), and turn this into a formal system. Fix a diagram Γ, and do an induction as in Definition 9.2.2 (p. 263).

Definition 9.4.1 (1) We distinguish $a \Rightarrow b$ and $a \Rightarrow_x b$, where the intuition of $a \Rightarrow_x b$ is: we know with strength x that *a*s are *b*s, and of $a \Rightarrow b$ that it has been decided taking all information into consideration that $a \Rightarrow b$ holds.
(We introduce this notation to point informally to our idea of information strength, and beyond, to logical systems with varying strengths of implications.)
(2) $a \rightarrow b$ implies $a \Rightarrow_a b$, likewise $a \not\rightarrow b$ implies $a \Rightarrow_a \neg b$.
(3) $a \Rightarrow_a b$ implies $a \Rightarrow b$, likewise $a \Rightarrow_a \neg b$ implies $a \Rightarrow \neg b$. This expresses the fact that direct arrows are uncontested.
(4) $a \Rightarrow b$ and $b \Rightarrow_b c$ imply $a \Rightarrow_b c$, likewise for $b \Rightarrow_b \neg c$. This expresses concatenation – but without deciding if it is accepted! Note that we cannot make ($a \Rightarrow b$ and $b \Rightarrow c$ imply $a \Rightarrow_b c$) a rule, as this would make concatenation of two composed paths possible.
(5) We decide acceptance of composed paths as in Definition 9.2.3 (p. 265), where preclusion uses accepted paths for deciding.

Note that we reason in this system not only with, but also about relative strength of truth values, which are just nodes. This is then, of course, used in the acceptance condition, in preclusion more precisely.

9.4.4 Inheritance as Reasoning with Prototypes

Some of the issues we discuss here apply also to the more general picture of information and its transfer. We present them here for motivational reasons: it seems easier to discuss them in the (somewhat!) more concrete setting of prototypes than in the very general situation of information handling. These issues will be indicated.

It seems natural to see information in inheritance networks as information about prototypes. (We do not claim that our use of the word "prototype" has more than a vague relation to the use in psychology. We do not try to explain the usefulness of prototypes either, one possibility is that there are reasons why birds fly, and why penguins don't, etc.) In the Tweety diagram, we will thus say that prototypical birds will fly, prototypical penguins will not fly. More precisely, the property "fly" is part of the bird prototype, is the property "$\neg fly$" is part of the penguin prototype. Thus, the information is given for some node, which defines its application or domain (bird or penguin in our example) – beyond this node, the property is not defined (unless inherited, of course). It might very well be that no element of the domain has ALL the properties of the prototype; every bird may be exceptional in some sense. This again shows that we are very far from the ideal picture of small and big subsets as used in systems P and R. (This, of course, goes beyond the problem of prototypes.)

Of course, we will want to "inherit" properties of prototypes; for instance, in Diagram 9.4.1 (p. 277), a "should" inherit the property e from b, and the property $\neg d$ from c. Informally, we will argue as follows: Prototypical as have property b and prototypical bs have property e; so it seems reasonable to assume that prototypical as also have property e – unless there is better information to the contrary. A plausible solution is then to use upward chaining inheritance as described above to find all relevant information, and then compose the prototype.

We now discuss three points whose importance goes beyond the treatment of prototypes:

(1) Using upward chaining has an additional intuitive appeal: We consider information at a the best, so we begin with b (and c), and only then, tentatively, add information e from b. Thus, we begin with strongest information, and add weaker information successively – this seems good reasoning policy.
(2) In upward chaining, we also collect information at the source (the end of the path), and do not use information which was already filtered by going down – thus the information we collect has no history, and we cannot encounter problems of iterated revision, which are problems of history of change. (In downward chaining, we only store the reasons why something holds, but not why something does not hold, so we cannot erase this negative information when the reason is not valid anymore. This is an asymmetry apparently not much noted before. Consider Diagram 9.1.2 (p. 259). Here, the reason why u does not accept y as information, but $\neg y$, is the preclusion via x. But from z, this preclusion is not valid anymore, so the reason why y was rejected is not valid anymore, and y can now be accepted.)
(3) We come back to the question of extensions vs. direct scepticism. Consider the Nixon Diamond, Diagram 9.1.1 (p. 257). Suppose Nixon were a subclass of Republican and Quaker. Then the extensions approach reasons as follows: Either the Nixon class prototype has the pacifist property, or the hawk property, and we consider these two possibilities. But this is not sufficient: The Nixon class prototype might have neither property – they are normally neither pacifists, nor hawks, but some are this, some are that. So the conceptual basis for the extensions approach does not hold: "Tertium non datur" just simply does not hold – as in intuitionist logic, where we may have a proof neither for ϕ, nor for $\neg\phi$.

Once we fixed this decision, i.e. how to find the relevant information, we can still look upward or downward in the net and investigate the changes between the prototypes in going upward or downward, as follows: For example, in the above example, we can look at the node a and its prototype, and then at the change going from a to b, or, conversely, look at b and its prototype, and then at the change going from b to a. The problem of finding the information, and this dynamics of information change have to be clearly separated.

In both cases, we see the following:

(1) The language is kept small, and thus efficient.
For instance, when we go from a to b, information about c is lost, and "c" does not figure anymore in the language, but f is added. When we go from b to a, f is lost, and c is won. In our simple picture, information is independent, and contradictions are always between two bits of information.

(2) Changes are kept small and need a reason to be effective. Contradictory, stronger information will override the old one, but no other information, except in the following case: Making new information (in-) accessible will cause indirect changes, i.e. information now made (in-) accessible via the new node. This is similar to formalisms of causation: if a reason is not there anymore, its effects vanish too.

It is perhaps more natural when going downward also to consider "subsets" as follows: Consider Diagram 9.4.1 (p. 277). bs are ds, and cs are $\neg d$s, and cs are also bs. So it seems plausible to go beyond the language of inheritance nets, and conclude that bs which are not cs will be ds, in short to consider $(b - c)$s. It is obvious which such subsets to consider, and how to handle them: For instance, loosely speaking, in $b \cap d$ e will hold, in $b \cap c \cap d$ $\neg f$ will hold, in $b \cap d \cap Cc$ f will hold, etc. This is just putting the bits of information together.

We turn to another consideration, which will also transcend the prototype situation, and we will (partly) use the intuition that nodes stand for sets and arrows for (soft, i.e. possibly with exceptions) inclusion in a set or its complement.

In this reading, specificity stands for soft, i.e. possibly with exceptions, set inclusion. So, if b and c are visible from a, and there is a valid path from c to b (as in Diagram 9.4.1 (p. 277)), then a is a subset of both b and c, and c a subset of b, so $a \subseteq c \subseteq b$ (softly). But then a is closer to c than a is to b. Automatically, a will be closest to itself. This results in a partial, and not necessarily transitive relation between these distances.

When we go now from b to c, we lose information d and f, win information $\neg d$, but keep information e. Thus, this is minimal change: We give up (and win) only the necessary information, but keep the rest. As our language is very simple, we can use the Hamming distance between formula sets here. (We will make a remark on more general situations just below.)

When we look now again from a, we take the set-closest class (c), and use the information of c, which was won by minimal change (i.e. the Hamming closest) from information of b. So, we have the interplay of two distances, where the set distance certainly is not symmetrical, as we need valid paths for access and comparison. If there is no such valid path, it is reasonable to make the distance infinite.

We make now the promised remark on more general situations: In richer languages, we cannot count formulas to determine the Hamming distance between two situations (i.e. models or model sets), but have to take the difference in propositional variables. Consider, e.g. the language with two variables, p and q. The models (described by) $p \wedge q$ and $p \wedge \neg q$ have distance 1, whereas $p \wedge q$ and $\neg p \wedge \neg q$ have distance 2. Note that this distance is NOT robust under redefinition of the language. Let p' stand for $(p \wedge q) \vee (\neg p \wedge \neg q)$ and q' for q. Of course, p' and q' are equivalent descriptions of the same model set, as we can define all the old singletons also in the new language. Then the situations $p \wedge q$ and $\neg p \wedge \neg q$ have now distance 1, as one corresponds to $p' \wedge q'$, the other to $p' \wedge \neg q'$.

There might be misunderstandings about the use of the word "distance" here. The authors are fully aware that inheritance networks cannot be captured by distance semantics in the sense of preferential structures. But we do NOT think here of distances from one fixed ideal point, but of relativized distances: Every prototype is the origin of measurements. For example, the bird prototype is defined by "flying, laying eggs, having feathers ...". So, we presume that all birds have these properties of the prototype, i.e. distance 0 from the prototype. When we see that penguins do not fly, we move as little as possible from the bird prototype, so we give up "flying", but not the rest. Thus, penguins (better: the penguin prototype) will have distance 1 from the bird prototype (just one property has changed). So there is a new prototype for penguins, and considering penguins, we will not measure from the bird prototype, but from the penguin prototype, so the point of reference changes. This is exactly as in distance semantics for theory revision, introduced in [LMS01], only the point of reference is not the old theory T, but the old prototype, and the distance is a very special one, counting properties assumed to be independent. (The picture is a little bit more complicated, as the loss of one property (flying) may cause other modifications, but the simple picture suffices for this informal argument.)

We conclude this section with a remark on prototypes. Realistic prototypical reasoning will probably neither always be upward nor always be downward. A medical doctor will not begin with the patient's most specific category (name and birthday or so), nor will he begin with all he knows about general objects. Therefore, it seems reasonable to investigate upward and downward reasoning here.

9.5 Detailed Translation of Inheritance to Modified Systems of Small Sets

For background material on abstract size semantics, the reader is referred to Chap. 3 (p. 53).

9.5.1 Normality

As we saw already in Sect. 9.4.2 (p. 276), normality in inheritance (and Reiter defaults, etc.) is relative, and as much normality as possible is preserved. There is no set of absolute normal cases of X, which we might denote $N(X)$, but only for ϕ a set $N(X, \phi)$, elements of X, which behave normally with respect to ϕ. Moreover, $N(X, \phi)$ might be defined, but not $N(X, \psi)$ for different ϕ and ψ. Normality in the sense of preferential structures is absolute: If x is not in $N(X)$ ($= \mu(X)$ in preferential reading), we do not know anything beyond classical logic. This is the dark Swedes' problem: Even dark Swedes should probably be tall. Inheritance systems are different: If birds usually lay eggs, then penguins, though abnormal with respect to flying, will still usually lay eggs. Penguins are fly-abnormal birds, but will continue to be egg-normal birds – unless we have again information to the contrary. So the absolute, simple $N(X)$ of preferential structures splits up into many, by default independent, normalities. This corresponds to intuition: There are no absolutely normal birds, each one is particular in some sense, so $\bigcap\{N(X,\phi) : \phi \in \mathcal{L}\}$ may well be empty, even if each single $N(X,\phi)$ is almost all birds.

What are the laws of relative normality? $N(X,\phi)$ and $N(X,\psi)$ will be largely independent (except for trivial situations, where $\phi \leftrightarrow \psi$, ϕ is a tautology, etc.). $N(X,\phi)$ might be defined and $N(X,\psi)$ not. Connections between the different normalities will be established only by valid paths. Thus, if there is no arrow, or no path, between X and Y, then $N(X,Y)$ and $N(Y,X)$ – where X, Y are also properties – need not be defined. This will get rid of the unwanted connections found with absolute normalities, as illustrated by Fact 9.5.1 (p. 286).

We interpret now "normal" by "big set", i.e. essentially "ϕ holds normally in X" iff "there is a big subset of X, where ϕ holds". This will, of course, be modified.

9.5.2 Small Sets

The main interest of this section is perhaps to show the adaptations of the concept of small and big subsets necessary for a more "real-life" situation, where we have to relativize. The amount of changes illustrates the problems and what can be done, but also perhaps what should not be done, as the concept is stretched too far. For more background, see Chap. 3 (p. 53).

As said, the usual informal way of speaking about inheritance networks (plus other considerations) motivates an interpretation by sets and soft set inclusion – $A \rightarrow B$ means that "most As are Bs". Just as with normality, the "most" will have to be relativized, i.e. there is a B-normal part of A, and a B-abnormal one, and the first is B-bigger than the second – where "bigger" is relative to B, too. A further motivation for this set interpretation is the often evoked specificity argument for preclusion. Thus, we will now translate our remarks about normality into the language of big and small subsets.

Consider now the system P (with cumulativity), see Definition 2.3 (p. 35). Recall from Remark 3.2.1 (p. 58) that small sets (see Sect. 3.2.2.6 (p. 58)) are used in two conceptually very distinct ways: $\alpha \mid\!\sim \beta$ iff the set of $\alpha \wedge \neg\beta$-cases is a small subset (in the absolute sense, there is just one system of big subsets of the α-cases) of the set of α-cases. The second use is in information transfer, used in cumulativity, or cautious monotony, more precisely: If the set of $\alpha \wedge \neg\gamma$-cases is a small subset of the set of α-cases, then $\alpha \mid\!\sim \beta$ carries over to $\alpha \wedge \gamma : \alpha \wedge \gamma \mid\!\sim \beta$. (See Chap. 3 (p. 53) and also the discussion in [Sch04], (p. 86), after Definition 2.3.6.) It is this transfer which we will consider here, and not things like AND, which connect different $N(X,\phi)$ for different ϕ.

Before we go into details, we will show that, e.g. the system P is too strong to model inheritance systems, and that, e.g. the system R is too weak for this purpose. Thus, preferential systems are really quite different from inheritance systems.

Fact 9.5.1

(a) System P is too strong to capture inheritance.
(b) System R is too weak to capture inheritance.

Proof (a) Consider the Tweety diagram, Diagram 9.2.1 (p. 256). $c \to b \to d$, $c \not\to d$. There is no arrow $b \not\to c$, and we will see that P forces one to be there. For this, we take the natural translation, i.e. $X \to Y$ will be "$X \cap Y$ is a big subset of X". We show that $c \cap b$ is a small subset of b, which we write $c \cap b < b$. $c \cap b = (c \cap b \cap d) \cup (c \cap b \cap \mathbf{C}d)$. $c \cap b \cap \mathbf{C}d \subseteq b \cap \mathbf{C}d < b$, the latter by $b \to d$, thus $c \cap b \cap \mathbf{C}d < b$, essentially by right weakening. Set now $X := c \cap b \cap d$. As $c \not\to d$, $X := c \cap b \cap d \subseteq c \cap d < c$, and by the same reasoning as above $X < c$. It remains to show $X < b$. We use now $c \to b$. As $c \cap \mathbf{C}b < c$ and $c \cap X < c$, by cumulativity $X = c \cap X \cap b < c \cap b$; so essentially by OR $X = c \cap X \cap b < b$. Using the filter property, we see that $c \cap b < b$.

(b) Second, even R is too weak: In the diagram $X \to Y \to Z$, we want to conclude that most of X is in Z, but as X might also be a small subset of Y, we cannot transfer the information "most Ys are in Z" to X. □

We have to distinguish direct information or arrows from inherited information or valid paths. In the language of big and small sets, it is easy to do this by two types of big subsets: big ones and very big ones. We will denote the first big, the second BIG. This corresponds to the distinction between $a \Rightarrow b$ and $a \Rightarrow_a b$ in Definition 9.4.1 (p. 281).

We will have the implications $BIG \to big$ and $SMALL \to small$, so we have nested systems. Such systems were discussed in [Sch95-1], see also [Sch97-1]. This distinction seems to be necessary to prevent arbitrary concatenation of valid paths to valid paths, which would lead to contradictions. Consider, e.g. $a \to b \to c \to d$, $a \to e \not\to d$, $e \to c$. Then concatenating $a \to b$ with $b \to c \to d$, both valid, would lead to a simple contradiction with $a \to e \not\to d$, and not to preclusion, as it should be – see below.

For the situation $X \to Y \to Z$, we will then conclude that:

If $Y \cap Z$ is a Z-BIG subset of Y and $X \cap Y$ is a Y-big subset of X, then $X \cap Z$ is a Z-big subset of X. (We generalize already to the case where there is a valid path from X to Y.)

We call this procedure information transfer.

$Y \to Z$ expresses the direct information in this context, so $Y \cap Z$ has to be a Z-BIG subset of Y. $X \to Y$ can be direct information, but it is used here as channel of information flow. In particular, it might be a composite valid path, so in our context, $X \cap Y$ is a Y-big subset of X. $X \cap Z$ is a Z-big subset of X: this can only be big, and not BIG, as we have a composite path.

The translation into big and small subsets and their modifications is now quite complicated: We seem to have to relativize, and we seem to need two types of big and small. This casts, of course, a doubt on the enterprise of translation. The future will tell if any of the ideas can be used in other contexts.

We investigate this situation now in more detail, first without conflicts. The way we cut the problem is not the only possible one. We were guided by the idea that we should stay close to usual argumentation about big and small sets, should proceed carefully, i.e. step by step, and should take a general approach.

Note that we start without any X-big subsets defined, so X is not even a X-big subset of itself.

(A) The simple case of two arrows, and no conflicts.
(In slight generalization:) If information ϕ is appended at Y, and Y is accessible from X (and there is no better information about ϕ available), ϕ will be valid at X. For simplicity, suppose there is a direct positive link from X to Y, written sloppily $X \to Y \models \phi$. In the big subset reading, we will interpret this as: $Y \wedge \phi$ is a ϕ-BIG subset of Y. It is important that this is now direct information, so we have "BIG" and not "big". We read now $X \to Y$ also as: $X \cap Y$ is a Y-big subset of X – this is the channel, so just "big". We want to conclude by transfer that $X \cap \phi$ is a ϕ-big subset of X.

We do this in two steps: First, we conclude that $X \cap Y \cap \phi$ is a ϕ-big subset of $X \cap Y$, and then, as $X \cap Y$ is a Y-big subset of X, $X \cap \phi$ itself is a ϕ-big subset of X. We do NOT conclude that $(X - Y) \cap \phi$ is a ϕ-big subset of $X - Y$. This is very important, as we want to preserve the reason of being ϕ-big subsets – and this goes via Y! The transition from "BIG" to "big" should be at the first step, where we conclude that $X \cap Y \cap \phi$ is a ϕ-big (and not ϕ-BIG) subset of $X \cap Y$, as it is really here where things happen, i.e. transfer of information from Y to arbitrary subsets $X \cap Y$.

We summarize the two steps in a slightly modified notation, corresponding to the diagram $X \to Y \to Z$:

(1) If $Y \cap Z$ is a Z-BIG subset of Y (by $Y \to Z$) and $X \cap Y$ is a Y-big subset of X (by $X \to Y$), then $X \cap Y \cap Z$ is a Z-big subset of $X \cap Y$.
(2) If $X \cap Y \cap Z$ is a Z-big subset of $X \cap Y$ and $X \cap Y$ is a Y-big subset of X (by $X \to Y$) again, then $X \cap Z$ is a Z-big subset of X, so $X \ldots \to Z$.

Note that (1) is very different from cumulativity or even rational monotony, as we do not say anything about X in comparison to Y: X need not be any big or medium size subset of Y.

Seen as strict rules, this will not work, as it would result in transitivity, and thus monotony: We have to admit exceptions, as there might just be a negative arrow $X \not\to Z$ in the diagram. We will discuss such situations below in (C), where we will modify our approach slightly, and obtain a clean analysis.

(Here and in what follows, we are very cautious and relativize all normalities. We could perhaps obtain our objective with a more daring approach, using absolute normality here and there. But this would be a purely technical trick (interesting in its own right), and we look here more for a conceptual analysis, and, as long as we do not find good conceptual reasons why to be absolute here and not there, we will just be relative everywhere.)

We now try to give justifications for the two (defeasible) rules. They will be philosophical and can certainly be contested and/or improved.

For (1): We look at Y. By $X \rightarrow Y$, Ys information is accessible at X, so, as Z-BIG is defined for Y, Z-big will be defined for $Y \cap X$. Moreover, there is a priori nothing which prevents X from being independent from Y, i.e. $Y \cap X$ to behave like Y with respect to Z – by default: of course, there could be a negative arrow $X \not\rightarrow Z$, which would prevent this. Thus, as $Y \cap Z$ is a Z-BIG subset of Y, $Y \cap X \cap Z$ should be a Z-big subset of $Y \cap X$. By the same argument (independence), we should also conclude that $(Y - X) \cap Z$ is a Z-big subset of $Y - X$. The definition of Z-big for $Y - X$ seems, however, less clear.

To summarize, $Y \cap X$ and $Y - X$ behave by default with respect to Z as Y does, i.e. $Y \cap X \cap Z$ is a Z-big subset of $Y \cap X$ and $(Y - X) \cap Z$ is a Z-big subset of $Y - X$. The reasoning is downward, from supersets to subsets, and symmetrical to $Y \cap X$ and $Y - X$. If the default is violated, we need a reason for it. This default is an assumption about the adequacy of the language. Things do not change wildly from one concept to another (or, better: from Y to $Y \wedge X$), they might change, but then we are told so – by a corresponding negative link in the case of diagrams.

For (2): By $X \rightarrow Y$, X and Y are related, and we assume that X behaves as $Y \cap X$ does with respect to Z. This is upward reasoning, from subset to superset and it is NOT symmetrical: There is no reason to suppose that $X - Y$ behaves the same way as X or $Y \cap X$ do with respect to Z, as the only reason for Z we have, Y, does not apply. Note that putting relativity aside (which can also be considered as being big and small in various, per-default independent dimensions) this is close to the reasoning with absolutely big and small sets: $X \cap Y - (X \cap Y \cap Z)$ is small in $X \cap Y$, so a fortiori small in X, and $X - (X \cap Y)$ is small in X, so $(X - (X \cap Y)) \cup (X \cap Y - (X \cap Y \cap Z))$ is small in X by the filter property, so $X \cap Y \cap Z$ is big in X, so a fortiori $X \cap Z$ is big in X.

Thus, in summary, we conclude by default that

(3) if $Y \cap Z$ is a Z-BIG subset of Y and $X \cap Y$ is a Y-big subset of X, then $X \cap Z$ is a Z-big subset of X.

(B) The case with longer valid paths, but without conflicts.
Treatment of longer paths: Suppose we have a valid, composed path from X to Y, $X \ldots \rightarrow Y$, and not any longer a direct link $X \rightarrow Y$. By induction, i.e.

upward chaining, we argue – using directly (3) – that $X \cap Y$ is a Y-big subset of X, and conclude by (3) again that $X \cap Z$ is a Z-big subset of X.

(C) Treatment of multiple and perhaps conflicting information.

Consider Diagram 9.5.1 (p. 291).

We want to analyse the situation and argue that, e.g. X is mostly not in Z.

First, all arguments about X and Z go via the Ys. The arrows from X to Ys, and from Y' to Y could also be valid paths. We look at information which concerns Z (thus U is not considered), and which is accessible (thus Y'' is not considered). We can be slightly more general and consider all possible combinations of accessible information, not only those used in the diagram by X. Instead of arguing on the level of X, we will argue one level above, on the Ys and their intersections, respecting specificity and unresolved conflicts.

(Note that in more general situations, with arbitrary information appended, the problem is more complicated, as we have to check which information is relevant for some ϕ – conclusions can be arrived at by complicated means, just as in ordinary logic. In such cases, it might be better to look first at all accessible information for a fixed X, then at the truth values and their relation, and calculate closure of the remaining information.)

We then have (using the obvious language: "most As are Bs" for "$A \cap B$ is a big subset of A", and "MOST As are Bs" for "$A \cap B$ is a BIG subset of A"):

In Y, Y'', and $Y \cap Y''$, we have that MOST cases are in Z. In Y' and $Y \cap Y'$, we have that MOST cases are not in Z (= are in $\mathbf{C}Z$). In $Y' \cap Y''$ and $Y \cap Y' \cap Y''$, we are UNDECIDED about Z.

Thus

$Y \cap Z$ will be a Z-BIG subset of Y, $Y'' \cap Z$ will be a Z-BIG subset of Y'', $Y \cap Y'' \cap Z$ will be a Z-BIG subset of $Y \cap Y''$.

$Y' \cap \mathbf{C}Z$ will be a Z-BIG subset of Y', $Y \cap Y' \cap \mathbf{C}Z$ will be a Z-BIG subset of $Y \cap Y'$.

$Y' \cap Y'' \cap Z$ will be a Z-MEDIUM subset of $Y' \cap Y''$, $Y \cap Y' \cap Y'' \cap Z$ will be a Z-MEDIUM subset of $Y \cap Y' \cap Y''$.

This is just simple arithmetic of truth values, using specificity and unresolved conflicts, and the nonmonotonicity is pushed into the fact that subsets need not preserve the properties of supersets.

In more complicated situations, we implement, e.g. the general principle (P2.2) from Definition 9.2.2 (p. 263), to calculate the truth values. This will use in our case specificity for conflict resolution, but it is an abstract procedure, based on an arbitrary relation $<$.

This will result in the "correct" truth value for the intersections, i.e. the one corresponding to the other approaches.

It remains to do two things: (C.1) We have to assure that X "sees" the correct information, i.e. the correct intersection and, (C.2), that X "sees" the accepted Ys, i.e. those through which valid paths go, in order to construct not only the result, but also the correct paths.

(Note that by split validity preclusion, if there is valid path from A through B to C, $\sigma : A \cdots \to B, B \to C$, and $\sigma' : A \cdots \to B$ is another valid path from A to B, then $\sigma' \circ B \to C$ will also be a valid path. Proof: If not, then $\sigma' \circ B \to C$ is precluded, but the same preclusion will also preclude $\sigma \circ B \to C$ by split validity preclusion, or it is contradicted, and a similar argument applies again. This is the same argument as the one for the simplified definition of preclusion – see Remark 9.2.1 (p. 267), (4).)

(C.1) Finding and inheriting the correct information:

X has access to Z-information from Y and Y', so we have to consider them. Most of X is in Y, most of X is in Y', i.e. $X \cap Y$ is a Y-big subset of X, $X \cap Y'$ is a Y'-big subset of X, so $X \cap Y \cap Y'$ is a $Y \cap Y'$-big subset of X; thus most of X is in $Y \cap Y'$.

We thus have Y, Y', and $Y \cap Y'$ as possible reference classes, and use specificity to choose $Y \cap Y'$ as reference class. We do not know anything, e.g. about $Y \cap Y' \cap Y''$; so this is not a possible reference class.

Thus, we use specificity twice, on the Y's-level (to decide that $Y \cap Y'$ is mostly not in Z), and on X's-level (the choice of the reference class), but this is good policy, as, after all, much of nonmonotonicity is about specificity.

We should emphasize that nonmonotonicity lies in the behaviour of the subsets, determined by truth values and comparisons thereof, and the choice of the reference class by specificity. But both are straightforward now and local procedures, using information already decided before. There is no complicated issue here like determining extensions.

We now use above argument, described in the simple case, but with more detail, speaking in particular about the most specific reference class for information about Z, $Y \cap Y'$ in our example – this is used essentially in (1.4), where the "real" information transfer happens, and where we go from BIG to big.

(1.1) By $X \to Y$ and $X \to Y'$ (and there are no other Z-relevant information sources), we have to consider $Y \cap Y'$ as reference class.

(1.2) $X \cap Y$ is a Y-big subset of X (by $X \to Y$) (it is even Y-BIG, but we are immediately more general to treat valid paths), $X \cap Y'$ is a Y'-big subset of X (by $X \to Y'$). So $X \cap Y \cap Y'$ is a $Y \cap Y'$-big subset of X.

(1.3) $Y \cap Z$ is a Z-BIG subset of Y (by $Y \to Z$), $Y' \cap \mathbf{C}Z$ is a Z-BIG subset of Y' (by $Y' \not\to Z$), so by preclusion $Y \cap Y' \cap \mathbf{C}Z$ is a Z-BIG subset of $Y \cap Y'$.

(1.4) $Y \cap Y' \cap \mathbf{C}Z$ is a Z-BIG subset of $Y \cap Y'$, and $X \cap Y \cap Y'$ is a $Y \cap Y'$-big subset of X, so $X \cap Y \cap Y' \cap \mathbf{C}Z$ is a Z-big subset of $X \cap Y \cap Y'$.

This cannot be a strict rule without the reference class, as it would then apply to $Y \cap Z$, too, leading to a contradiction.

(2) If $X \cap Y \cap Y' \cap \mathbf{C}Z$ is a Z-big subset of $X \cap Y \cap Y'$, and $X \cap Y \cap Y'$ is a $Y \cap Y'$-big subset of X, so $X \cap \mathbf{C}Z$ is a Z-big subset of X.

We make this now more formal.

We define for all nodes X, Y two sets: $B(X, Y)$ and $b(X, Y)$, where $B(X, Y)$ is the set of Y-BIG subsets of X and $b(X, Y)$ is the set of Y-big subsets of X. (To distinguish undefined from medium/MEDIUM-size, we will also have to define $M(X, Y)$ and $m(X, Y)$, but we omit this here for simplicity.)

Diagram 9.5.1 Multiple and conflicting information

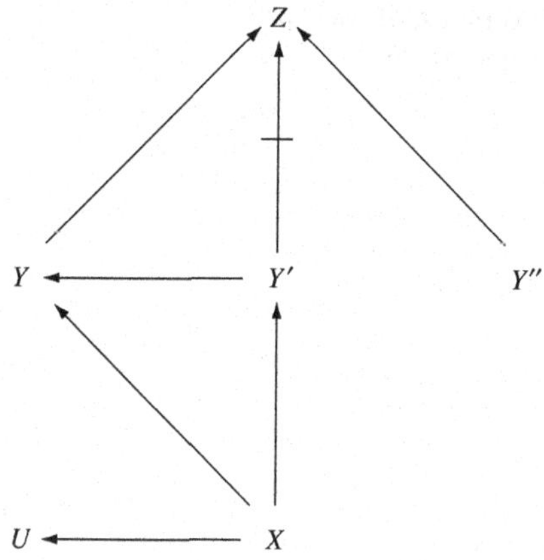

The translations are then:

(1.2′) $X \cap Y \in b(X, Y)$ and $X \cap Y' \in b(X, Y') \Rightarrow X \cap Y \cap Y' \in b(X, Y \cap Y')$;
(1.3′) $Y \cap Z \in B(Y, Z)$ and $Y' \cap \mathbf{C}Z \in B(Y', Z) \Rightarrow Y \cap Y' \cap \mathbf{C}Z \in B(Y \cap Y', Z)$ by preclusion;
(1.4′) $Y \cap Y' \cap \mathbf{C}Z \in B(Y \cap Y', Z)$ and $X \cap Y \cap Y' \in b(X, Y \cap Y') \Rightarrow X \cap Y \cap Y' \cap \mathbf{C}Z \in b(X \cap Y \cap Y', Z)$ as $Y \cap Y'$ is the most specific reference class;
(2′) $X \cap Y \cap Y' \cap \mathbf{C}Z \in b(X \cap Y \cap Y', Z)$ and $X \cap Y \cap Y' \in b(X, Y \cap Y') \Rightarrow X \cap \mathbf{C}Z \in b(X, Z)$.
Finally,
(3′) $A \in B(X, Y) \to A \in b(X, Y)$, etc.

Note that we used, in addition to the set rules, preclusion, and the correct choice of the reference class.

(C.2) Finding the correct paths:
Idea:

(1) If we come to no conclusion, then no path is valid, this is trivial.
(2) If we have a conclusion:
(2.1) All contradictory paths are out: e.g. $Y \cap Z$ will be Z-big, but $Y \cap Y' \cap \mathbf{C}Z$ will be Z-big. So there is no valid path via Y.
(2.2) Thus, not all paths supporting the same conclusion are valid.

Consider the following Diagram 9.5.2 (p. 292):

There might be a positive path through Y, a negative one through Y', a positive one through Y'' again, with $Y'' \to Y' \to Y$, so Y will be out, and only Y'' in. We can see this, as there is a subset, $\{Y, Y'\}$ which shows a change: $Y' \cap Z$ is Z-BIG, $Y' \cap \mathbf{C}Z$ is Z-BIG, $Y'' \cap Z$ is Z-BIG, and $Y \cap Y' \cap \mathbf{C}Z$ is Z-BIG, and the latter can only happen if there is a preclusion between Y' and Y, where Y looses. Thus, we can see this situation by looking only at the sets.

We now show equivalence with the inheritance formalism given in Definition 9.2.3 (p. 265).

Diagram 9.5.2 Valid paths or conclusions

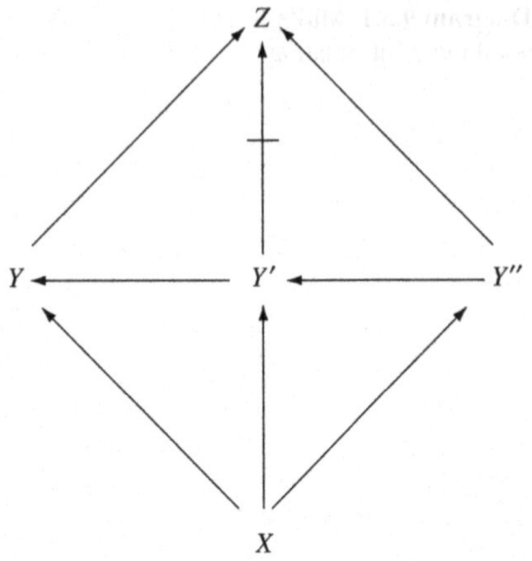

Fact 9.5.2 The above definition and the one outlined in Definition 9.2.3 (p. 265) correspond.

Proof By induction on the length of the deduction that $X \cap Z$ (or $X \cap \boldsymbol{C}Z$) is a Z-big subset of X. (Outline)

It is a corollary of the proof that we have to consider only subpaths and information of all generalized paths between X and Z. Make all sets (i.e. one for every node) sufficiently different, i.e. all sets and Boolean combinations of sets differ by infinitely many elements, e.g. $A \cap B \cap C$ will have infinitely many less elements than $A \cap B$, etc. (Infinite is far too many, we just choose it by laziness to have room for the $B(X, Y)$ and the $b(X, Y)$. Put in $X \cap Y \in B(X, Y)$ for all $X \to Y$, and $X \cap \boldsymbol{C}Y \in B(X, Y)$ for all $X \not\to Y$ as base theory.

Length $= 1$: Then big must be BIG, and, if $X \cap Z$ is a Z-BIG subset of X, then $X \to Z$, likewise for $X \cap \boldsymbol{C}Z$.

We stay close now to above Diagram 9.5.1 (p. 292), so we argue for the negative case. Suppose that we have deduced $X \cap \boldsymbol{C}Z \in b(X, Z)$, we show that there must be a valid negative path from X to Z. (The other direction is easy.) Suppose for simplicity that there is no negative link from X to Z – otherwise we are finished. As we can distinguish intersections from elementary sets (by the starting hypothesis about sizes), this can only be deduced using (2′). So there must be some suitable $\{Y_i : i \in I\}$ and we must have deduced $X \cap \bigcap Y_i \in b(X, \bigcap Y_i)$, the second hypothesis of (2′). If I is a singleton, then we have the induction hypothesis, so there is a valid path from X to Y. So suppose I is not a singleton. Then the deduction of $X \cap \bigcap Y_i \in b(X, \bigcap Y_i)$ can only be done by (1.2′), as this is the only rule having in the conclusion an elementary set on the left in $b(\ldots)$, and a true intersection on the right. Going back along (1.2′), we find $X \cap Y_i \in b(X, Y_i)$, and by the induction hypothesis, there are valid paths from X to Y_i.

The first hypothesis of (2′), $X \cap \bigcap Y_i \cap \boldsymbol{C}Z \in b(X \cap \bigcap Y_i, Z)$ can be obtained by (1.3′) or (1.4′). If it was obtained by (1.3′), then X is one of the Y_i, but then there is a direct link from X to Z (due to the "B", BIG). As a direct link always wins by specificity, the link must be negative, and we have a valid, negative path from X to Z. If it was obtained by $(1.4')$, then its first hypothesis $\bigcap Y_i \cap CZ \in B(\bigcap Y_i, Z)$ must have been deduced, which can only be by $(1.3')$, but the set of Y_i there was chosen to take all Y_i into account for which there is a valid path from X to Y_i and arrows from Y_i to Z (the rule was only present for the most specific reference class with respect to X and Z!), and we are done by the definition of valid paths in Sect. 9.2 (p. 258). □

We summarize our ingredients.

Inheritance was done essentially by (1) and (2) of (A) above and its elaborations $(1.i)$, (2) and $(1.i')$, $(2')$. It consisted of a mixture of bold and careful (in comparison to systems P and R) manipulation of big subsets. We had to be bolder than the systems P and R are, as we have to transfer information also to perhaps small subsets. We had to be more careful, as P and R would have introduced far more connections than are present. We also saw that we are forced to loose the paradise of absolute small and big subsets, and have to work with relative size.

We then have a plug-in decision what to do with contradictions. This is a plug-in, as it is one (among many possible) solutions to the much more general question of how to deal with contradictory information in the presence of a (partial, not necessarily transitive) relation which compares strength. At the same place in our procedure, we can plug-in other solutions, so our approach is truly modular in this aspect. The comparing relation is defined by the existence of valid paths, i.e. by specificity. This decision is inherited downward using again the specificity criterion.

Perhaps, the deepest part of the analysis can be described as follows: Relevance is coded by positive arrows, and valid positive paths, and thus is similar to Kripke structures for modality, where the arrows code dependencies between the situations for the evaluation of the modal quantifiers. In our picture, information at A can become relevant only to node B iff there is a valid positive path from B to A. But, relevance (in this reading, which is closely related to causation) is profoundly nonmonotonic, and any purely monotonic treatment of relevance would be insufficient. This seems to correspond to intuition. Relevance is then expressed in our translation to small and big sets formally by the possibility of combining different small and big sets in information transfer. This is, of course, a special form of relevance, and there might be other forms.

Bibliography

[AA00] O. Arieli, A. Avron, "General patterns for nonmononic reasoning: From basic entailment to plausible relations", Logic Journal of the Interest Group in Pure and Applied Logics, Vol. 8, No. 2, pp. 119–148, 2000

[ABK07] A. Avron, J. Ben-Naim, B. Konikowska, "Cut-free ordinary sequent calculi for logics having generalized finite-valued semantics", Journal Logica Universalis, Vol. 1, pp. 41–69, 2006

[AGM85] C. Alchourron, P. Gardenfors, D. Makinson, "On the logic of theory change: Partial meet contraction and revision functions", Journal of Symbolic Logic, Vol. 50, pp. 510–530, 1985

[Ant97] A. Antonelli, "Defeasible inheritance on cyclic networks", Artificial Intelligence, Vol. 92, pp. 1–23, 1997

[Ant99] A. Antonelli, "A directly cautious theory of defeasible consequence for default logic via the notion of general extensions", Artificial Intelligence, Vol. 109, pp. 71–109, 1999

[Ant05] A. Antonelli, "Grounded Consequence for Defeasible Reasoning", Cambridge University Press, 2005

[Aqv00] L. Aqvist, "Three characterizability problems in deontic logic", Nordic Journal of Philosophical Logic, Vol. 5, No. 2, pp. 65–82, 2000

[BB94] S. Ben-David, R. Ben-Eliyahu, "A modal logic for subjective default reasoning", Proceedings LICS-94, 1994

[BCMG06] R. Booth, S. Chopra, T. Meyer, A. Ghose, "A unifying semantics for belief change", ECAI, pp. 793–797, 2004

[BGH95] J. Barwise, D. Gabbay, C. Hartonas, "On the logic of information flow", Journal of the IGPL, Vol. 3, No. 1, pp. 7–50, 1995

[BGW05] H. Barringer, D.M. Gabbay, J. Woods, "Temporal dynamics of support and attack networks: From argumentation to zoology", Mechanizing Mathematical Reasoning, Volume dedicated to Joerg Siekmann, D. Hutter, W. Stephan (eds.), Springer Lecture Notes in Computer Science 2605, pp. 59–98, 2005

[BLS99] S. Berger, D. Lehmann, K. Schlechta, "Preferred history semantics for iterated updates", Journal of Logic and Computation, Vol. 9, No. 6, pp.817–833, 1999

[BMP97] H. Bezzazi, D. Makinson, R.P. Perez, "Beyond rational monotony: Some strong non-Horn rules for nonmonotonic inference relations", Journal of Logic and Computation, Vol. 7, No. 5, pp. 605–631, 1997

[BN06] R. Booth, A. Nitka, "Reconstructing an agent's epistemic state from observations about its beliefs and non-beliefs", Journal of Logic and Computation, Vol. 18, No. 5, pp. 755–782, 2008

[BS85] G. Bossu, P. Siegel, "Saturation, nonmonotonic reasoning and the closed-world assumption", Artificial Intelligence, Vol. 25, pp. 13–63, 1985

[Bou90a] C. Boutilier, "Conditional logics of normality as modal systems", AAAI, Boston, p. 594, 1990

D.M. Gabbay, K. Schlechta, *Logical Tools for Handling Change in Agent-Based Systems*, Cognitive Technologies, DOI 10.1007/978-3-642-04407-6,

[CJ02] J. Carmo, A.J.I. Jones, "Deontic logic and contrary-to-duties", in: D. Gabbay, F. Guenthner (eds.), "Handbook of Philosophical Logic", Vol. 8, pp. 265–343, Kluwer, Dordrecht, Holland, 2002

[CP00] S. Chopra, R. Parikh, "Relevance sensitive belief structures", Annals of Mathematics and Artificial Intelligence, Vol. 28, No. 1–4, pp. 259–285, 2000

[DP94] A. Darwiche, J. Pearl, "On the logic of iterated belief revision", in: R. Fagin (ed.), "Proceedings of the Fifth Conference on Theoretical Aspects of Reasoning about Knowledge", Morgan Kaufman, Pacific Grove, CA, pp. 5–23, 1994

[DP97] A. Darwiche, J. Pearl, "On the logic of iterated belief revision", Journal of Artificial Intelligence, Vol. 89, No. 1–2, pp. 1–29, 1997

[Del87] J.P. Delgrande, "A first-order conditional logic for prototypical properties", Artificial Intelligence, Vol. 33, pp. 105–130, 1987

[Dun95] P.M. Dung, "On the acceptability of arguments and its fundamental role in nonmonotonic reasoning, logic programming and n-person games", Artificial Intelligence, Vol. 77, pp. 321–357, 1995

[FH96] N. Friedman, J. Halpern, "Plausibility measures and default reasoning", Journal of the ACM, Vol. 48, pp. 1297–1304, 1996

[Gab85] D.M. Gabbay, "Theoretical foundations for non-monotonic reasoning in expert systems", in: K.R. Apt (ed.), "Logics and Models of Concurrent Systems", Springer, Berlin, pp. 439–457, 1989

[Gab98] D.M. Gabbay, "Fibring Logics", Oxford University Press, New York, 1998

[Gab04] D.M. Gabbay, "Reactive Kripke semantics and arc accessibility", in: W. Carnielli, F.M. Dionesio, P. Mateus (eds.), "Proceedings CombLog04", Centre of Logic and Computation, University of Lisbon, pp. 7–20, 2004

[Gab08a] D.M. Gabbay, "Reactive Kripke models and contrary to duty obligations", in: R.v.d. Meyden, L.v.d. Torre (eds.), "DEON-2008", Deontic Logic in Computer Science, LNAI 5076, Springer, pp. 155–173, 2008

[Gab08b] D.M. Gabbay, "Reactive Kripke models and contrary to duty obligations", Journal of Applied Logic (special issue on Deon-2008) (to appear)

[Gab08c] D.M. Gabbay,"Reactive Kripke semantics and arc accessibility", Pillars of Computer Science: Essays Dedicated to Boris (Boaz) Trakhtenbrot on the Occasion of his 85th Birthday, A. Avron, N. Dershowitz, A. Rabinovich (eds.), LNCS, Springer, Berlin, Vol. 4800, pp. 292–341, 2008

[Gab08d] D.M. Gabbay, "Logical modes of attack in argumentation networks", (in preparation)

[GS08a] D. Gabbay, K. Schlechta, "Cumulativity without closure of the domain under finite unions", hal-00311938, arXiv 0808.3077, Review of Symbolic Logic, Vol 1, No. 3, pp. 372–392, 2008

[GS08b] D. Gabbay, K. Schlechta, "Reactive preferential structures and nonmonotonic consequence", hal-00311940, arXiv 0808.3075, Review of Symbolic Logic, Vol 2, No. 2, pp. 414–450

[GS08c] D. Gabbay, K. Schlechta, "Roadmap for preferential logics", hal-00311941, arXiv 0808.3073, Journal of Applied Nonclassical Logic, Vol. 19, No. 1, pp. 43–95, 2009

[GS08d] D. Gabbay, K. Schlechta, "A theory of hierarchical consequence and conditionals", hal-00311937, arXiv 0808.3072, Journal of Logic, Language and Information (to appear)

[GS08e] D. Gabbay, K. Schlechta, "Defeasible inheritance systems and reactive diagrams" (hal-00336105, arXiv 0811.0075), Logic Journal of the IGPL, Vol. 17, pp. 1–54, 2009

[GS08f] D. Gabbay, K. Schlechta, "Logical tools for handling change in agent-based systems", hal-00336103, arXiv 0811.0074 (to appear with Springer)

[GS08g] D. Gabbay, K. Schlechta, "A semantics for obligations" (hal-00339393, arXiv 0811.2754), (submitted)

[GS08h] D. Gabbay, K. Schlechta, "A comment on work by Booth and Nitka on revision" (hal-00336103, arXiv 0811.0074), (submitted)

[GS09a] D. Gabbay, K. Schlechta, "Size and logic" (arXiv 0903.1367), Review of Symbolic Logic, Vol. 2, No. 2, pp. 396–413, 2009

[GS09b] D. Gabbay, K. Schlechta, "Independence – revision and defaults", hal-00375421 and arXiv 0904.2199, Studia Logica, Vol. 92, pp. 381–394, 2009 (to appear)

[Gol03] R. Goldblatt, "Mathematical modal logic: A view of its evolution", Journal Applied Logic, Vol. 1, No. 5–6, pp. 309–392, 2003

[Han69] B. Hansson, "An analysis of some deontic logics", Nous, Vol. 3, pp. 373–398, Reprinted in: R. Hilpinen (ed.), "Deontic Logic: Introductory and Systematic Readings", Reidel, Dordrecht, pp. 121–147, 1971

[Haw07] J. Hawthorne, "Nonmonotonic conditionals that behave like conditional probabilities above a threshold", Journal of Applied Logic, Vol. 5, No. 4, pp. 625–637, 2007

[Haw96] J. Hawthorne, "On the logic of nonmonotonic conditionals and conditional probabilities", Journal of Philosophical Logic, Vol. 25, No. 2, pp. 185–218, 1996

[HM07] J. Hawthorne, D. Makinson, "The quantitative/qualitative watershed for rules of uncertain inference", Studia Logica, Vol. 86, No. 2, pp. 247–297, 2007

[Ker99] G. Kern-Isberner, "Postulates for conditional belief revision", in: T. Dean (ed.), "Proceedings IJCAI 99", Morgan Kaufmann, San Mateo, CA, pp. 186–191, 1999

[KLM90] S. Kraus, D. Lehmann, M. Magidor, "Nonmonotonic reasoning, preferential models and cumulative logics", Artificial Intelligence, Vol. 44, No. 1–2, pp.167–207, July 1990

[Kri59] S. Kripke, "A completeness theorem in modal logic", Journal of Symbolic Logic, Vol. 24, No. 1, pp. 1–14, 1959

[LM92] D. Lehmann, M. Magidor, "What does a conditional knowledge base entail?", Artificial Intelligence, Vol. 55, No. 1, pp. 1–60, May 1992

[Leh92a] D. Lehmann, "Plausibility logic", Proceedings CSL91, E. Boerger, G. Jaeger, H. Kleine-Buening, M.M. Richter (eds.), pp. 227–241, Springer, New York, 1992

[Leh92b] D. Lehmann, "Plausibility logic", Technical Report TR-92-3, Hebrew University, Jerusalem 91904, Israel, February 1992

[LMS01] D. Lehmann, M. Magidor, K. Schlechta, "Distance semantics for belief revision", Journal of Symbolic Logic, Vol. 66, No. 1, pp. 295–317, March 2001

[Lew73] D. Lewis, "Counterfactuals", Blackwell, Oxford, 1973

[Mak94] D. Makinson, "General patterns in nonmonotonic reasoning", in: D. Gabbay, C. Hogger, J. Robinson (eds.), "Handbook of Logic in Artificial Intelligence and Logic Programming", Vol. III: "Nonmonotonic and Uncertain Reasoning", Oxford University Press, Oxford, 1994, pp. 35–110

[Mak09] D. Makinson, "Propositional relevance through letter-sharing" Journal of Applied Logic, Vol. 195, pp. 377–387, 2009

[MG91] D. Makinson, P. Gardenfors, "Relations between the logic of theory change and nonmonotonic logic", in: A. Fuhrmann, M. Morreau (eds.), "The Logic of Theory Change", Springer, Berlin, pp. 185–205, 1991

[McC80] J. McCarthy, "Circumscription – A form of non-monotonic reasoning", Artificial Intelligence, Vol. 13, pp. 27–39, 1980

[MDW94] J.J.Ch. Meyer, F.P.M. Dignum, R.J. Wieringa, "The paradoxes of deontic logic revisited: A computer science perspective", University of Utrecht, NL, Department of Computer Science, Technical Report, 1994

[Mon70] R. Montague, "Universal grammar", Theoria, Vol. 36, pp. 373–98, 1970

[Mor98] L. Morgenstern, "Inheritance comes of age: Applying nonmonotonic techniques to problems in industry", Artificial Intelligence, Vol. 103, pp. 1–34, 1998

[Pac07] O. Pacheco, "Neighbourhood semantics for deontic and agency logics" CIC007, October 2007

[Rod97] O.T. Rodrigues, "A methodology for iterated information change", PhD thesis, Imperial College, London, 1997

[Sch90] K. Schlechta, "Semantics for defeasible inheritance", in: L.G. Aiello (ed.), "Proceedings ECAI 90", London, pp. 594–597, 1990

[Sch92] K. Schlechta, "Some results on classical preferential models", Journal of Logic and Computation, Vol. 2, No. 6, pp. 675–686, 1992

[Sch93] K. Schlechta, "Directly sceptical inheritance cannot capture the intersection of extensions", Journal of Logic and Computation, Vol. 3, No. 5, pp. 455–467, 1995

[Sch95-1] K. Schlechta, "Defaults as generalized quantifiers", Journal of Logic and Computation, Vol. 5, No. 4, pp. 473–494, 1995

[Sch95-2] K. Schlechta, "Logic, topology, and integration", Journal of Automated Reasoning, Vol. 14, 353–381, 1995

[Sch96-1] K. Schlechta, "Some completeness results for stoppered and ranked classical preferential models", Journal of Logic and Computation, Vol. 6, No. 4, pp. 599–622, 1996

[Sch96-3] K. Schlechta, "Completeness and incompleteness for plausibility logic", Journal of Logic, Language and Information, Vol. 5, No. 2, pp. 177–192, 1996

[Sch97-1] K. Schlechta, "Nonmonotonic Logics: Basic Concepts, Results, and Techniques" Springer Lecture Notes Series, LNAI 1187, 243pp., January 1997

[Sch04] K. Schlechta, "Coherent Systems", Elsevier, Amsterdam, 2004

[Sch07a] K. Schlechta, "Factorization", HAL, arXiv.org 0712.4360v1, 2007

[SGMRT00] K. Schlechta, L. Gourmelen, S. Motre, O. Rolland, B. Tahar, "A new approach to preferential structures", Fundamenta Informaticae, Vol. 42, No. 3–4, pp. 391–410, 2000

[SM94] K. Schlechta, D. Makinson, "Local and global metrics for the semantics of counterfactual conditionals", Journal of Applied Non-Classical Logics, Vol. 4, No. 2, pp. 129–140, 1994, also LIM Research Report RR 37, 09/94

[Sco70] D. Scott, "Advice in modal logic", in: K. Lambert (ed.), "Philosophical Problems in Logic", Reidel, Dordrecht, 1970

[Sho87b] Yoav Shoham, "A semantical approach to nonmonotonic logics", Proceeding of the Logics in Computer Science, pp. 275–279, Ithaca, New York, 1987, IEEE Computer Society, and Proceedings of the IJCAI, Vol. 87, pp. 388–392

[SL89] B. Selman, H. Levesque, "The tractability of path-based inheritance", Proceedings of the IJCAI, pp. 1140–1145, 1989

[Spo88] W. Spohn, "Ordinal conditional functions: A dynamic theory of epistemic states", in: W.L. Harper and B. Skyrms, (eds.), "Causation in Decision, Belief Change, and Statistics", Reidel, Dordrecht, Vol. 2, pp. 105–134, 1988

[Sta68] R. Stalnaker, "A theory of conditionals", in: N. Rescher (ed.), Studies in Logical Theory, Blackwell, Oxford, pp. 98–112, 1968

[Tou84] D.S. Touretzky, "Implicit ordering of defaults in inheritance systems", Proceedings of the AAAI, Vol. 84, pp. 322–325, 1984

[Tou86] D.S. Touretzky, "The Mathematics of Inheritance Systems", Los Altos, London, 1986

[THT87] D.S. Touretzky, J.F. Horty, R.H. Thomason, "A clash of intuitions: The current state of nonmonotonic multiple inheritance systems", Proceedings of the IJCAI, pp. 476–482, 1987

Index

D

E